QUAN*toons*

QUANtoons

Metaphysical Illustrations by Tomas Bunk

Physical Explanations by Arthur Eisenkraft and Larry D. Kirkpatrick

Claire Reinburg, Director
Judy Cusick, Senior Editor
J. Andrew Cocke, Associate Editor
Betty Smith, Associate Editor

Printing and Production Catherine Lorrain-Hale, Director
Nguyet Tran, Assistant Production Manager
Jack Parker, Desktop Publishing Specialist
Tim Weber, Technical Assistance

National Science Teachers Association
Gerald F. Wheeler, Executive Director
David Beacom, Publisher

08 07 06 4 3 2 1

Library of Congress Cataloging-in-Publication Data

Bunk, Tomas, 1945-
Quantoons : metaphysical illustrations by Tomas Bunk, physical explanations by Arthur Eisenkraft and Larry D. Kirkpatrick / by Tomas Bunk, Arthur Eisenkraft, and Larry D. Kirkpatrick.
p. cm.
ISBN 0-87355-265-2
1. Science—Problems, exercises, etc. 2. Physics—Problems, exercises, etc. 3. Mathematics—Problems, exercises, etc. 4. Science—Competitions. 5. Physics—Competitions. 6. Mathematics—Competitions. I. Eisenkraft, Arthur. II. Kirkpatrick, Larry D., 1941- III. Title.
Q182.B855 2006
507'.1—dc22

2005024051 2002153474

The problems and artwork in this book were originally published in different form in *Quantum* magazine, Vol. 2, No. 1 (September/October 1991), through Vol. 11, No. 6 (July/August 2001).

NSTA is committed to publishing quality materials that promote the best in inquiry-based science education. However, conditions of actual use may vary and the safety procedures and practices described in this book are intended to serve only as a guide. Additional precautionary measures may be required. NSTA and the authors do not warrant or represent that the procedures and practices in this book meet any safety code or standard or federal, state, or local regulations. NSTA and the author(s) disclaim any liability for personal injury or damage to property arising out of or relating to the use of this book including any of the recommendations, instructions, or materials contained therein.

QUANtoons

Introduction

QUANTOONS BRINGS PHYSics to you through masterful illustrations, quotes, text and challenging problems. The simple classic physics problem of crossing a raging river and determining where you land on the other shore turns into a metaphor of traversing the river of life from birth to death in a cartoon illustration filled with humor and poignancy. The Heraclitus quote "You can never step into the same river twice" adds another aspect of appreciation to the physics story of relative motion, movement in two dimensions, and calculations of least time.

The colors in a rainbow, the sounds of rustling leaves, and the splattering of waves crashing into rocks communicate to us through our senses and our imaginations. Physics can add to our appreciation and understanding of these natural occurrences if the insights of physics are made accessible.

Quantoons is a compilation of Contest Problems that were published in *Quantum* magazine during its 11-year run (1990–2001). *Quantum* was a collaborative effort of the United States and Russia. Published by the National Science Teachers Association (NSTA), this semimonthly magazine was targeted at

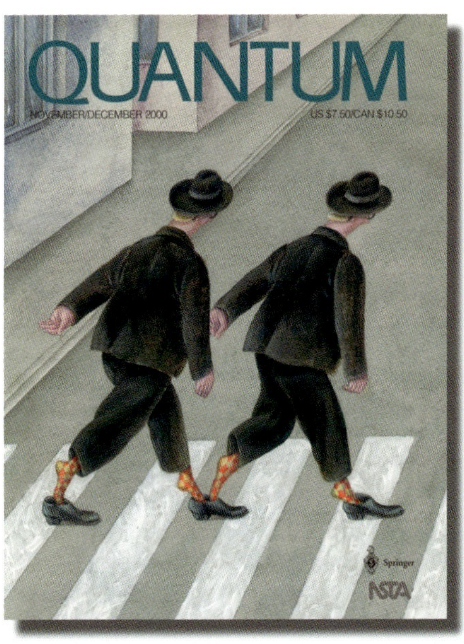

students and teachers and anyone else interested in science and mathematics. Borrowing from original notes and articles in the Russian publication *Kvant*, *Quantum* added articles and problems from American authors. The Contest Problem was one such addition. In every issue, a physics problem was presented and interested readers were invited to submit solutions. The best of these solutions were acknowledged in a subsequent issue, and often used as the basis for the published solution. The Contest Problems were also intended to be descriptions of physics enhanced by the creative cartoons that accompanied them.

In the first issue of *Quantum*, Bill Aldridge quotes the great Russian scientist and poet Mikhail Lomonosov as he viewed the Northern Lights: "Nature, where are your laws? The dawn appears from the dark northern climes! Does not the sun there set up its throne? Are not the ice-bound seas emitting fire?" For 11 years, *Quantum* attempted to answer some of these questions while illuminating the minds of so many who will one day provide us with other glimpses into the wonder of the universe.

Bill Thurston, in that same first issue, reflected on the beginning of his illustrious career as a mathematician. "As a child, I often hated arithmetic and mathematics in school. Pages of exercises were tedious and dull ... I stared out the window and let my mind wander. Sometimes I tried to puzzle something out .... Might the square root of 2 eventually be periodic if you write it out in base 12 instead of base 10? How many ways are there to fold a map into sixteenths, in quarters each way?" This spirit of inquiry pervaded the many issues of *Quantum* and stimulated readers to invent their own questions and allow their minds to explore.

The history of *Quantum* is worth recounting briefly. Arthur Eisenkraft was first introduced to the Russian publication, *Kvant*, by his friend Sergey Krotov of the former Soviet Union during their years as Academic Directors for their respective Physics Olympiad teams. The problems, articles, and humor in *Kvant* seemed like something that could be imported and massaged for United States audiences. Lots of interested people stepped up to the plate. Bill Aldridge, then Executive Director of NSTA, led the charge to create a magazine of "the highest quality." Bill came through on his commitment. He enlisted the help of Sheldon Glashow, a Nobel Laureate in physics, William Thurston, a Fields medalist in mathematics, and Yuri Ossipyan, vice-president of the Academy of Sciences of the USSR, to launch the magazine. Edward Lozansky served as an international consultant and Tim Weber took on the responsibility of managing editor. NSTA, under Bill Aldridge's leadership, with financial support from the National Science Foundation (NSF), committed resources to insuring that *Quantum* met the needs of our intended audience. He also brought the American Association of Physics Teachers (AAPT) and the National Council of Teachers of Mathematics (NCTM) on board. Larry Kirkpatrick and Mark Saul became the field editors for physics and mathematics, respectively.

Arthur Eisenkraft and Larry Kirkpatrick collaborated from the beginning on the physics Contest Problems. The physics was great, as were the literary quotes accompanying each, but better illustrations were needed. Tomas Bunk, a professional cartoonist with credits including *MAD* magazine and Garbage Pail Kids, was approached. His first reaction was "But I don't know physics." Arthur responded, "Well, that might be an asset. The next article is about light. How would you draw light?" Tomas replied, "I guess like a super-hero because it travels so fast." And so the collaboration began. Of course, as Tomas learned enough physics to illustrate that first picture, Arthur learned about art. The illustration had light at the circus moving through hoops, ricocheting off mirrors, and darting through a water-filled aquarium. Arthur loved the sketch but the light did not always obey the laws of physics. What to do? Tomas explained that to shift the path of light would require a new illustration because the balance and structure of the art would be off in this portrayal. Arthur decided it would be better to state below the illustration, "Light is misbehaving in this picture. Can you find where?" The cartoon became another dimension to the Contest Problem. As the Contest Problems evolved, so did Tomas's illustrations. They began to take on political commentary, historical ideas, and larger issues of philosophy while always providing insights into the physics with whimsy and humor.

Quantoons adds a feature that was not present in the original *Quantum* series. Each illustration now has a brief commentary by Tomas Bunk. This peek at the creative mind of a visual artist not only provides insight into how Tomas views the world, but also how people who are not trained in physics can appreciate the world of science and make it their own.

Some readers will move right to the cartoons while others will begin with the quotes. These readers may then be curious enough to check out the physics text that introduces the topic. Still others will be intrigued by the complexity of some of the physics problems, be tempted to invent a solution, and then may reward themselves with an investigation of the illustration. Our hope is that you will find your own, personal way to enjoy *Quantoons* and better appreciate the world we share.

—Arthur Eisenkraft and
Larry D. Kirkpatrick

This book is dedicated to . . .

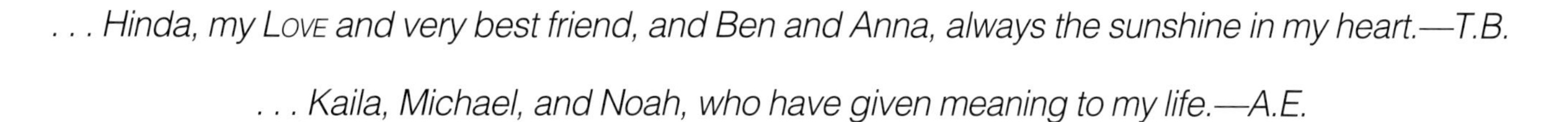

. . . Hinda, my LOVE and very best friend, and Ben and Anna, always the sunshine in my heart.—T.B.

. . . Kaila, Michael, and Noah, who have given meaning to my life.—A.E.

. . . Karen, Jennifer, Monica, and Peter, who have taught me how to view the world.—L.D.K.

QUANtoons

A snail that moves like light

"The cause is hidden but the effect is known."
—*Ovid,* Metamorphoses

MOST OF OUR READERS know that light bouncing off a mirror travels along a path that can be adequately described as "the angle of incidence equals the angle of reflection." Light traveling from a point in air to a point in water is certainly more complicated. In this case, the light bends (refracts) at the boundary between the two surfaces. The amount of bending is a property of the water and the color of the light. Light entering other transparent substances, like quartz or diamond, refract by different amounts. Willebrord Snell in 1621 was able to give a mathematical description of the behavior of light, which is now known as Snell's law:

$$n_1 \sin \theta_1 = n_2 \sin \theta_2,$$

where n_1 and n_2 are the indices of refraction. We can see that if the light enters water ($n = 1.33$) from air ($n = 1.00$) at an angle of 30°, the angle in water would be 22°:

$$n_1 \sin \theta_1 = n_2 \sin \theta_2,$$
$$1.00 \sin 30° = 1.33 \sin \theta_2,$$
$$\theta_2 = 22°.$$

Measuring the angle of refraction is one way to tell whether that's a diamond or a piece of glass in that ring you bought.

What fascinates many people about the study of physics is the alternative ways of explaining phenomena. The great mathematician Pierre de Fermat recognized (in 1657) that the path of light is the path that requires the least time.[1] If you try all possible paths from the light source *A* to the object *B* after they hit the mirror, you'll find that the shortest path, and so the quickest, is the path through point *D* (fig. 1 on the next page), where the angle of incidence equals the angle of reflection.

[1]The "extremum path."

Light with its human face is having a super blast speeding in a flash through space and objects, breaking mirrors and here and there some rules, shooting through a fishbowl—surprising the wet inhabitants—and through a solid glass prism without getting a headache. This spectacular high-speed performance is taking place in a cosmic circus tent where the alligator and his dog assistant are in charge of the modest flashlight source. In the audience we see the professional science community verifying the correct course and angles of light but not being very impressed by the seriousness of the performance. But that does not bother our lightweight space cadet, who thinks that coming from a flashlight he is Flash Gordon himself.

—T.B.

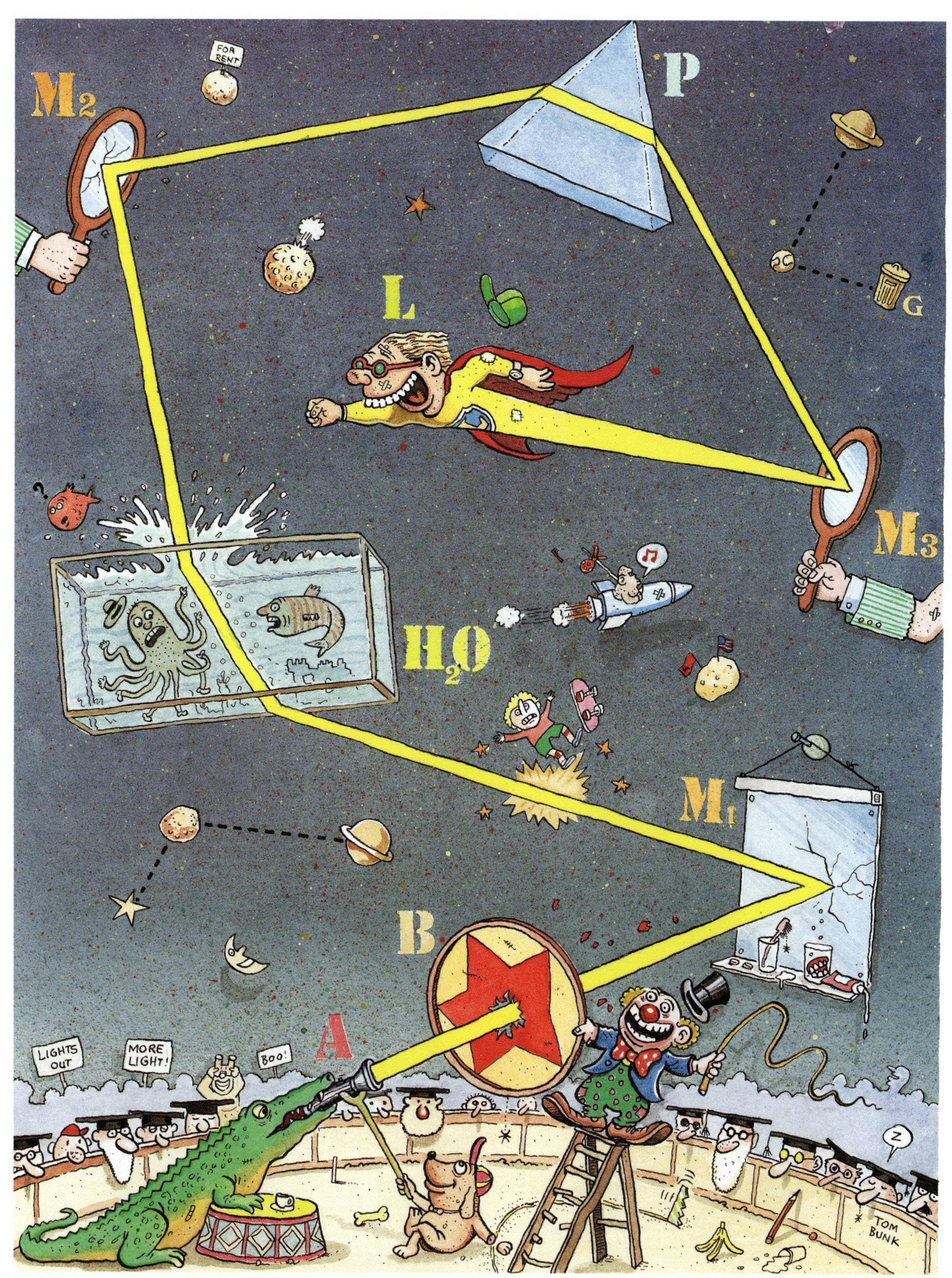

Light is bending the rules a bit here. (Can you see where?)

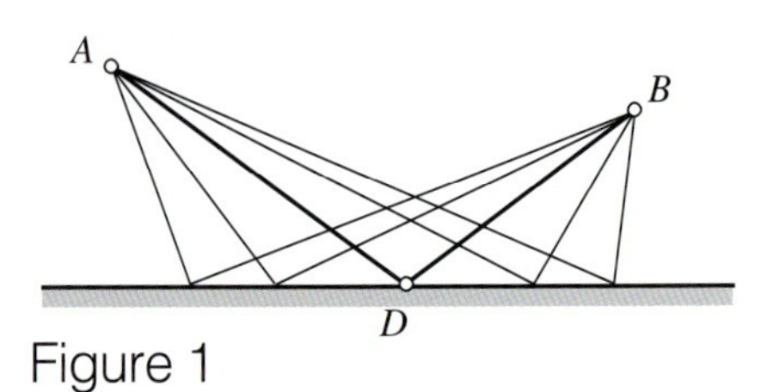

Figure 1

You can demonstrate this for yourself by drawing lots of paths and measuring them. You can also prove it with some simple geometry or by using some calculus.

Fermat's theorem is also valid for refraction: the path light takes when it passes from air to water must be the path requiring the least time. In this case least time is not identical to least distance, since light travels more slowly in water that in air. The speed of light in a substance is equal to the speed of light in a vacuum divided by the substance's index of refraction n.

Proving that the path of the light is the quickest one takes some ingenuity. You can draw lots of paths of light traveling from point A in air to point B in water (fig. 2). You can then measure the lengths of the lines in air and water. But Fermat's theorem states that the path should take the least *time*, not the least distance. We can multiply the lengths in water by 1.33, since the

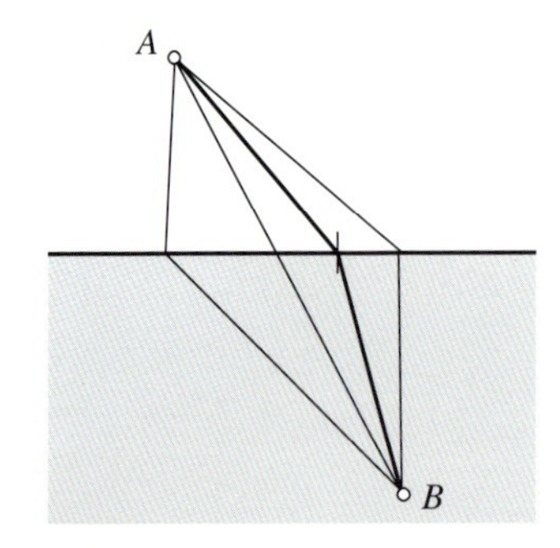

Figure 2

light takes longer to travel in water by a factor of 1.33. Then add this distance to the distance in air. The path that minimizes this sum is the path the light takes. And—guess what? It's the same path described by Snell's law! Those of you who have some calculus background can prove it mathematically.

Leaving light behind, we enter the world of slow-moving mollusks to find our contest problem. A snail must get from one corner of a room (dimensions 5 m × 10 m × 15 m) to the diagonally opposite corner in the least time. The snail can walk on any of the four walls but may not walk on the floor or ceiling. What is the path that the snail should take? In part B of the contest problem, for our more advanced readers, the snail finds that the 15 meter wall that must be traveled is sticky—

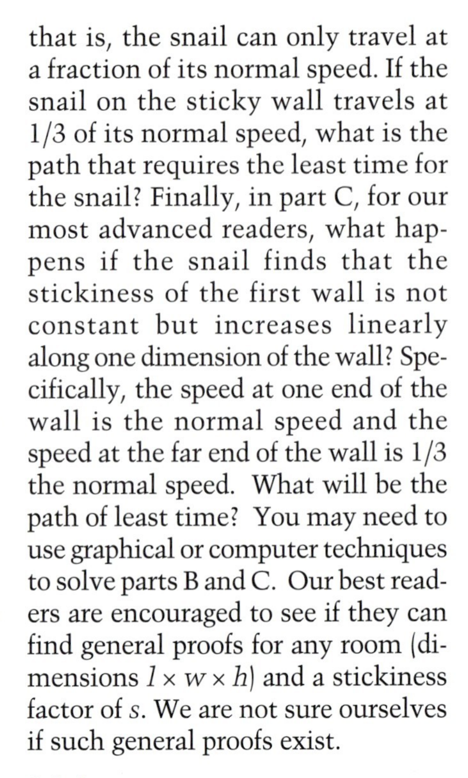

that is, the snail can only travel at a fraction of its normal speed. If the snail on the sticky wall travels at 1/3 of its normal speed, what is the path that requires the least time for the snail? Finally, in part C, for our most advanced readers, what happens if the snail finds that the stickiness of the first wall is not constant but increases linearly along one dimension of the wall? Specifically, the speed at one end of the wall is the normal speed and the speed at the far end of the wall is 1/3 the normal speed. What will be the path of least time? You may need to use graphical or computer techniques to solve parts B and C. Our best readers are encouraged to see if they can find general proofs for any room (dimensions $l \times w \times h$) and a stickiness factor of s. We are not sure ourselves if such general proofs exist.

Solution

You were asked to help a snail find the quickest path from one corner of a room to a diagonally opposite corner.

In the first case, in which all walls were identical and the dimensions of the room were 5 × 10 × 15, there are at least three ways to solve the problem. The first is to choose different crossover points at the edge between the two walls and calculate the total distance that the snail travels. This numerical method may appear to be tedious, but it will actually converge on the correct solution quickly. A second method is to call the height of the crossover point x, write the total distance traveled in terms of x, and differentiate. By setting the derivative equal to zero, the minimum distance will be revealed as the solution to the equation. The third method is the elegant solution. In this case, the wall is opened up. The room is now a large rectangle of dimensions 25 × 5. The shortest distance will be the diagonal connecting the two corners of the rectangle. If the snail starts at the lower corner of the 15-meter wall, the crossover point can be found by using similar triangles. The crossover point is

$$\frac{x}{15} = \frac{5 - x}{10},$$
$$x = 3.$$

In the second case, one of the walls was declared "sticky," meaning that the snail could travel at only 1/3 of its speed on this wall. Unlike the first case, the shortest distance is no longer the shortest time! Since the snail travels at different speeds on the two walls, the quickest path will be the one where the snail travels a greater distance on the faster wall. Once again, the straightforward but tedious solution would be to assign the variable x to the crossover point, write an equation that describes all paths in terms of x, and the minimum time will be revealed.

The more elegant solution in this case is to realize that light always takes the least time to travel, and that this snail traveling on a sticky wall is like light traveling in a slower medium. We then recognize that the solution will be Snell's law (or, if you'll forgive us, "Snail's law"). Even with this knowledge, we are faced with a fourth-order equation, which we choose to solve by numerical techniques:

$$n_1 \sin \theta_1 = n_2 \sin \theta_2.$$

Since the stickiness factor is 3, then $n_1 = 3$ and $n_2 = 1$, and it follows that

$$3\frac{x}{\sqrt{15^2 + x^2}} = \frac{5 - x}{\sqrt{(5 - x)^2 + 10^2}}.$$

We'll try different values of x and see if the value of the left side of the equation is equal to the value of the right side.

x	left side	right side
1	0.1996	0.3714
2	0.3965	0.2873
1.5	0.2985	0.3304
1.7	0.3378	0.3134
1.6	0.3182	0.3219
1.63	0.3241	0.3194
1.62	0.3221	0.3202

This method can give us any accuracy we desire. It would certainly be easier to plug the equations into a spreadsheet program and have all values given "instantly."

The third part of the problem, to solve for a wall whose stickiness varies along one dimension, was solved by Jason Jacobs of Harvard University. We will leave this problem as a tease.

The leaky pendulum

"Would you like to swing on a star,
Carry moonbeams home in a jar . . .?"
—Johnny Burke[1]

THE MOTION OF SIMPLE pendulums has played an interesting role in physics and technology. Galileo is reported to have made an important discovery about the motion of a pendulum while watching the swinging of a chandelier during a church service. He discovered that the period of a simple pendulum is (almost) independent of the amplitude of its swing. This led to the use of pendulums to measure time intervals and the development of accurate clocks.

Any mass hanging from a pivot is a pendulum. An orangutan swinging by one arm from a branch is one example. Your leg pivoting from your hip as you walk is another. Both of these examples are fairly complicated because the objects aren't rigid structures and the mass is distributed in such a complicated fashion. As physics students, you should practice looking at complicated situations and finding ways to simplify them. What do you imagine would be the simplest pendulum? Certainly not the chandelier that so intrigued Galileo.

The simplest pendulum would probably be a compact mass attached to a long string. Physicists call this "the simple pendulum." If we let a simple pendulum swing through small angles, we find that its period T (the time to complete one cycle) is given by $T = 2\pi\sqrt{L/g}$, where L is the length from the pivot to the center of mass of the pendulum bob and g is the acceleration due to gravity. This is similar to the solution of many other problems in physics where objects repeat their motion over and over again. Notice that the period doesn't depend on the mass of the bob. Does this surprise you? It should! And we recommend that you plan on devoting

[1]From "Swinging on a Star," music by Jimmy Van Heusen, which was the Academy Award–winning song in 1944 (sung by Bing Crosby in *Going My Way*).

Archimedes is taking a nice hot bath, hoping to get enlightened and inspired, finding brilliant solutions as he usually does in his hot tub, and when it occurs we find him running full of excitement through nighttime alleys shouting, "Eureka!!!" This time his bathtub is swinging like a pendulum from point A to point B, demolishing both a bit. A number of observers hide safely behind the entrance, among them Einstein showing his mischievous tongue. The swinging tub is leaking, causing the tub to loose mass and create a minor flood on the floor beneath. Already the cleaning lady is busy mopping up the master's mess and the plumber is rushing by, while higher up Icarus is getting ready to land and cool off his wings before he takes off again to reach the sun.

—T.B.

Y
A
B
H2O
TOM
BUNK

some quiet time to wondering about the insignificance of the mass of a pendulum.

In this month's contest problem, we'll study the period of a simple pendulum that is slowly losing mass—the so-called leaky pendulum. The pendulum bob is a cubical container of negligible mass with an edge length of $2a$. It is initially filled with a fluid of mass M_0. The cube is tied to a very light string to form a simple pendulum of length L_0. The fluid flows through a small hole in the bottom of the cube at a constant rate r. At any time t the level of the fluid in the container is l and the length L of the pendulum is measured relative to the instantaneous center of mass.

Part A: Find the period as a function of time for small angular displacements.

Part B: Sketch a graph of the period as a function of time, being sure to label the endpoints of the graph.

Part C: How do your answers change if the mass of the container is also M_0 (the same as the initial mass of the fluid) and the center of mass of the container is located at its geometrical center?

Solution

Very good solutions to this problem were submitted by Ben Davenport (Charlotte, North Carolina), Samuel Dorsett (Mitchell, Indiana), and Jesse Tseng (Little Rock, Arkansas). We also received a packet of solutions from Campbell High School in Campbell, Missouri.

We begin by calculating the length of time t_{max} it takes for the fluid to run out of the container. This is given by the initial mass M_0 of the fluid divided by the rate r at which the fluid runs out. Therefore,

$$t_{max} = \frac{M_0}{r} = \frac{8a^3\rho}{r},$$

where ρ is the density of the fluid. In order to avoid writing this set of constants repeatedly, let's define a new dimensionless, timelike variable τ by the relationship

$$\tau = \frac{t}{t_{max}}, \quad 0 \le \tau \le 1.$$

In terms of τ, the mass M remaining as a function of time is given by

$$M(t) = M_0(1 - \tau)$$

and the height l of the fluid as a function of time is given by

$$l(t) = 2a(1 - \tau).$$

Since the center of mass of the fluid is located at its geometrical center, the length of the pendulum is given by

$$L(t) = L_0 + a - \frac{l(t)}{2} = L_0 + a\tau.$$

This expression could have been written down directly by realizing that the center of mass moves from L_0 to $L_0 + a$ at a uniform rate during the "time" τ.

Under the simplifying assumptions of this problem, the period of the pendulum is

$$T = 2\pi\sqrt{\frac{L_0 + a\tau}{g}} = T_0\sqrt{1 + \frac{a\tau}{L_0}},$$

where $T = 2\pi\sqrt{L_0/g}$ is the initial period of the pendulum. The graph of the period versus time is shown as the upper curve in figure 2 for the case $L_0 = 2a$. Although the curve appears to be straight, it actually has a slight curve due to the square root. Note that the period is not defined after the fluid has all run out as the pendulum no longer has any mass.

When the container has a mass M_0, we must calculate the combined center of mass x_{cm} of the remaining fluid and the mass of the container

$$x_{cm} = \frac{m_c x_c + m_f x_f}{m_c + m_f},$$

where the subscripts "c" and "f" refer to the container and fluid, respectively. If we choose to calculate the center of mass x_{cm} relative to the center of the container, we have $x_c = 0$ and

$$x_{cm} = \frac{M_0(1-\tau)a\tau}{M_0 + M_0(1-\tau)} = a\frac{\tau(1-\tau)}{2-\tau}.$$

Therefore, the period of the pendulum as a function of time is

$$T = T_0\sqrt{1 + \frac{a\tau(1-\tau)}{L_0(2-\tau)}}.$$

This function is shown by the lower curve in figure 2. Note that the maximum period occurs about 60% of the way through the time period and that the maximum period is less than for the case with the massless container. Note also that the period returns to its initial value when the fluid has completely run out. You probably anticipated this because the center of mass must

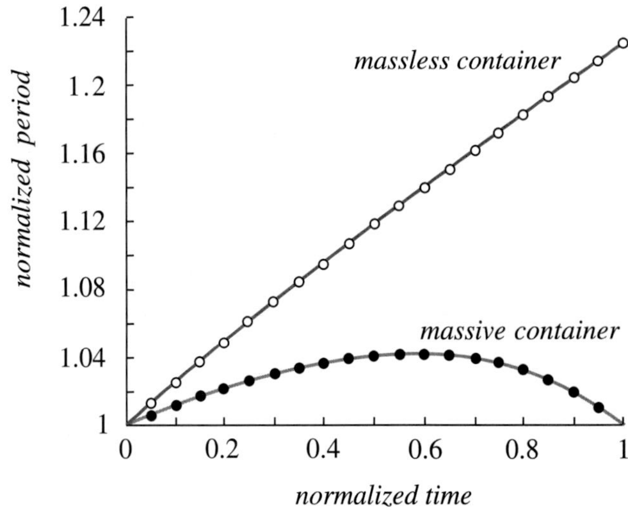

Figure 2

return to the center of the container.

Those of you who know some calculus can differentiate the expression for the center of mass to find that it reaches its maximum value when

$$\tau = 2 - \sqrt{2} = 0.586.$$

Notice that this maximum occurs 58.6% of the way through the time period independent of the rate of flow. This value can now be substituted back into the expression for the period to find its maximum value

$$T_{max} = T_0\sqrt{1 + \frac{\left(3 - 2\sqrt{2}\right)a}{L_0}}.$$

We hope you enjoyed thinking about the leaky pendulum. ◘

What goes up . . .

"Whether this agent [of gravity] be material or immaterial, I have left to the consideration of my readers."
—Isaac Newton

EVERY SCHOOL-AGE KID HAS heard the prediction "What goes up must come down." Many kids challenge it with a question relating to helium balloons. Some inquisitive and persistent kids have wondered what would happen if that object thrown up was going terrifically fast. Would it still come down?

The first step that a physicist will take in analyzing what happens to an object thrown straight up at a high speed is to analyze what happens if it is thrown at a low speed. The object goes up, reaches a peak, and descends. If we increase the throwing speed, the object goes higher, stays in the air longer, and then returns to the ground.

Now we have to refine this description. One way to do that is to quantify it. We owe our ability to write these motion equations to Galileo Galilei. Galileo was the first person to make measurements of an object's motion and record them. He was also the first person to describe this rising and falling object in a mathematical language. What Galileo found, and you can also discover, is that the falling object changes its speed. When the object descends it goes faster and faster. Well, everybody knows that.

But not everybody knows this: if I look at the object every 2 meters, does its speed change by the same amount? Specifically, if the speed increases to 6 m/s in the first 2 meters, will it increase to 12 m/s in the next 2 meters? The answer is no. So, how about this: if I look at the object every 2 seconds, does its speed change by the same amount? The answer is yes. Galileo defined this change in speed for every unit of time as the acceleration of an object. Cars accelerate when they change their speed. A runner changes her speed as she races. An object accelerates when it's dropped. What's special about the falling object is that the acceleration

A huge sophisticated mechanical construction, operated by scientists, clowns, and technicians, was specially created to kick a ball as high into space as possible to observe how long it will take the ball to get back to Earth. This ball is shooting all the way to Mars, so it doesn't look like it's coming down any time soon. Riding on the ball is a soccer player listening to the radio. The excited audience in the flying box is enthusiastic about the mighty kick, while above the image we see the divine creator wondering what creature is shooting a ball all the way into his cosmic playground. There seems to be no place safe from human curiosity. On the Moon they are already playng golf while another creature is dusting off planet Mars, a horse is operating a satelite, and a diver is hunting stars. By the way, the ball was never seen again.

—T.B.

G
MARS
M
B
A
Y
Tom Bunk

is a constant. Very close to our Earth, an object falling will change its speed by 9.8 m/s each and every second. It's very hard for a car to have a constant acceleration. It's almost impossible for a human to have a constant acceleration. But every object dropped close to the Earth (and not appreciably affected by the air—no feathers, please) has a perfectly constant acceleration of 9.8 m/s every second.

This acceleration of 9.8 m/s every second works on the way up as well as on the way down. If you throw a ball up at 50 m/s (that's about 100 miles per hour), its speed will be only 40 m/s after the first second (it's really 40.2 m/s but we're going to use the approximation of 10 m/s every second instead of 9.8 m/s every second because we can subtract 10 faster than we can subtract 9.8). After the second second its speed is only 30 m/s; after the third second, 20 m/s; after the fourth second, 10 m/s; and after the fifth second, its speed is 0 m/s. Zero m/s is not moving at all. The ball is stopped in midair. In the sixth second, its speed still decreases by 10 m/s, so its new speed will be –10 m/s. The minus sign is the mathematical way of informing us that the ball is moving down.

From this simple analysis we should conclude that all objects that go up must come down. No matter what speed the ball starts with, it will eventually reach 0 m/s and start heading down. Oh, but that's only if the acceleration is constant at 9.8 m/s every second. Galileo found this to be true *near* the Earth. It's not true as we move far enough away from the Earth. As an object moves farther and farther from the Earth, the acceleration toward the Earth gets smaller and smaller. The acceleration of the Moon toward the Earth is only 0.0027 m/s every second. That's just the acceleration needed for the Moon to stay in a fixed orbit about the Earth. In this case, the change of 0.0027 m/s every second isn't a change in speed but a change in the direction of the velocity. If the Moon had no change in velocity, it would continue traveling on a straight path away from the Earth. The change in velocity keeps the Moon in orbit about the Earth. Pick up a physics textbook for a more complete (and mathematical) treatment of this.

If the acceleration decreases as we increase the distance from the Earth, then there is a possibility that the resulting decreases in speed will never slow the object down to zero. That means that the object will never return to the Earth. The object escapes! The speed that an object needs to escape from the Earth can be derived most simply by using the conservation of energy.

The more challenging contest problems that follow are taken from the 1985 International Physics Olympiad, which was held in Portoroz, Yugoslavia. To simplify the problem, assume that all the planets revolve in circles about the Sun. Neglect air resistance and the rotation of the Earth on its axis. In parts B–D, neglect the energy used in escaping from the Earth's gravitational field. Assume that the orbital velocity of the Earth is 30 km/s and that the ratio of the distances of the Earth and Mars from the Sun is 2/3.

A. For our beginning physics students, derive the equations and values for the velocity required for a satellite to escape from Earth.

B. For those of you who wish to attempt a bit more, derive the equations and values for the minimum velocity required for a satellite launched from Earth to escape the solar system.

C. For our more advanced readers, suppose that the satellite has been launched with a velocity less than this escape velocity. Determine the velocity of the satellite when it crosses the orbit of Mars. Assume that Mars is not near the point of crossing.

D. Finally, let the satellite enter the gravitational field of Mars. Find the minimum velocity needed for the satellite to be launched from the Earth and escape from the solar system using the gravitational field of Mars. (This is often described as using Mars as a "slingshot.")

Solution

A. In order for a satellite probe to escape from the Earth, the sum of the kinetic energy and potential energy must be greater than or equal to zero:

$$\frac{1}{2}m_p v_p^2 - \frac{Gm_p m_E}{R_O} = 0,$$

$$v_p = \sqrt{\frac{2Gm_E}{R_O}},$$

where m_p is the mass of the probe, v_p is its velocity, m_E is the mass of the Earth, and R_O is the radius of the Earth.

B. The condition for the probe to escape from the solar system is similar to that of escaping the Earth. In this case, however, the potential energy function is related to the mass of the Sun:

$$v_a = \sqrt{\frac{2Gm_S}{R_E}},$$

where m_S is the mass of the Sun, R_E is the Earth–Sun distance, and v_a is the probe's velocity relative to the Sun. The Earth's velocity about the Sun can be determined by recognizing that the gravitational attraction of the Sun holds the Earth in orbit:

$$\frac{m_E v_E^2}{R_E} = \frac{Gm_E m_S}{R_E^2},$$

$$v_E = \sqrt{\frac{Gm_S}{R_E}}.$$

The escape velocity can now be written

$$v_a = v_E\sqrt{2}.$$

Since the probe can be shot in the direction of the Earth's orbital velocity, the required velocity is diminished by the value of the Earth's velocity:

$$v_a' = v_E\left(\sqrt{2}-1\right) = 12.3 \text{ km/s},$$

where v_a' is the probe's velocity relative to the Earth.

C. Let v_b and v_b' be the velocities of launching the probe in the Sun's

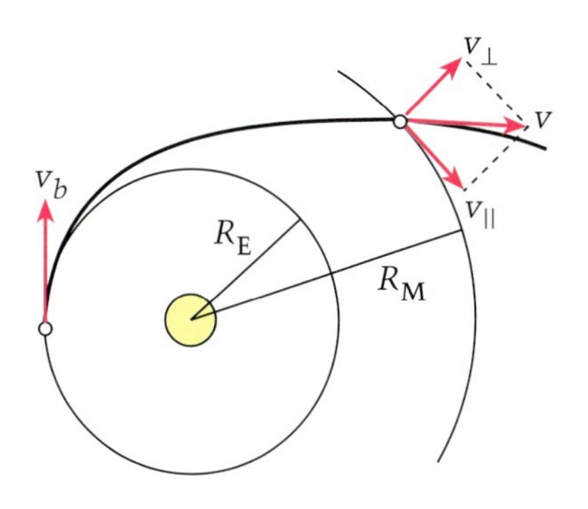

Figure 1

and Earth's system of reference, respectively. Then $v_b = v_b' + v_E$ (see part B). From the conservation of angular momentum of the probe

$$m_p v_b R_E = m_p v_{||} R_M$$

(R_M is the Mars–Sun distance) and the conservation of energy

$$\frac{m_p v_b^2}{2} - \frac{G m_p m_S}{R_E} = \frac{m_p\left(v_{||}^2 + v_\perp^2\right)}{2} - \frac{G m_p m_S}{R_M},$$

we get, for the parallel component of the velocity (fig. 1),

$$v_{||} = (v_b' + v_E) r$$

and, for the perpendicular component,

$$v_\perp = \sqrt{\left(v_b' + v_E\right)^2\left(1 - r^2\right) - 2v_E^2(1 - r)},$$

where $r = R_E/R_M$.

D. The minimum velocity of the probe in the Mars system of reference to escape from the solar system is $v_p'' = v_M\left(\sqrt{2} - 1\right)$ in the direction parallel to the orbit of Mars (v_M is the velocity of Mars around the Sun). The role of Mars is thus to change the velocity of the probe so that it leaves its gravitational field with this velocity.

In the Mars system, the energy of the probe is conserved. That is not true, however, in the Sun's system, in which this encounter can be considered an elastic collision between Mars and the probe. The velocity of the probe before it enters the gravitational field of Mars is therefore, in the Mars system, equal to the velocity with which the probe leaves its gravitational field. The components of the former velocity are $v_\perp'' = v_\perp$ and $v_{||}'' = v_{||} - v_M$, so

$$\begin{aligned} v'' &= \sqrt{v_{||}''^2 + v_\perp''^2} \\ &= \sqrt{v_\perp^2 + \left(v_{||} - v_M\right)^2} \\ &= v_S''. \end{aligned}$$

Using the expressions for $v_{||}$ and $v_\perp$ from part C, we can now find the relation between the launching velocity v_b' from the Earth and the velocity $v_p'' = v_M\left(\sqrt{2} - 1\right)$:

$$\begin{aligned} &\left(v_b' + v_E\right)^2\left(1 - r^2\right) - 2v_E^2(1 - r) + v_M^2 \\ &\quad + \left(v_b' + v_E\right)^2 r^2 - 2v_M\left(v_b' + v_E\right) r \\ &= v_M^2\left(3 - 2\sqrt{2}\right). \end{aligned}$$

The velocity of Mars around the Sun is

$$v_M = \sqrt{\frac{G m_M}{R_M}} = \sqrt{r}\, v_E,$$

and the equation for v_b' takes the form

$$\begin{aligned} &\left(v_b' + v_E\right)^2 - 2r\sqrt{r}\, v_E\left(v_b' + v_E\right) \\ &\quad + \left(2\sqrt{2}r - 2\right)v_E^2 = 0. \end{aligned}$$

The physically relevant solution is

$$\begin{aligned} v_b' &= v_E\left(r\sqrt{r} - 1 + \sqrt{r^3 + 2 - 2\sqrt{2}r}\right) \\ &= 5.5 \text{ km/s}. \end{aligned}$$

The clamshell mirrors

"Mirror facing facing mirror
Doubles—exquisite effect;
Between them in the shadow stands a
Crystal cube which will reflect."
—Johann Wolfgang von Goethe

WHAT IS IT THAT CAUSES our fascination of worlds within worlds? A picture in a picture in a picture absorbs us as we wonder what the different worlds are all about. Readers are encouraged to spend a delightful afternoon looking at the drawings of M. C. Escher or reading Lewis Carroll's *Alice's Adventures in Wonderland* to see how two people were also intrigued by these ideas.

A simpler set of worlds occurs with two plane mirrors. Can you remember your last visit to the beauty salon or barber shop, where you observed the infinite number of images of yourself formed by mirrors facing each other? Each mirror produced images of everything in front of it—including the images formed by the other mirror. In this case the images were all of the same size. They just appeared to get smaller and smaller because they were farther and farther away from you. What happened when you moved between the mirrors? How did the images move? How does the number of images vary with the angle between the mirrors? What does the speed of light have to do with how the images move when you raise your hand?

Although you may have less experience with curved mirrors, you probably expect that similar things would happen. A very popular physics toy consists of two concave spherical mirrors facing each other like the shells of a clam, as shown in the photograph on page 16. If a coin or button is placed on the surface of the lower mirror, its image appears in the hole at the center of the top mirror. The image looks so real that people will try to push the button or pick up the coin. It's fun to watch their faces as they discover that there is actually nothing there!

The separation of the mirrors in this toy has been carefully chosen so that the real image appears in the hole as if it were sitting on a clear portion of the upper mirror. The image is the same size as the object.

A. Begin by finding the separation of the mirrors that corresponds

The show put on by the grand escape artist is taking place outside the city gates—for safety reasons the police have closed the roads. The audience around the event doesn't seem particularly interested in the show itself—they're looking into the mirrors. While the seriously chained artist performs with utter elegance his escape with the help of the concave mirror, through a tiny hole on top of another concave mirror, we see another person running through a regular mirror into another reality. An old man who does not like his mirror image solves his problem with a hammer, and a cubist woman improves her looks, while our hero flies up into the air, enjoying his freedom. There is nothing like a nice illusion instead of being chained down to reality.

—T.B.

I
G
M2
O
66
Tom Bunk
M1

Photo courtesy of Larry Kirkpatrick

to the photograph. Draw a ray diagram to show how the light reflects from each surface to form the image.

B. If you haven't already discovered a second separation that also produces a real image in the hole with unit magnification, find it and draw the corresponding ray diagram. Are there any differences in the images produced in the two cases?

C. In fact, there are an infinite number of solutions. Our more advanced readers may want to find a few of them to discover the general method for obtaining additional solutions. As this can be done in at least two different ways, this problem should give you a challenge and, we hope, much enjoyment.

Solution

In this problem we asked you to find the separation of two spherical mirrors that would produce a real image in a hole in the upper mirror. Furthermore, the image was to be the same size as the object. If you want to experiment with the mirrors, they are available from Edmund Scientifics (www.scientificsonline.com) as the Optic Mirage (item 3072381).

Since both mirrors are required to obtain the image, we know that the image is formed by two reflections—one from the upper mirror followed by one from the lower mirror. Since the only dimension in the problem is the focal length of the mirrors, let's choose to express all other distances in units of the focal length. If we assume that the mirrors are separated by nf, the object distance s_1 for the image produced by the upper mirror (mirror 1) is also nf. The image distance s_1' can be found from the mirror formula:

$$\frac{1}{s_1} + \frac{1}{s_1'} = \frac{1}{f}.$$

Therefore,

$$s_1' = \frac{nf}{n-1}.$$

For $n > 1$, the image will be real and located below the surface of the lower mirror (mirror 2). The image acts as an object for the lower mirror with an object distance

$$s_2 = nf - s_1' = \frac{n(n-2)}{n-1} f.$$

Using the mirror formula a second time, we locate the image formed by the lower mirror:

$$s_2' = f \frac{n(n-2)}{n^2 - 3n + 1}.$$

However, the conditions of the problem require that

$$s_2' = nf.$$

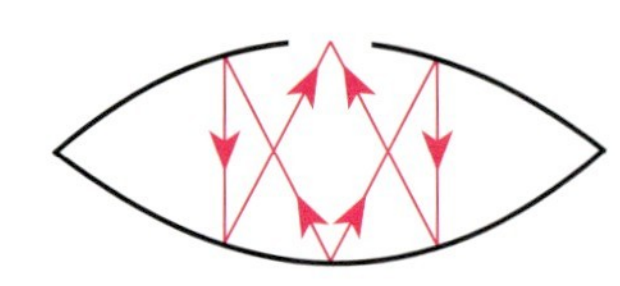

Figure 1

This yields a quadratic equation in n with the roots $n = 1$ and $n = 3$.

The solution $n = 1$ is the one actually used in the Optic Mirage. If we go back and look at our mathematics for this case, we find that the image produced by the upper mirror is located at infinity. Therefore, the rays forming the image leave the upper mirror parallel to each other, as shown in figure 1. These parallel rays are focused by the lower mirror to form an image at its focal point, which is located in the hole in the upper mirror. Notice the symmetry of the problem. Interchanging the image and the object has no effect.

If we take a few liberties with infinities, we can use the formula for the magnification of the image

$$m = \frac{-s'}{s}$$

to find the magnification at each stage and then multiply them together to get the overall magnification. For this case, we obtain $m = -1$, indicating that the image is inverted. Observation of the image shows that this is correct.

The solution $n = 3$ is a surprise to many students. The ray diagram for this case is shown in figure 2. The upper mirror forms a real image midway between the mirrors that has a magnification of $-1/2$. The

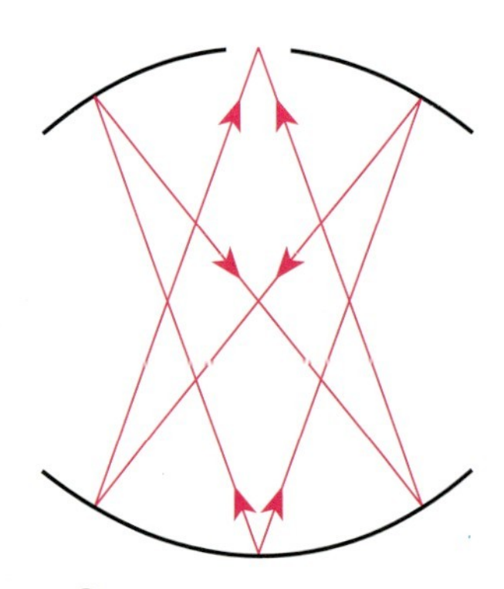

Figure 2

lower mirror then forms an image in the hole with a magnification of –2. Therefore, the overall magnification is +1 and the image is erect. Notice once again the symmetry of the problem. The upper half of figure 2 is just a mirror reflection of the lower half. This must be the case for all solutions to this problem.

Additional solutions can be found by letting the light reflect two, three, four, or more times from each mirror. In each case, symmetry requires that the rays be parallel, forming an image at infinity, or they form an image midway between the mirrors. For more information about these solutions, we refer you to the article by Andrzej Sieradzan in *The Physics Teacher* (November 1990, page 534).

Shake, rattle, and roll

"She stood in silence, listning to the voices of the ground . . ."
—William Blake, "The Book of Thel"

ONE PERSON DESCRIBED how the bedroom wall moved across the room. Another watched as a huge wave of concrete traveled along the highway. We all saw the massive destruction when one bridge roadway collapsed on top of another. The earthquake in the San Francisco area that coincided with the 1989 World Series gave us a glimpse of the power and energy in our planet.

In the fury of the destruction, the Earth is whispering secrets about its composition. The Earth is not solid rock. The Earth is not of uniform density. Longitudinal and transverse waves, called *P* and *S* waves, travel through the Earth as a result of the quake. The differences in *P* and *S* wave behaviors can give us clues about the structure of the Earth while also allowing us to locate the epicenter of the quake.

Although the speeds of the *P* and *S* waves vary within the Earth, the *P* waves always travel faster than the *S* waves. This fact gives us the ability to locate the epicenter of the quake. By knowing the relative speeds of the *P* and *S* waves and measuring the delay in the arrival of the *S* waves, we can determine the distance from the epicenter. Here's an analogy. If you run at 3 m/s and a friend walks at 1 m/s, you will always arrive at a given location before your friend. If you arrive 10 seconds earlier, the distance traveled was 15 meters. If you arrive 20 seconds earlier, the distance traveled was 30 meters.

Let's assume that the epicenter is near the Earth's surface and that the *P* and *S* waves have constant but unequal velocities. If at one location on the Earth the waves arrive with a time difference of 2 minutes, we know that the epicenter of the quake must be situated a specified distance from this location. But in which direction? We don't know. We therefore trace the circumference of a circle on the surface with

Here we are looking at the high drama of a tremendous earthquake, causing tsunamis to splash enormous waves in all directions, and a gargantuan volcanic eruption, spitting out even the devil himself above a town surrounded by a calm agricultural landscape. In front we have numerous scientists with their top-notch equipment trying to determine the Earth's innermost secrets and where the epicenter of the quake might be. Some measure the S waves and others the P waves, some take a look at the fiery inside and some listen to it. In the very front is the boss in charge, looking like some Old Testament prophet investigating with his assistants the seismographic chart, which looks like a Torah. Whether Mother Earth will reveal her deepest secrets is not clear yet, but the quest to learn where we come from and what we are made of will keep the curious among us on our toes and off the streets for a long time.

—T.B.

THE WORLD
THE EARTH
YOU ARE HERE!
THE EARTH
CRUST
LITHOSPHERE
ASTHENOSPHERE
MANTLE
OUTER CORE
INNER CORE
TOM BUNK

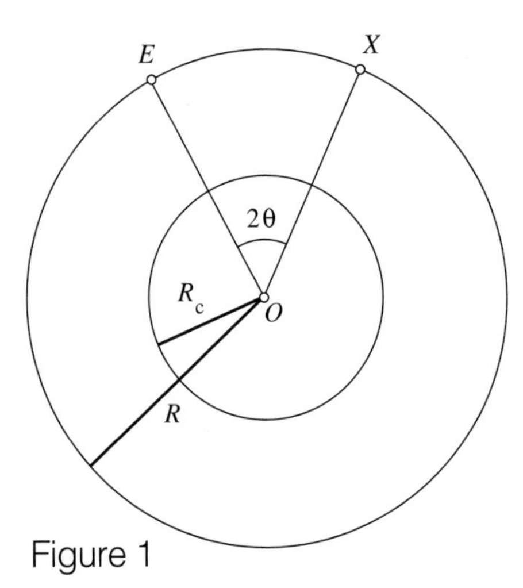

Figure 1

a radius specified by this time delay. The epicenter can be located on any part of this circumference. If we have a second location with a (different) time delay, this will provide us with a second circle. A third location and a third circle will uniquely determine the actual point location of the epicenter.

The *P* waves are able to travel through solids and liquids, while the *S* waves travel only through solids. The *P* waves arrive at locations on the opposite side of the Earth; the *S* waves do not. This information leads us to conclude that a portion of the interior of the Earth is liquid. By carefully observing where the *P* waves travel and where the *S* waves do not, we can infer more about the size of this liquid core of the Earth.

More curious is the observation that there are positions on the Earth where neither the *P* nor the *S* waves arrive. These shadow zones are somehow protected from disturbances at some locations. What could cause such a shadow region? One explanation is that the *P* waves travel at a different speed within the liquid core. A *P* wave traveling from a solid mantle into a liquid core will change speeds and change direction (that is, they will refract). The result of this refraction is the creation of the shadow region.

Professor Cyril Isenberg, academic leader of the British Physics Olympiad Team, challenged students worldwide in the 1986 International Physics Olympiad with a problem about the *P* and *S* waves of an earthquake. We present parts of that problem as a challenge to our readers.

Let's assume that the Earth is composed of a central liquid spherical core of radius R_c that is surrounded by a solid, homogeneous mantle of radius R. The velocities of the *S* and *P* waves through the mantle are v_S and v_P, respectively. An earthquake occurs at point *E* on the surface of the Earth and produces *P* and *S* seismic waves. A seismologist observes the waves at location *X*. The angular separation between *E* and *X* measured from the center of the Earth *O* is 2θ, as shown in figure 1.

A. Our beginning physics students should try to show that the seismic waves that travel through the mantle in a straight line arrive at *X* at a time t (the travel time after the earthquake) given by $t = 2R \sin\theta/v$ for $\theta \le \arccos(R_c/R)$, where $v = v_P$ for the *P* waves and v_S for the *S* waves.

B. After an earthquake an observer measures the time delay between the arrival of the *S* wave and the *P* wave as 2 minutes, 11 seconds. Deduce the angular separation of the earthquake from the observer using these data:

$$R = 6{,}370 \text{ km}$$
$$R_c = 3{,}470 \text{ km}$$
$$v_P = 10.85 \text{ km/s}$$
$$v_S = 6.31 \text{ km/s}$$

C. The observer in part B notices that at some time after the arrival of the *P* and *S* waves, there are two further recordings on the seismometer separated by a time interval of 6 minutes, 37 seconds. Explain this result and verify that it is indeed associated with the angular separation determined in part B.

D. For those of you who wish to plunge deeper, draw the path of a seismic *P* wave that arrives at an observer, where $\theta \le \arccos(R_c/R)$, after two refractions at the mantle–core interface. Obtain a relation for *P* waves between θ and i, the angle of incidence of the seismic *P* wave at the mantle–core interface.

E. For our advanced problem solvers, using the data above and the additional fact that the speed of the *P* waves in the liquid core is 9.02 km/s, draw a graph of θ versus i. Comment on the physical consequences of the form of this graph for observers stationed at different points on the Earth's surface.

F. Sketch the variation of the travel time taken by the *P* and *S* waves as a function of θ for $0 \le \theta \le 90$ degrees.

Solution

A. In part A, you were asked to calculate the time it would take for *P* or *S* waves emanating from the earthquake location *E* to reach an observation point *X*. From figure 2 we can see that

$$EX = 2R \sin\theta.$$

Therefore,

$$t = \frac{2R\sin\theta}{v},$$

where $v = v_P$ for *P* waves and $v = v_S$ for *S* waves. This is valid provided that *X* is at an angular separation less than or equal to *X′*, defined by the tangential ray to the liquid core. From figure 2, *X′* has an angular separation given by

$$2\phi = 2\cos^{-1}\left(\frac{R_c}{R}\right).$$

B. Given the delay time between *P* and *S* waves, you were next asked to deduce the angular separation of *E* and *X*. Using the result from part A,

$$t = \frac{2R\sin\theta}{v},$$

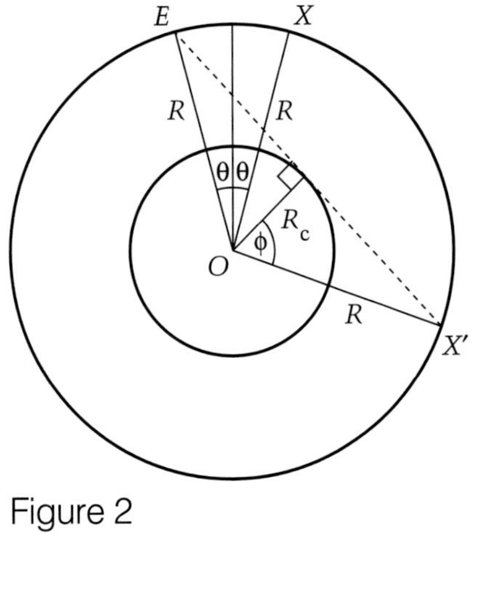

Figure 2

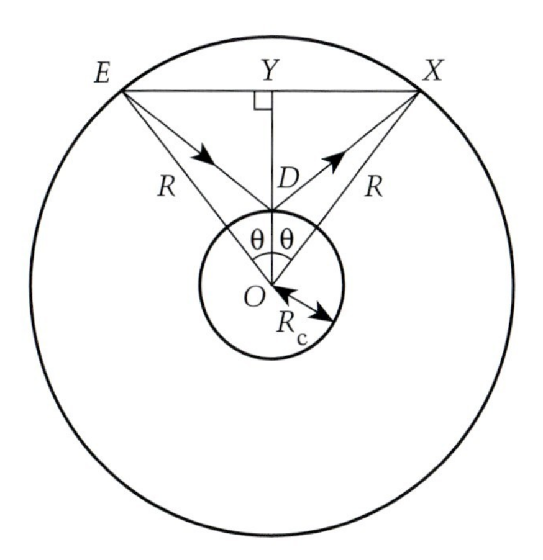

Figure 3

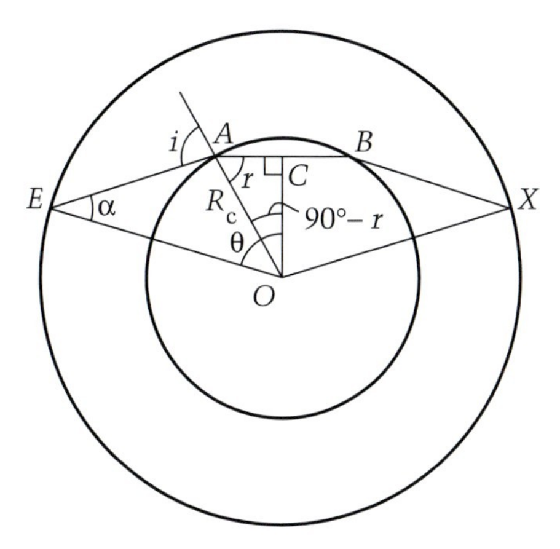

Figure 4

we can express the time delay as

$$\Delta t = 2R\sin\theta\left(\frac{1}{v_S} - \frac{1}{v_P}\right). \quad (1)$$

Substituting the data given, we get

$$131 = 2(6370)\left(\frac{1}{6.31} - \frac{1}{10.85}\right)\sin\theta.$$

Therefore, the angular separation of E and X is

$$2\theta = 17.84°.$$

This result is less than

$$2\cos^{-1}\left(\frac{R_c}{R}\right) = 2\cos^{-1}\left(\frac{3470}{6370}\right) = 114°,$$

and consequently the seismic wave is not refracted through the core.

C. If a second set of P and S waves had a longer delay, readers first had to hypothesize an explanation for the second set of delayed waves and then see if the result is consistent with the time delay given in part B.

The observations are most likely due to reflections from the mantle–core interface. Using the symbols in figure 3, we can express the time delay $\Delta t'$ as

$$\Delta t' = (ED + DX)\left(\frac{1}{v_S} - \frac{1}{v_P}\right) = 2ED\left(\frac{1}{v_S} - \frac{1}{v_P}\right).$$

In triangle EYD,

$$(ED)^2 = (R\sin\theta)^2 + (R\cos\theta - R_c)^2 = R^2 + R_c^2 - 2RR_c\cos\theta,$$

since $\sin^2\theta + \cos^2\theta = 1$. Therefore,

$$\Delta t' = 2\sqrt{R^2 + R_c^2 - 2RR_c\cos\theta}\left(\frac{1}{v_S} - \frac{1}{v_P}\right).$$

Using equation (1), we get

$$\Delta t' = \frac{\Delta t\sqrt{R^2 + R_c^2 - 2RR_c\cos\theta}}{R\sin\theta} = 396.7\text{ s} = 6\text{ min } 37\text{ s}.$$

Thus, the subsequent interval produced by reflection of seismic waves at the mantle–core interface is consistent with an angular separation of 17.84°.

D. Since a P wave is able to travel through the core, you were asked to draw the path of the refracted P waves and derive the relation between the angle of incidence and the angular separation of E and X.

From figure 4, we get

$$\theta = \angle AOC + \angle EOA = (90 - r) + (i - \alpha). \quad (2)$$

The law of refraction (Snell's law) gives

$$\frac{\sin i}{\sin r} = \frac{v_P}{v_{cP}}. \quad (3)$$

From triangle EAO and the law of sines, we get

$$\frac{R_c}{\sin\alpha} = \frac{R}{\sin i}. \quad (4)$$

Substituting equations (3) and (4) into (2) yields

$$\theta = 90 - \sin^{-1}\left(\frac{v_{cP}}{v_P}\right)\sin i + 1 - \sin^{-1}\left(\frac{R_c}{R}\right)\sin i. \quad (5)$$

E. Our most talented readers were then asked to draw a graph of the relationship expressed in part D and to comment on the physical consequences.

Substituting $i = 0°$ into equation (5) gives $\theta = 90°$; $i = 90°$ gives $\theta = 90.8°$. Substituting numerical values for $i = 0°$ to $i = 90°$, one finds a minimum value at 55° and the corresponding minimum value of θ: $\theta_{min} = 75.8°$ (fig. 5). As θ has a minimum value of 75.8°, observers at positions for which $2\theta < 151.6°$ will not observe the earthquake as seismic waves. For $2\theta < 114°$, however, the direct, nonrefracted waves will reach the observer.

F. Finally, readers were asked to sketch a comparison of the travel times for P and S waves for all angles. In this sketch (fig. 6) we can get a better sense of the "shadow" region where no earthquake waves will be observed.

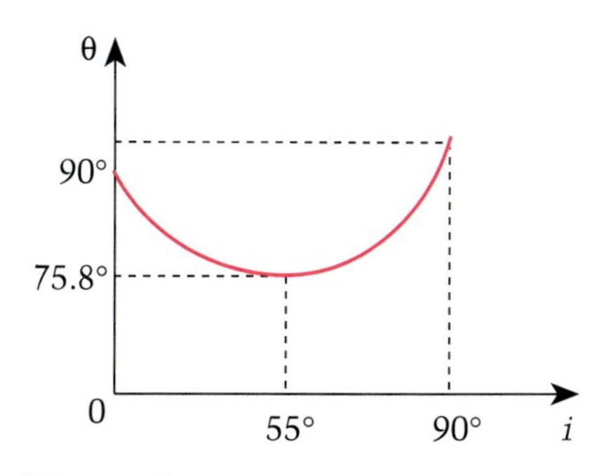

Figure 5

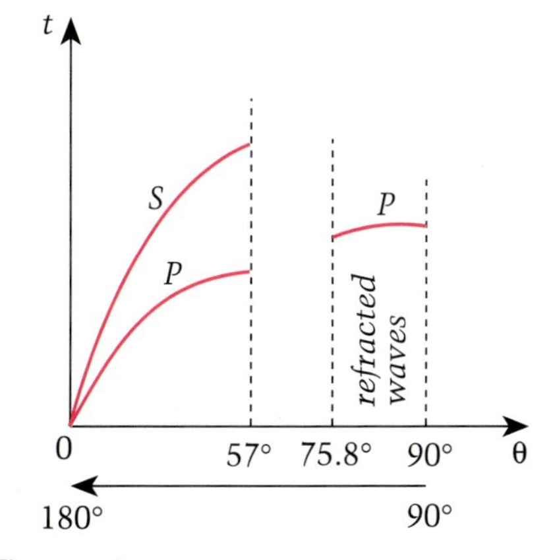

Figure 6

Sources, sinks, and gaussian spheres

"I do not perceive in any part of space, whether vacant or filled with matter, anything but forces and the lines in which they are exerted."
—Michael Faraday

WHEN YOU VISIT AN ART gallery, you can gain an enhanced appreciation of a sculpture by comparing views from different angles. Likewise, it's often useful in physics to compare two different physical systems with the same mathematics in order to develop a better physical feeling for both. A prime example of this occurs with electrostatics and hydrodynamics.

Let's consider the steady-state flow of an incompressible fluid. Water is a very good approximation. Conservation of mass requires that the flow of mass into the volume be equal to the flow of mass out of the volume. Another way of stating this is that the net flow of mass—or flux—through the surface must be zero.

This is true unless there are sources or sinks inside the volume. If there are sources, the net flow of fluid through the surface must be positive—that is, there must be a net flow out of the volume. On the other hand, if the region contains a sink, the net flow will be inward, or negative. In general, the net flow will just be the sum of the positive contributions due to the sources and the negative contributions due to the sinks. If we let ρ represent the density of the fluid and $\mathbf{v}$ the velocity of the fluid at each point in space, this can be written in mathematical terms as

$$\sum \rho v_n \Delta A = \sum (\text{sources} + \text{sinks}),$$

where ΔA is a small piece of the surface, v_n is the component of the velocity normal to the surface element (v_n is positive when it points out of the volume). Only the sources and sinks *inside* the volume are included on the right side. In the notation of calculus we get

$$\int \rho v_n dA = \int (\text{sources} + \text{sinks}) dV.$$

Even though nothing is flowing in the case of the electrostatic field surrounding a distribution of electric charge, the mathematical equation is the same. We can think of the electric

High up in the snow-covered mountains we find a scientific inquiry taking place.The crew has launched a charged water ball high up into the air to find out the conservation of electric charge. While some tasks require sophisticated generators, others are simple, like the one on top where the divine finger charges the first human with life and ejects his brains, or like Ben Franklin, who is drying his laundry with three lightning strikes. Another scientist is determining how effective a shave would be in electrostatic-hydrodynamic terms when plugged into the charged water ball. Others are more on the receiving end of the lightning and suffer minor burns or become a well-functioning toaster. The inquiry is still going on.

—T.B.

ATTENTION
DANGER
HIGH
VOLTAGE
ELECTRIC
BILL
STAFF
tom bunk

field E as the analogue of the flow of mass ($\rho\mathbf{v}$) in the fluid. In the electric case, the sources and sinks are charges; positive charges are sources of the electric field, and negative charges are sinks. This is all stated in Gauss's law:

$$\sum E_n dA = \frac{q_{enc}}{\varepsilon_0},$$

where ε_0 is a proportionality constant that depends on the units and q_{enc} is the net charge *inside* the volume. Note that it is easy to understand that only the charges inside the volume affect the sum by appealing to the case of the fluid. By analogy with fluid flow, the sum on the left side of the equation is called the "flux."

In cases of high symmetry, Gauss's law is very useful for finding the electric field due to a collection of point charges or a distribution of charge. As an example, let's calculate the electric field surrounding a positive point charge q located at the origin. Let's choose our surface to be a sphere of radius r centered on the origin to match the symmetry of the charge distribution. Because of the symmetry, we expect that the electric field will point radially outward and have the same value at all points on the surface. Therefore, the sum on the left side is easy to calculate:

$$\sum E_n \Delta A = E\sum \Delta A = EA = E4\pi r^2.$$

Setting this equal to the right side and solving for E, we get

$$E = \frac{1}{4\pi\varepsilon_0}\frac{q}{r^2},$$

which we recognize as Coulomb's law.

For our contest problem, we apply Gauss's law to find the electric field for several different cases with spherical symmetry. The key to all but the last part is figuring out how much charge is enclosed in the gaussian sphere.

A. Find the electric field at all points outside a sphere of radius a that contains a uniform density of charge ρ and show that it has the same form as Coulomb's law.

B. Find the electric field at all points inside this sphere. Does the value at the surface agree with the value found in part A?

C. Now assume that a spherical region of radius b has been removed from the center of the sphere. What is the electric field at all points in space?

D. As a final challenge, find the electric field at all points inside the hole when the hole is moved off center. Assume that it is moved a distance c in the $+x$ direction, but not so far that the hole penetrates the surface. (Hint: consider the superposition of a complete sphere with charge density ρ and a "hole" formed by a smaller sphere with charge density $-\rho$.)

Solution

A completely correct solution was submitted by Eric Joanis of Gatineau, Quebec.

A. Because we have a spherically symmetric distribution of charge, we choose a spherical gaussian surface of radius r centered on the spherical charge distribution. We can then use the result given in the statement of the problem:

$$E4\pi r^2 = \frac{q_{enc}}{\varepsilon_0}, \quad (1)$$

where

$$q_{enc} = \rho V = \rho\frac{4}{3}\pi a^3. \quad (2)$$

Combining these two equations and solving for the electric field E, we get

$$E_o = \frac{a^3\rho}{3\varepsilon_0 r^2}, \quad (3)$$

where we have added the subscript o to indicate that this is the value outside the sphere—that is, for $r \geq a$.

We can show that this has the same form as Coulomb's law by substituting the value of [r] from equation (2) into equation (3).

B. We can use the same technique to find the electric field E_i inside the sphere ($r \leq a$) if we remember that only the charge *inside* the gaussian sphere contributes to the field. Therefore, the enclosed charge is given by

$$q_{enc} = \rho\frac{4}{3}\pi r^3,$$

where we use the radius r of the gaussian sphere instead of the radius a of the complete sphere as before. Therefore,

$$E_i = \frac{r\rho}{3\varepsilon_0}. \quad (4)$$

At the surface of the sphere $r = a$ and

$$E_o = E_i = \frac{a\rho}{3\varepsilon_0},$$

as expected.

C. The only complication caused by the spherical hole at the center of the sphere is in the calculation of the enclosed charge. We can calculate this charge by taking the charge of the complete sphere and subtracting off the contribution due to the hole. There are three regions. Outside the sphere ($r \geq a$) we have

$$\begin{aligned} q_{enc} &= \frac{4}{3}\pi a^3\rho - \frac{4}{3}\pi b^3\rho \\ &= \frac{4}{3}\pi\rho\left(a^3 - b^3\right), \end{aligned}$$

with the resulting electric field

$$E = \frac{\rho\left(a^3 - b^3\right)}{3\varepsilon_0 r^2}.$$

In the spherical shell ($b \leq r \leq a$) we have

$$\begin{aligned} q_{enc} &= \frac{4}{3}\pi r^3\rho - \frac{4}{3}\pi b^3\rho \\ &= \frac{4}{3}\pi\rho\left(r^3 - b^3\right). \end{aligned}$$

Inside the spherical hole ($r \leq a$) $E = 0$, since there is no enclosed charge.

D. When the hole is moved off-center, we must be careful to remember the vector nature of electric fields. Inside the bigger sphere (assumed to be completely filled with charge ρ), the electric field E_B is given by equation (4), but now we write it in vector form:

$$E_B = \frac{\rho}{3\varepsilon_0}\mathbf{r},$$

where **r** is a radial vector from the origin to the point of interest. We now express this in terms of rectangular coordinates:

$$E_B = \frac{\rho}{3\varepsilon_0}(x, y, z).$$

We can do the same thing for the smaller sphere of negative charge, but we must remember that the center of this sphere has been shifted to $x = c$:

$$E_S = \frac{-\rho}{3\varepsilon_0}(x - c, y, z).$$

When we add the two contributions to the field, we find that the y- and z-components cancel, and we are left with a constant x-component:

$$E = E_B + E_S = \frac{\rho}{3\varepsilon_0}(c, 0, 0).$$

Notice the surprising result that the electric field inside the hole has a constant value independent of the size of the hole, the size of the larger sphere, and the location within the hole. It depends only on the amount of offset and the charge density.

The tip of the iceberg

"And ice, mast-high, came floating by,
As green as emerald."
—Samuel Taylor Coleridge,
"The Rime of the Ancient Mariner"

CLOUDS FLOAT IN THE SKY. Rocks plunge to the bottom of lakes. Children gaze at the sky as a helium balloon rises and rises, wondering what its future will be. Physicists also wonder about such things. And as they wonder, they think about gravity, buoyancy, sinking, and floating.

One of the first people to ponder floating and sinking was Archimedes. It's difficult to believe that Archi-medes was once so consumed by his king's challenge of determining the constituency of the royal crown that in his burst of insight, he jumped out of the bathtub and ran through the town screaming "Eureka!" If you heard such a commotion outside and then observed a naked man running down the street screaming a Greek word or even its English equivalent ("I found it!"), what would you do? Your first thought probably involves locking doors or calling police. When you found out that this was the reaction of a wise man after figuring out whether the king's crown was pure gold or an alloy, I can't imagine many people unlocking their doors or hanging up the phone: "It's okay, dear—just a wise man discovering a new law of physics."

Archimedes's law was a great achievement. Everybody knew that an object dropped in water made the water level rise (that is, it displaced some water). But Archimedes was the first to recognize that the amount of water displaced is related to the object's volume in a very precise way!

You can try to "discover" Archimedes's law, and we promise you won't have to take a bath or run through your town. Gather 100 pennies. Fill a glass with water. Carefully place the pennies in the water. What happens when 50 pennies are placed in water? What happens when 100 pennies are placed in water? Does the water rise? Does the water spill out of the glass? How

How much will the water rise if you sink the Titanic in your fish tank? That's what Archimedes is trying to find out, sitting in his bathtub and observing the sinking steamer. First thing he sees is the captain of the ship running away, and the upper-class passengers are looking for safety by climbing on floating pencils, which in some cases conveniently become rockets flying out of the tank and to the Moon, while Columbus is arriving with his Spanish pencil on a Japanese wave more than 400 years too late to discover the safe haven of the Titanic. The liner is being pushed down by the divine finger that not only creates life but also extinguishes it. It is not clear if Archimedes went running through the streets shouting "Eureka!" on this particular night.

—T.B.

TITANIC
TOM BUNK

much water? Can you find the relationship between the water and the pennies? That was part of Archimedes's challenge.

We know that when you try to submerge a table tennis ball in water, it will find its way to the surface. A well-designed boat will float. The same boat, with lots of cargo, may sink. Why is it that some people find floating in water easier than others?

One helpful way of analyzing floating and sinking is to note that in a container of water, the water at each level of the container is, in effect, floating at that level. Physicists would say that the water at each depth is in equilibrium. The buoyant force on any piece of water pushing it up must be exactly equal to the gravitational force or weight of the piece of water pulling it down. We then conclude that the pressure (force per area) of the water increases with depth. The bottom of a submerged object experiences a higher pressure than its top. The difference in pressure pushes the object up. Gravity is pulling the object down. Since the piece of water moves neither up nor down, the force due to the pressure differences must exactly equal the weight.

The pressure at the top of the piece of water is equal to p and the pressure at the bottom of the piece of water is equal to $p + \Delta p$. This difference in pressure provides the upward force that must be equal to the weight of the piece of water:

$$(p + \Delta p)A - pA = m_w g = \rho_w V g.$$

If we assume that the piece of water is a slab of length l, width w, and height Δh, then

$$(p + \Delta p)A - pA = \rho_w A \Delta h g,$$
$$\Delta p = \rho_w g \Delta h.$$

Since the density of the water and the acceleration due to gravity are constant, we can conclude that the pressure change is proportional to the change in depth of the water.

If we now replace our piece of water with an identically shaped piece of wood, we know that the upward force on the wood is identical to the upward force on the piece of water. This is due to the pressure difference between the top of the wood and the bottom of the wood. The weight of the wood may be different than the weight of the piece of water. If the wood weighs less than the equivalent volume of water, the buoyant force prevails and pushes the wood to the surface. If the wood weighs more than the equivalent volume of water, the force of gravity prevails and the wood sinks to the bottom.

And so we have theoretically derived Archimedes's principle: *A body in water (or any fluid) will have a buoyant force equal to the weight of the fluid that it displaces.* When Archimedes sat in his bathtub and displaced that bath water, he reached this same conclusion. Anybody feel like shouting "Eureka"?

Have you ever wondered how much of an ice cube is beneath the water? Or, maybe, how much of an iceberg is below the surface? The density of ice is 0.92 g/cm³. The density of ocean water is 1.04 g/cm³. Assume that the weight of the iceberg is W_i, its volume is V_i, and its density is ρ_i; the volume of the displaced water is V_w; and the density of the displaced water is ρ_w. Then

$$W_i = \rho_i V_i g.$$

The buoyant force equals

$$F_B = \rho_w V_w g.$$

Since the buoyant force equals the weight,

$$\rho_i V_i g = \rho_w V_w g,$$
$$\frac{V_w}{V_i} = \frac{\rho_i}{\rho_w}.$$

Assuming the densities of ice and water given above,

$$\frac{V_w}{V_i} = \frac{0.92}{1.04} = 0.88.$$

We conclude that 88% of the iceberg is below the surface, or that we see only the "tip of the iceberg."

Our problem relates to buoyancy and the tendency of a tall object floating on the surface to tip over. If the mast of a ship tilts due to a strong wind or a wave, the submerged part of the ship will no longer be directly below the center of mass. The buoyant force and the weight will exert torques on the ship that could cause it to capsize.

We can simulate this tilting ship by considering what happens if you hold a pencil at its tip and lower the other end into a large pail of water. At first we expect that the pencil will remain vertical. As more of the pencil enters the water, the pencil will have a tendency to tilt in the water. If the pivot is at a given height above the water, the pencil will choose a specific stable angle. Don't believe us—try it!

Our problem has three parts:

A. What is the relationship between the angle of the pencil and the height of the pivot above the water?

B. Why is this position stable?

C. What happens to the fraction of the submerged pencil as the pivot is lowered?

Solution

Ben Davenport of the North Carolina School for Science and Mathematics submitted a correct solution to this problem and then provided an interesting extension that we'll tempt you with later.

A. The three forces acting on the pencil are the force of gravity, the force of buoyancy, and the force of the pivot. If we choose to take torques about the point at the top, the force of the pivot produces no torque and can be ignored (see figure 1):

$$\Sigma \tau_p = 0,$$
$$F_g = -\rho_r L A g,$$
$$F_b = \rho_w \left(L - \frac{H}{\cos\theta} \right) A g.$$

Using the moment arms for the torques about P, we get equation 1

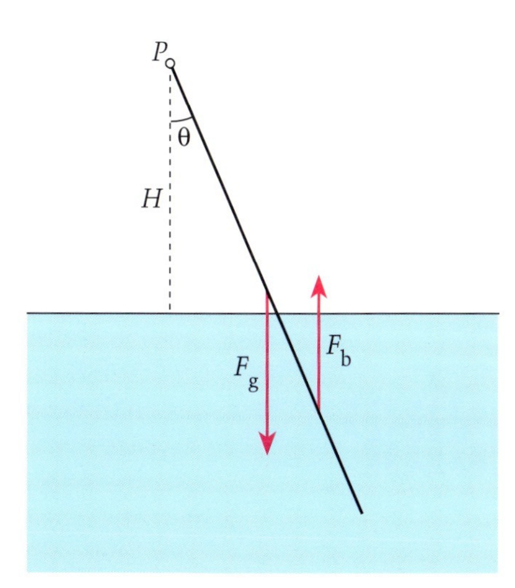

Figure 1

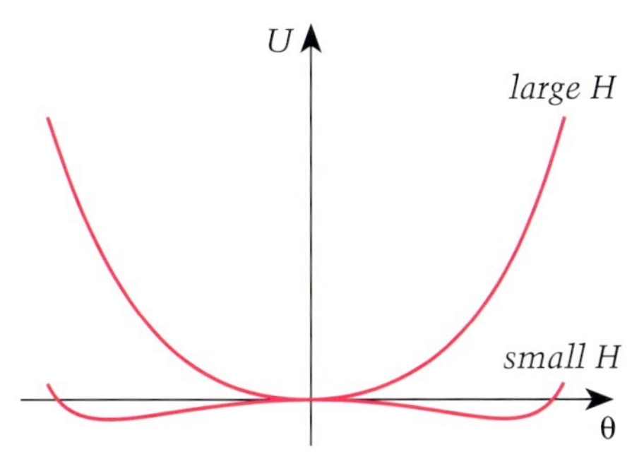

Figure 2

in the box below. You can divide each side of equation 1 by sin θ, but you then must realize that sin θ = 0 is a solution to the equation and, therefore, θ may be equal to 0.

The second solution is

$$\cos\theta = \left(\frac{H}{L\sqrt{1-\rho_r/\rho_w}}\right),$$

where $H \le L\sqrt{1-\rho_r/\rho_w}$, since cos θ must be less than or equal to 1.

B. In order to sketch a graph of potential energy versus theta, it's useful to derive the equation for potential energy—equation 2 in the box. The extrema of potential energy represent equilibria. You can get a sense of this by thinking about stable points for a roller coaster. For our pencil in water, we plot the potential energy versus theta on a spreadsheet for small H and for large H (fig. 2). From the graphs we can see that for large H, zero degrees is a stable equilibrium. For small H, zero degrees is an unstable equilibrium, and the angle given by the relation in part A is the stable equilibrium.

C. To solve the last part of the contest problem, let the submerged part of the pencil be referred to as S. Then

$$S = L - \frac{H}{\cos\theta},$$

where

$$\cos\theta = \frac{H}{L\sqrt{1-\rho_r/\rho_w}}.$$

Therefore,

$$S = L\left(1-\sqrt{1-\rho_r/\rho_w}\right).$$

We then conclude that the submerged part of the stick is constant—that is, independent of H. You can verify this by actually performing the experiment.

As we mentioned above, Ben Davenport challenged himself (and now we challenge the rest of our readers) with what happens if H remains constant but the pivot point of the pencil is lowered.

The details of this problem and the experimental verification of the solution can be found in an article by Joseph Priest and David F. Griffing in the April 1990 issue of *The Physics Teacher* (pp. 210–13).

1. $$\Sigma\tau = -\left(\tfrac{1}{2}\sin\theta\right)(\rho_r LAG) + \left[L - \left(\frac{L-H/\cos\theta}{2}\right)\right](\sin\theta)\rho_w\left(L-\frac{H}{\cos\theta}\right)Ag = 0$$

2. $$\Delta U = -\int_0^\theta \tau d\theta$$

$$= -\int_0^\theta \frac{AgL^2\rho_r}{2}\left(-1+\frac{\rho_w}{\rho_r}\left[1-\left(\frac{H}{L\cos\theta}\right)^2\right]\right)\sin\theta d\theta$$

$$= \frac{AgL^2\rho_r}{2}(1-\cos\theta)\left[1-\frac{\rho_w}{\rho_r}+\frac{\rho_w}{\rho_r}\left(\frac{H}{L}\right)^2\frac{1}{\cos\theta}\right]$$

A topless roller coaster

"And in their motions harmony divine . . ."
—John Milton, Paradise Lost

WHY IS IT THAT AMUSEment park rides are such fun for some and a horror for others? Why will some of us rush to a park to try the newest, most frightening ride while others will only begrudgingly accept a ride on a ferris wheel? One of us (LDK) enjoys roller coasters so much that he spent 1.5 hours navigating on the New York subway system to take a 1.8-minute ride on the Cyclone at Coney Island. And this was an old-fashioned, wooden roller coaster!

The amusement park rides can take on an added dimension of interest and fun when you try to understand the physics inherent in the designs. Many amusement parks now have activities created by members of the American Association of Physics Teachers and the National Science Teachers Association that serve as guides for field trips for physics classes across the county. In these activities, people build accelerometers, predict and measure speeds, and record heart beat changes as you spin around, flip upside down, or watch the floor drop from under you. The rides provide many opportunities for testing the ideas of physics in "real-world" situations.

In July 1993 the International Physics Olympiad was held in Williamsburg, Virginia. During a respite from the grueling five-hour exams, the best high school students from approximately 40 countries spent a day at Busch Gardens enjoying themselves as they rode and tried to explain the physics in the park.

Designing an amusement park ride is quite a challenge. The ride must be entertaining *and* safe. Modern roller coasters have added a new dimension to roller coaster riding by adding such things as loops and corkscrews. These require our readers, as future ride designers, to apply knowledge of circular motion and centripetal acceleration in addition

If life is sometimes compared to a roller coaster, here is one that would fit the comparison. It is full of surprises, sprinkled with surrealistic elements. Science alone cannot explain the hair-raising thrill of riding a roller coaster or the unbearable horror of it, or the surprise when suddenly during the speedy ride we are confronted with a cow in front of us or with a steamship puffing and panting up the curve, or a skeleton riding in a coffin, or an airplane that is trying to find its way out of the chaotic construction—or the surprise when we suddenly find ourselves riding in the wrong direction and see other riders coming toward us. This roller coaster has endless surprises like real life has, even if it is realistically not very probable.

—T.B.

SOOORY, OUT OF ORDER!
Tom Bunk

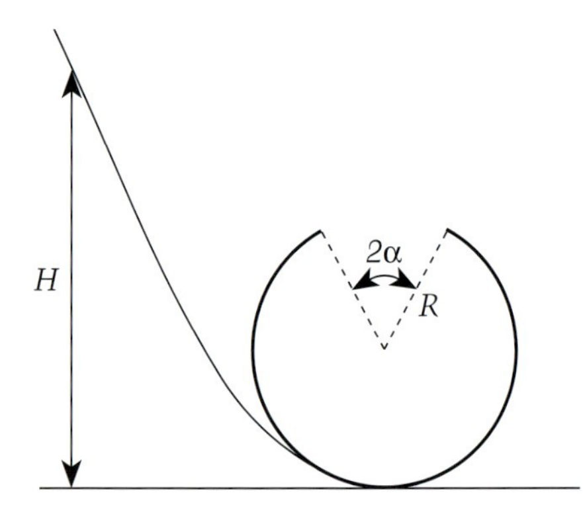

Figure 1

to the conservation of mechanical energy in analyzing the rides. Of course, in order to do this in a simple way we can make a number of assumptions such as (1) there are no frictional forces (including air resistance), (2) the kinetic energy of the wheels can be neglected, (3) the train of roller coaster cars stays on the track without the safety rail, and (4) the train is a point mass. The last assumption allows us to neglect such things as the rotational kinetic energy, the angular momentum, and the orientation of the train.

Up to now, all of the roller coasters of the world use a continuous track. But that does not restrict our imagination. Let's imagine that we remove the top portion of the track in a vertical loop, creating the so-called topless roller coaster. This allows us to combine the physics of circular motion in a gravitational field with that of projectile motion.

Assume that the vertical loop is a circle with a radius R and that the portion that is missing has an angle 2α centered about the top of the loop as shown in figure 1.

A. From what height H must a car be released so that it will leave the loop at one side of the gap and still arrive at the other side of the gap to continue the trip? Check your answer by setting $\alpha = 0$ to see if you get the expected result for a complete roller coaster.

B. It is useful to analyze the problem using the dimensionless ratio H/R because this sets the scale for the roller coaster. Draw a graph of H/R versus α.

C. For what range of angle α is this possible if the height is restricted to $H < 3R$?

D. Discuss the motion for the case $H = 5R/2$.

E. Discuss the motion for the case when H is a minimum.

Solution

This problem was used on the semifinal exam to select the 1992 US Physics Team that competed in the XXIII International Physics Olympiad held in Helsinki, Finland, during July 1992. Our solution is modeled after the one submitted by Ben Davenport of the North Carolina School of Science and Mathematics in Durham.

When the point-mass train leaves the track, it becomes a projectile subject only to the force of gravity. Because of the symmetry of the problem, we know that the train will meet the track on the other side with the correct speed and angle if the parabola describing the motion is also symmetric. We can, therefore, begin by remembering (or deriving) the formula for the range L of a projectile over flat ground in the absence of air resistance:

$$L = \frac{2v^2 \sin\alpha\cos\alpha}{g}, \qquad (1)$$

where v is the speed of the train and g is the usual acceleration due to gravity. Note that the angle of the projectile with respect to the ground is α.

The range of the train must be equal to the horizontal distance across the opening in the track, which we can get from trigonometry:

$$L = 2R \sin \alpha. \qquad (2)$$

Equating equations (1) and (2) leads to the following condition on the speed of the train at the time it leaves the track:

$$v^2 = \frac{gR}{\cos\alpha}. \qquad (3)$$

Since there is no friction, conservation of mechanical energy requires that the sum of the kinetic energy and the gravitational potential energy be the same at height H and at the height of the end of the track. Therefore,

$$mgH = mgR(1 + \cos \alpha) + \tfrac{1}{2}mv^2.$$

After canceling the common factor m, we substitute in the condition on v^2 from equation (3):

$$gH = gR(1+\cos\alpha) + \frac{gR}{2\cos\alpha}.$$

Defining $k = H/R$, the equation simplifies to

$$k = 1 + \cos\alpha + \frac{1}{2\cos\alpha}. \qquad (4)$$

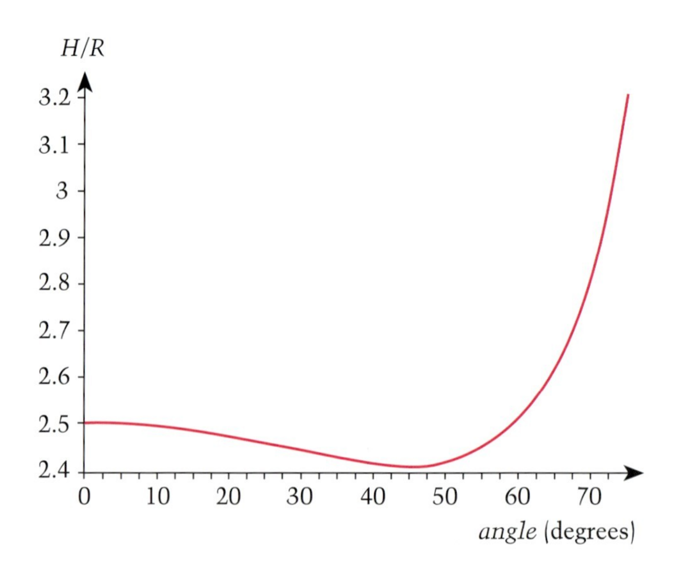

Figure 2

Substituting $\alpha = 0$, we find $k = 5/2$, which is the answer we expect to get from a complete loop by setting the gravitational force equal to the centripetal force required for the train to execute the loop.

The graph of H/R versus α is given in figure 2. We have not plotted the graph beyond 75° so that we can see the details of the graph in the range of interest.

To find the value of α for the case of $k = 3$, we solve the quadratic equation in cos α to obtain

$$\cos\alpha = \frac{(k-1) \pm \sqrt{(k-1)^2 - 2}}{2}. \quad (5)$$

Plugging in the value of $k = 3$, we get

$$\cos\alpha = 1 \pm \frac{1}{\sqrt{2}}.$$

Throwing out the positive root because $|\cos\alpha| \le 1$, we obtain a maximum angle of 73°, which can be verified on the graph in figure 4. The graph also tells us that the angle α can be as small as we want.

The case $k = 2.5$ is an interesting one, since this corresponds to the minimum height for which the car can complete the closed loop without falling off the top. We can also see from the graph that there are two possible values of α for this case, since the curve is multivalued in this range. We could have anticipated this because there are two angles, $45° \pm \theta$, that produce the same range in projectile motion. Plugging $k = 5/2$ into equation (5) yields

$$\cos\alpha = \frac{3}{4} \pm \frac{1}{4}$$

or

$$\alpha = 0°, 60°.$$

Notice that this is also the maximum height for which there are two possible angles.

The minimum value of k is obtained by setting the argument of the radical in equation (5) equal to zero and solving for k:

$$k = 1 + \sqrt{2} \cong 2.414,$$

which corresponds to an angle of 45°. Alternatively, we could set the derivative of equation (4) with respect to cos α (or α) equal to zero.

Ben states that "this makes sense physically, since 45° is the optimum angle for maximum range. That is, we get the most out of our velocity if the projectile leaves the track at this angle." Even in this case, the maximum height of the train exceeds $2R$. ◘

Row, row, row your boat

"It is not possible to step twice into the same river."
—Heraclitus

ONE OF THE EARLY ARGUMENTS against a spinning Earth held that objects would not fall straight down. Given that we now know (with the help of Eratosthenes) that our home planet has a diameter of 6,500 km (4,000 miles), then parts of our planet must be moving at 1,700 km/hr (1,000 mph). If the critics had been correct and you drop an object that takes 0.5 s to reach the ground, the object would land 240 m behind you. Along comes Galileo to refute what appears to be common sense.

Galileo proposes that a person climb the mast of a ship. If the ship is not moving and a ball is dropped, it will certainly fall straight down. The defenders of the stationary Earth would then predict that if the ship were moving, the ball dropped from the mast would land toward the rear of the boat. This is because, they would say, the boat glides forward while the ball is descending. Galileo suggested the correct behavior. The ball maintains the original horizontal motion of the ship and lands in the identical location as when the ship stood still.

If this works for a ship, it should also work for the Earth. The vertical motion is independent of the horizontal motion. A ball on a stationary ship or a gliding ship will land in the same place whether the ship is moving or stationary. This may seem obvious to some of our readers, but it is quite subtle and still confounds some people.

Imagine hopping aboard a rowboat and paddling from one shore to the opposite shore with no current. The trip takes you 15 minutes. If you return to the river and venture across again, paddling to the opposite shore with the same strokes, but with a stiff current dragging you downstream, will you arrive at the opposite shore in less time, in more time, or in the same time? You probably recognize that your would land further downstream on this second

Through the image streams a river with two very different shores on each side. The one symbolizes life, the other death. The life side is the sunny side, people are enjoying themselves under the tree of life, eating, drinking, writing, painting. The opposite side is the dark side, where the trees are chopped down and Time stands still. Our hero has to paddle across the river from point A to point B, but God forbid life would be that easy, the river is wild and treacherous and he gets dragged downstream, undergoing several life-threatening misadventures, until he reaches the other shore as an old man. Now he has to drag himself up to get to point B as planned, where he apparently bites the dust, though we see him right after that ressurrected as a young boy crossing the river to the shore of the living again. Some people never get enough.

—T.B.

A
B
E=mc2
Tom Bunk

journey. Since you traveled further, maybe it should take more time. But your velocity is actually the sum (the vector sum!) of your paddling velocity and the velocity of the current. With this faster speed, maybe the journey should take less time. Or perhaps, the longer distance is exactly compensated by the greater speed and you arrive at the opposite shore in the same time. Our readers can use the fact that the motion across the river and the motion downstream are perpendicular to each other and are therefore unaffected by each other. The time is determined by the motion across the river independent of the speed of the current. The current determines where the boat lands downstream, but does not change the time.

Once again, this is quite subtle, and our readers should attempt to explain the solution of this puzzle to people not accustomed to thinking the way physicists do. If you can convince someone of this, then you, as a teacher, must really understand it.

Let's complicate the situation. What happens if you don't paddle straight across the river, but rather choose to paddle at some angle? Now you'll find that there is a component of your velocity that helps you across the river and a component that takes you upstream or downstream. In this way, you can head upstream and end your journey directly across from where you embarked.

So here is our problem. Assume that you wish to end up directly across the river and that you were permitted to walk on the far shore if you land upstream or downstream. What path takes the least time? Let's add some specific numbers (suggested by Resnick and Halliday in their *Fundamentals of Physics*): the river is 500 m wide; your rowing speed is 3,000 m/h; the river flows at 2,000 m/h; and your walking speed on the opposite shore is 5,000 m/h.

A. Solve for the possible range of angles qualitatively.

B. Describe the path quantitatively.

C. Calculate this minimum length of time.

". . . Merrily, merrily, merrily, merrily, life is but a dream."

Solution

Responses to this problem arrived from high school students, college students, and physics professors. Geographically, they included solutions from all over the United States as well as from Canada and Great Britain.

The problem assumed that you wish to end up directly across the river and that you are permitted to walk on the far shore if you land upstream or downstream. What path takes the least time? Part A of the problem asked for a qualitative discussion of the range of plausible angles.

The first correct solution was submitted by Ben Davenport of the North Carolina School of Science and Mathematics, a steady reader of *Quantum*. The best angle to take lies between the angle that provides the minimum rowing time and the angle that provides the minimum walking time. For the minimum rowing time, the boat should head directly across the river, or at least at $\theta = 90°$, where θ is measured relative to the upstream bank as shown in the figure. The minimum possible walking time is zero—that is, the boat arrives exactly at point *B* across the river. In order for this to occur, the component of the rowing velocity parallel to the river must be equal and opposite to the current. Note that this is also the path of minimum distance. Call this minimum angle θ_w. If the angle is decreased below θ_w, it will take longer to cross the river and the walking time will become nonzero. If the angle is increased above 90°, it will take a longer time to cross the river, and the walking time will also increase. Therefore, we have two boundaries on the angle the boat should take across the river. Just for fun, let's solve for θ_w:

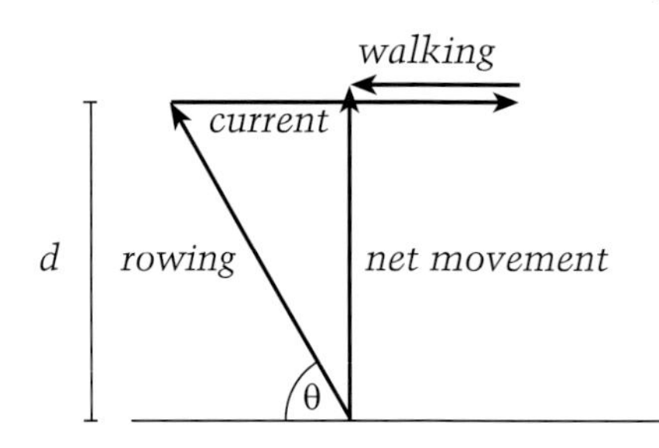

$$v_c = v_r \cos \theta_w,$$

where v_c is the speed of the current and v_r is the speed of the boat relative to the water. Then

$$\theta_w = \arccos \frac{v_c}{v_r} = 48°.$$

So we have limits on reasonable rowing angles:

$$48° < \theta < 90°.$$

Part B required readers to describe this path quantitatively. The rower's velocity component across the river is

$$v_{across} = v_r \sin \theta.$$

The rower's net velocity parallel to the river is

$$v_{along} = v_c - v_r \cos \theta.$$

The rower's walking velocity along the shore is given.

The total time is equal to the time spent rowing plus the time spent walking:

$$t = t_r + t_w.$$

The time spent rowing is the distance across the river *d* divided by the component of the rowing velocity directed perpendicular to the current:

$$t_r = \frac{d}{v_r \sin\theta}.$$

The time spend walking depends on the distance the boat lands from point *B* and the walking speed v_w. This distance depends on both the net velocity parallel to the river and the time spent rowing:

$$t_w = \frac{t_r\left(v_c - v_r \cos\theta\right)}{v_w},$$

$$t = t_r\left(\frac{v_c - v_r \cos\theta}{v_w} + 1\right).$$

$$\frac{dt}{d\theta} = \left(\frac{d}{v_w v_r}\right)\left[\frac{\sin\theta(v_r \sin\theta) - \cos\theta(v_w + v_c - v_r \cos\theta)}{\sin^{-2}\theta}\right] = 0$$

Plugging in for the time spent rowing from above, we get the following expression for the time to cross the river:

$$t = \frac{d(v_w + v_c - v_r \cos\theta)}{v_w v_r \sin\theta}.$$

One reader, W. Kenneth Beard of Cornwall, England, realized that this problem could also be solved by using Snell's Law, which he states "usually applies to waves experiencing a change of velocity on passing between two media of different density. The waves take the path giving the shortest time, which is precisely what is required here."

Part C required a calculation of the least time using the values provided for the speed of the current, the rowing speed, and the walking speed. Most readers differentiated their equation for the total time and set the derivative equal to zero to find the minimum-time path (see the equation in the box above). Since $\sin^2 \alpha + \cos^2 \alpha = 1$,

$$v_r - \cos\theta(v_w + v_c) = 0,$$

$$\theta = \cos^{-1}\left(\frac{v_r}{v_w + v_c}\right)$$

$$= \cos^{-1}\left(\frac{3}{7}\right) = 64.6°.$$

Thus, the rower should row 25.4° upstream of straight across for a minimum crossing time. At this angle, the rower arrives 132 m downstream and requires a total time of 12.6 minutes to complete the trip.

Andrew Menard from Saginaw, Michigan, went one step further in noting that "the width of the river does not matter. It will obviously affect the total time, but the optimal angle is not affected." Andrew also sent in a program for the TI-81 and TI-85 graphing calculators that shows the path and calculates the total time required for the journey. This numerical technique could also be used by our readers who have not yet encountered calculus.

How about a date?

"Time is what prevents everything from happening at once."
—John Archibald Wheeler

THE SEPTEMBER 1991 DIScovery of a frozen body in the Tirolean Alps has revived interest in radiocarbon dating. The body was that of a hunter (or possibly a shepherd). The discovery is particularly valuable to anthropologists because the body was virtually intact and fully clothed, and the hunter had been carrying various articles such as a bow and arrows. The date of the hunter's death was determined by carbon dating to be around 3300 B.C.

Modern techniques of carbon dating using mass spectroscopy have greatly reduced the size of the samples required for dating. The date for the death of the hunter was determined with a tissue sample the size of a tablet of artificial sweetener. Carbon dating techniques rely on the presence of two carbon isotopes in the atmosphere. The vast majority have a nucleus contain 6 protons and 6 neutrons. These are designated by the symbol ^{12}C, where the superscript denotes the total number of protons and neutrons. However, trace amounts of ^{14}C (containing two extra neutrons) are produced by collisions of cosmic rays with nitrogen atoms in the upper atmosphere. Currently the ratio of ^{14}C to ^{12}C in the atmosphere is 1.3×10^{-12} to 1.

All plant and animal life interact with the atmosphere, and the ratio of the two carbon isotopes in their bodies reaches equilibrium with the atmosphere. At the time of death, the ratio of ^{14}C to ^{12}C in a sample of the plant or animal is therefore equal to that in the atmosphere. After this the number of stable ^{12}C atoms in the sample remains constant. However, the radioactive decay of the ^{14}C causes their number to decrease exponentially according to the well-known decay law

$$N = N_0 e^{-\lambda t},$$

where N_0 is the original number of atoms in the sample, N is the num-

In order to have a date with someone, sometimes one has to go back a long way, sometimes millions of years. This illustration starts where we all more or less have started, in the jungle, living in trees. After descending from the tree, humans continue to follow all the evolutionary steps and missteps like making tools, painting rocks, hitting each other with clubs, inventing wheels, until one fine day 3300 years ago a man is crossing the Tyrolean Alps, where he freezes to death and disappears . . . until recently. Someone finds him extremely well preserved and ready for a date with the curious science community, which never sleeps. "Welcome back to life!" the scientist seems to say—and then they take him apart and put him in little boxes.

—T.B.

25
10 MILL.
5 MILL.
2 MILL.
1 MILL.
500,000 YEARS
300,000 YEARS
100,000
20,000
10,000
1
3
5
7
8
9
SNOW FROM YESTER YEAR
TOMAS BUNK

ber of atoms remaining after a time t, and the decay constant λ is a parameter that depends on how rapidly the atoms decay.

A useful measure of the decay rate is the half-life $t_{1/2}$. This is the time required for one half of any sample of radioactive atoms to decay and is independent of the size or age of the sample. For ^{14}C the half-life is measured to be 5,730 years = 1.81×10^{11} s. The decay constant and the half-life are related by

$$\lambda = \ln 2/t_{1/2}.$$

Instead of looking at the number of atoms remaining in a sample, we can focus on the rate R at which they decay. This can be obtained by differentiating the decay law to obtain

$$R = R_0 e^{-\lambda\tau},$$

where $R_0 = \lambda N_0$.

This brings us to our contest problem.

A. Assume that we have isolated a 1-g sample of carbon from a frozen animal and that the atmospheric ratio of the two carbon isotopes was the same when the animal died as it is now. What was the decay rate in decays per minute of the ^{14}C shortly after the animal died?

B. If the current decay rate is 1 decay per minute, how many years ago did the animal die?

Unfortunately, the ratio of the two isotopes of carbon has not been constant throughout time. It's possible to determine the dependence by dating samples from objects with well-determined ages. These could be such things as dated historical documents or tree rings. Let's look at two simplified scenarios to illustrate the problems associated with this variability.

C. How does the age of our sample change if the ratio varied linearly in the past? Assume that the ratio decreases by 0.1% of the current value for each century that we go back in time.

D. How does the age of our sample change if the atmospheric ratio has varied sinusoidally in the past? Assume a cosine dependence with an average value equal to the current value, an amplitude that is 5% of the current value, and a period of 628 years.

Solution

An excellent solution to this problem was submitted by Ben Davenport from the North Carolina School of Science and Mathematics. Ben was a semifinalist for the 1993 US Physics Team that competed in the International Physics Olympiad in Williamsburg.

A. We begin by calculating the value of the decay constant:

$$\lambda = \frac{\ln 2}{T_{1/2}}$$
$$= 1.21 \cdot 10^{-4}\ \text{year}^{-1}$$
$$= 3.83 \cdot 10^{-12}\ \text{s}^{-1}.$$

We can calculate the number N_0 of ^{14}C atoms from the ratio of the ^{14}C to ^{12}C atoms and the fact that 12 g of carbon contain Avogadro's number of atoms:

$$N_0 = \frac{\left(6.02 \cdot 10^{23}\ \text{atoms}\right)\left(1.30 \cdot 10^{-12}\right)}{12}$$
$$= 6.52 \cdot 10^{12}\ \text{atoms}.$$

Therefore, the decay rate for the 1-g sample of carbon shortly after the animal died was

$$R_0 = \lambda N_0 = 0.250\ \text{decays/s}$$
$$= 15\ \text{decays/min}.$$

B. Solving the equation for the change in the decay rate for t, we can obtain the time since the animal died:

$$t = \frac{\ln\left(R/R_0\right)}{-\lambda} = \frac{\ln(1/15)}{-\lambda}$$
$$= 7.07 \cdot 10^{11}\ \text{s}$$
$$= 22{,}400\ \text{years},$$

where R is the current decay rate of 1 decay/min.

C. With a decrease of 0.1% per century, we can use an iterative technique to find the age. As a first approximation to the decay rate R_0' at the time of the animal's death, let's use our age from part B. Then

$$R_0' = R_0\left(1 - \frac{0.001t}{100\ \text{years}}\right)$$
$$= 0.776(15\ \text{decays/min})$$
$$= 11.6\ \text{decays/min}$$

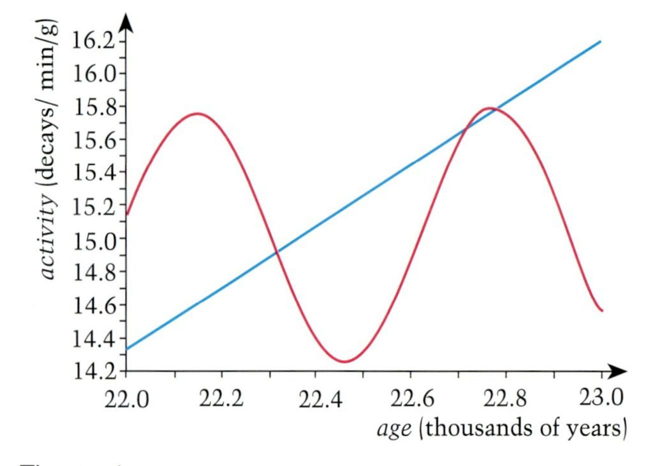

Figure 1
The ages assuming an increasing ratio at present.

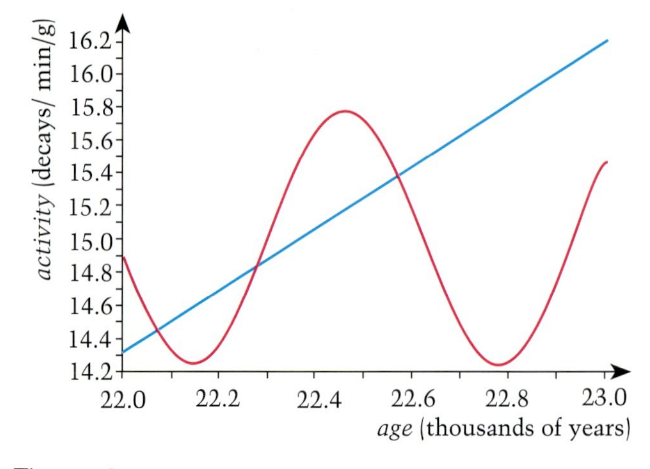

Figure 2
The ages assuming a decreasing ratio at present.

and

$$t = \frac{\ln\left(R/R_0'\right)}{-\lambda} = \frac{\ln(1/11.6)}{-\lambda}$$
$$= 6.40 \cdot 10^{11}\,\text{s}$$
$$= 20{,}300 \text{ years.}$$

If we iterate this procedure two more times, we settle in on an age of 20,500 years.

D. A graphical technique works very well for the sinusoidal dependence, where the historical decay rate is given by

$$R_0' = R_0\left[1 \pm 0.05\sin\left(\frac{2\pi t}{628 \text{ years}}\right)\right],$$

where we have used cos $(t - \pi/2) = \sin t$ to make R_0' have the current value when $t = 0$—that is, at the current time. If we let t represent time in the past, then the decay rate at any time in the past must be given by

$$R_0' = Re^{\lambda t}.$$

We get two different results (figures 1 and 2) depending on whether we assume that the ratio of the two carbon isotopes is increasing or decreasing at the current time. We see that in both cases we obtain three possible ages for the sample. About the best that we can say is that the age is 22,500 ± 300 years if the ratio is increasing and 22,300 ± 300 years if the ratio is decreasing. Ben observed that our solution to part A is equivalent to using the average atmospheric concentration and lies within both of these ranges.

Animal magnetism

"Ask the female Palme how shee
First did woo her husbands love;
Find the Magnet, ask how he
Doth th' obsequious iron move . . ."
—Thomas Stanley (1625–78)

THE MAGNETIC FORCE IS A strange beast indeed. It doesn't exist at all for neutral particles. And it only exists for charged particles if those particles are moving. Finally, the direction of the force isn't toward the magnetic field or away from the field but "sideways." This magnetic force protects us from cosmic rays by creating Van Allen belts of charged particles around the Earth; entertains us by creating pictures on our television sets; and provides a vital component of research in all areas of physics.

Let's begin to investigate this magnetic force by placing a charged particle (an electron) near a permanent magnet. If the particle is stationary, there is no force. If the electron is moving to the right and the magnet's north pole is behind this page (signified by dots in figure 1), then the force is toward the top of the page. But not for long! As the electron changes its velocity as a result of this force, the direction of the force changes as well. In fact, we observe that the electron moves in a circle. We conclude that the force must be a centripetal force. The magnetic force is always perpendicular to the velocity and to the magnetic field.

We have a way of describing this mathematically: the vector cross product. The magnetic force is

$$\mathbf{F} = q\mathbf{v} \times \mathbf{B},$$

where **F** is the magnetic force, q and **v** are the charge and velocity of the particle, and **B** is the strength of the external magnetic field. In fact, by measuring the force, charge, and velocity, this equation provides a definition for the magnetic field strength. Physicists have invented a

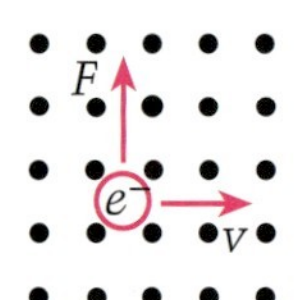

Figure 1

A depressed and tired electron is dragged into a charging machine to get some motivational skills and to start taking part in real life. After being charged, he enters another machine and he really gets a kick out of that, and suddenly he finds himself in the magnetic field where his life takes the right turn. Full of joy, he follows the rules of magnetism, without having any other choice. Completely involved now, he gets a racing car, flies a magnetic airplane, drives a luxury car, and after going broke he continues hitchhiking and then gets a job. During his mid-life crisis he becomes an easy rider and starts chasing the opposite sex. After marriage he finds himself out of breath, pulling his wife on water skis, but ahead of him are his two offspring on skates and scooter, starting anew the whole endless circle of life.

—T.B.

WE CHARGE
VISA
WE MAKE YOU MOVE!
MOVE IT!!
MAGNETIC FIELD ELECTRONS ONLY
ONE WAY
NO U TURN
BUK

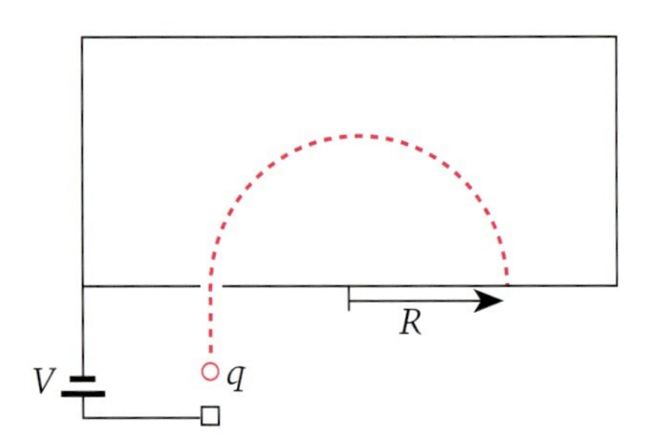

Figure 2

number of different rules to help remember the direction of a cross product. One such rule states that if you rotate the velocity vector **v** into the **B** field vector through the smaller angle between them in the same way that you would turn a screwdriver, the force is in the direction that the screw would move—for a positively charged particle. It would be opposite for a negatively charged particle. But remember, this is only true for velocity components that are perpendicular to the magnetic field. Charged particles moving parallel to the magnetic field experience no force whatsoever.

One way in which magnetic fields are used in research is in a mass spectrometer. A schematic of a mass spectrometer is shown in figure 2. Assume that we accelerate a particle with a positive charge q through a potential difference V. The particle gains kinetic energy

$$\tfrac{1}{2}mv^2 = qV.$$

Once the particle enters the magnetic field, the magnetic force provides the centripetal acceleration. Therefore,

$$qvB = \frac{mv^2}{R},$$

where R is the radius of the particle's circular path.

The mass spectrometer allows us to measure the radius of the particle and to determine its mass.

Parts of our contest problem were first given in the International Physics Olympiad (IPhO) in Czechoslovakia in 1977. This problem also appeared in the semifinal examination used to select the 1992 US team for the IPhO.

A. Show that the mass of the particle is given by

$$m = \frac{qB^2R^2}{2V}.$$

B. An electron gun accelerates through a potential difference V and emits them along the direction AB, as shown in figure 3. We want the particles to hit the target T located a distance d from the gun and at an angle α relative to AB. Find the strength of the uniform magnetic field B required for each of the following two situations: (a) the field is perpendicular to the plane defined by AB and AT; (b) the field is parallel to AT.

C. What are the numerical values of the magnetic field if V = 1,000 V, d = 5 cm, and α = 60°?

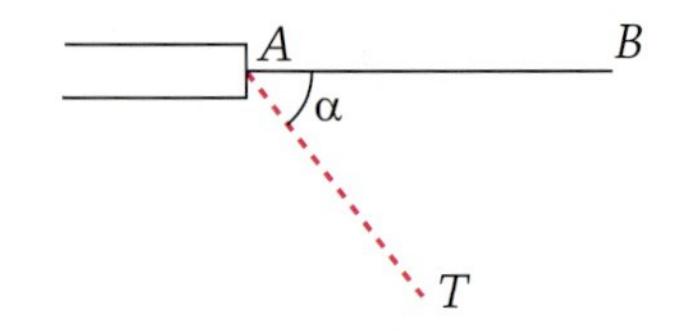

Figure 3

Solution

The best solution to this problem was submitted by Eric Joanis of Waterloo, Ontario.

In the problem we asked our readers to show that the mass of a particle in a mass spectrometer is given by

$$m = \frac{qB^2R^2}{2V}. \quad (1)$$

As explained in the problem, a particle with a mass m and charge q gains kinetic energy as it travels through a potential difference V according to

$$\tfrac{1}{2}mv^2 = qV. \quad (2)$$

Once the particle enters the magnetic field B, the magnetic force provides the centripetal acceleration

$$qvB = \frac{mv^2}{R}.$$

Combining these two equations yields equation (1) for the mass of the particle, which can be determined when the radius of its path can be measured.

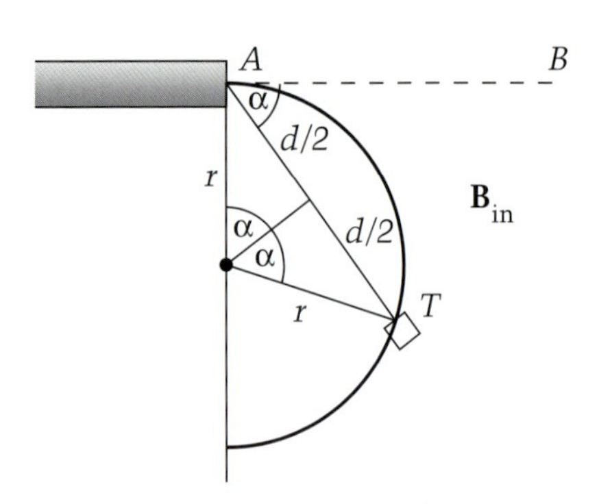

Figure 4

Part B of the problem involved some target practice with an electron in a magnetic field. If the field is perpendicular to the page, the particle travels in a circular path in the page. Since we know the mass of the electron, we can solve equation (1) for the magnetic field:

$$B = \frac{1}{r}\sqrt{\frac{2mV}{q}}.$$

From the geometry in figure 4, we see that

$$\frac{d}{2} = r\sin\alpha.$$

Therefore,

$$B = \frac{2\sin\alpha}{d}\sqrt{\frac{2mV}{q}}.$$

If the magnetic field is parallel to AT, the problem grows in complexity. There are now two components of the velocity. The component parallel to the field, $v\cos\alpha$, is unaffected by the field. The component perpendicular to the field, $v\sin\alpha$, causes the electron to move in a circle. The combined motion is that of a helix.

If the electron traveling along the helical path is going to hit the target T, the time it takes to travel a distance d (due to the parallel component) must equal the time it takes to complete one circle of the helix (due to the perpendicular component). The parallel time is given by

$$t_{\parallel} = \frac{d}{v\cos\alpha}. \quad (3)$$

The perpendicular time is

$$t_{\perp} = \frac{2\pi R}{v\sin\alpha}.$$

Because the radius R of the helix is determined by the component of the velocity perpendicular to the field, we have

$$R = \frac{mv\sin\alpha}{qB},$$

and the perpendicular time becomes

$$t_{\perp} = \frac{2\pi m}{qB}. \qquad (4)$$

Since the two times must be equal, we can equate equations (3) and (4) and solve for B to obtain

$$B = \frac{2\pi mv\cos\alpha}{qd}.$$

Solving equation (2) for v and substituting, we find that

$$B = \frac{2\pi\cos\alpha}{d}\sqrt{\frac{2mV}{q}}.$$

Note that the direction of the field does not matter.

The electron will also hit the target if it completes two circles or three circles or k circles before it travels the parallel distance to T. In that case we must modify the final equation to take this into account:

$$B = k\frac{2\pi\cos\alpha}{d}\sqrt{\frac{2mV}{q}}.$$

Part C of the contest problem asked readers to find the numerical values for the magnetic field given V = 1,000 V, d = 5 cm, and α = 60°. For the field perpendicular to the page, B = 3.7 mT; for the field parallel to the page, B = k(6.7 mT). ◘

Atwood's marvelous machines

"And your gravity fails
And negativity won't pull you through."
—Bob Dylan

YOU ARE STANDING ON THE fourth floor of a burning building and things aren't looking good. Suddenly you notice that there is a round beam sticking out of the wall with a rope draped over it. One end of the rope reaches to the ground and is tied to a sack of sand that you estimate has a mass of 45 kg. All those contrived physics problems you've solved in your lifetime have prepared you for this moment and you do not panic.

You decide to quickly calculate your chanced of surviving the ride to the ground using the rope. You start by drawing a free-body diagram like the one in figure 1 to determine all of the forces acting on the system. Since the beam is highly polished, you decide that you can ignore friction. (And it's just as well, since your physics class never got to the problems where friction was not neglected.) Letting M represent your mass, m the mass of the sack, g the acceleration due to gravity, T the tension in the rope, and a and A the accelerations of the sack and you, respectively, you apply Newton's second law to the problem at hand.

The beam must exert an upward force on the rope of $2T$ to balance the tension in each section of the rope. Let's choose the upward direction to be positive. The net force on you must be equal to your mass times your acceleration:

$$T - Mg = MA.$$

Likewise the net force on the sack

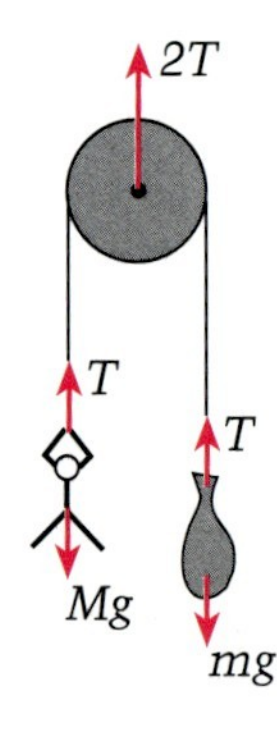

Figure 1

The old professor is sitting on his windowsill, thinking. His room is ablaze, the flames are already raging above his head. In his hand is one end of a rope, while the other end is attached to a bag of sand. To save his life he has to determine quickly the weight of the bag compared to his weight and all the other forces acting on the lifesaving action, like gravity, acceleration, friction, mass, etc. The other apartments below are also in flames, the tenants responding to the life-threatening situation each in their own way. At the same time a fat acrobat is jumping off the roof, his one leg attached to a rope, causing his skinny partner, who is holding the other end of the rope, to be pulled up and off the sidewalk. Who is going to bang his head first? Next to them is the lord of the jungle, hanging in the air, wondering what in the world he is doing up there.

—T.B.

TOMAS BUNK

must be equal to its mass times its acceleration:

$$T - mg = ma.$$

This gives us two equations in three unknowns. But there is a connection between the two accelerations. If you move downward a distance X, the sack must move upward a distance

$$x = -X$$

if the rope doesn't stretch. Since these displacements occur at the same time, the two velocities must have the same magnitudes and opposite directions:

$$v = -V.$$

Likewise the accelerations must be equal in magnitude and opposite in direction:

$$a = -A.$$

This result may seem obvious to you, but relationships that are a bit more complicated than this are often overlooked in solving physics problems like the last one we ask below.

We can substitute for a in equation (2) and reduce the problem to two equations in two unknowns. We can eliminate T by solving each equation for T and equating them to arrive at

$$A = g\frac{M - m}{M + m}.$$

If you have a mass of 90 kg, $A = g/3$, and you will hit the ground with a speed of

$$\begin{aligned} v &= \sqrt{2ax} \\ &= \sqrt{2 \cdot 3.3 \text{ m/s}^2 \cdot 15 \text{ m}} \\ &= 9.9 \text{ m/s}. \end{aligned}$$

Because this is equivalent to jumping from a height of 5 m, you have a good chance of escaping without injury. Of course, if you have less mass or are clever enough to wrap the rope around the beam to maximize the effects of friction, you'll have a safer fall.

This problem is an example of a classic physics problem known as Atwood's machine. In the lab it serves as a means of achieving a constant acceleration of any value less than g. Can you suggest a means of achieving a constant acceleration greater than g? Atwood's machine is often modified to test students' understanding of the applications of Newton's laws. For instance, the beam can be at the top of a wedge with the masses sliding on sloped, frictionless surfaces. And, of course, we could always include the effects of friction.

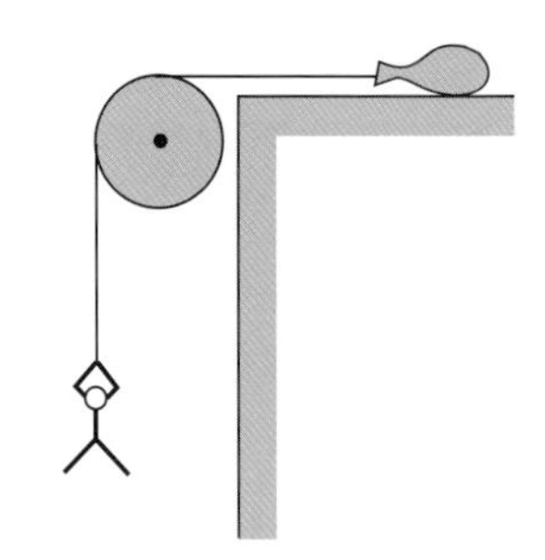

Figure 2

Our problem consists of two modifications of Atwood's machine. The second one (without the hints) appeared on the first of several examinations used in selecting the members of the U.S. Physics Team that competed in the 1993 International Physics Olympiad in Williamsburg, Virginia.

A. If the sack slides across the floor (assumed to be located at the height of the beam and frictionless) as shown in figure 2, what is your acceleration and the tension in the rope, assuming that your mass is 90 kg? This is sometimes called the "half Atwood's machine" and is often used in introductory physics labs to demonstrate Newton's second law. Be sure to check your answers in the limits of each mass going to zero to see if you get the expected results.

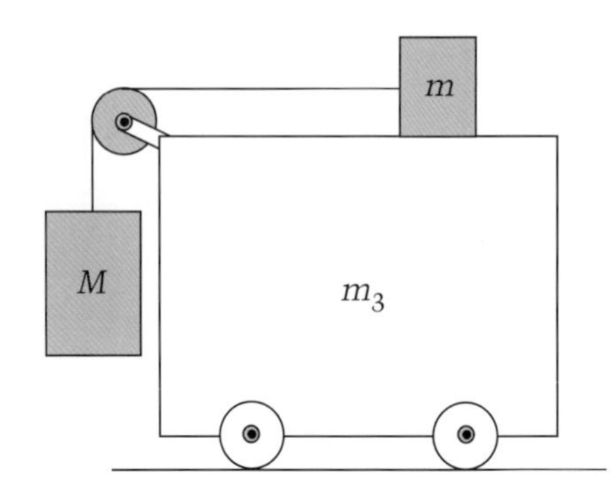

Figure 3

B. Let's assume that we have a scaled-down version of the situation in part A and that the half Atwood's machine is mounted on a car of mass m_3 that is free to move as shown in figure 3. What are the accelerations of all three masses and the tension in the rope just after the masses are released at rest?

If we let the system continue to move, the physics gets very complicated and requires advanced techniques for its solution. However, at time $t = 0$, the solution can be obtained with the techniques used to solve the normal Atwood's machine. Be sure that you draw free-body diagrams for all three masses and then write down Newton's second law for each mass. This will give you three equations in four unknowns (a_1, a_2, a_3, and T). If we assume that the rope does not stretch, you should be able to find a relationship between the three accelerations that will give you the fourth equation you need to solve the problem. As usual, be sure to include checks for the expected answers in the extreme cases.

Solution

Correct solutions to the problem were submitted by Jeff Dodson of Vista, California, and Scott Wiley of Weslaco, Texas.

Figure 4 shows the situation for part A. Let's choose a coordinate system in which down and left are positive. This means that both masses will have positive displacements when the system is released. Using the notation in figure 4 we can write down Newton's second law for each mass:

$$Mg - T = MA, \quad (1)$$
$$T = ma. \quad (2)$$

With our choice of coordinates, we can write the connection between the accelerations as

$$A = a. \quad (3)$$

Solving equations (1–3) yields

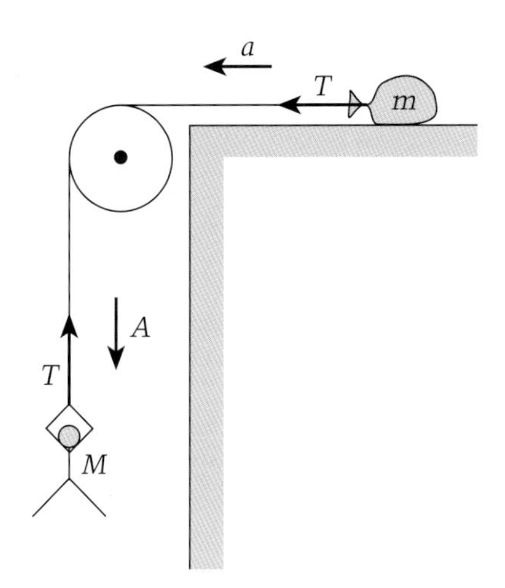

Figure 4

$$A = \left(\frac{M}{M+m}\right)g = \frac{2}{3}g,$$

$$T = \left(\frac{Mm}{M+m}\right)g = \frac{1}{3}Mg.$$

Notice that the acceleration of the system is larger than we calculated for the full Atwood's machine. This makes sense because the sack is no longer being retarded by gravity.

If we place the system in part A on a moving cart, we have the situation depicted in figure 5. Using the notation in figure 5 and our earlier choice of coordinates, we can once again write down Newton's second law for each mass:

$$Mg - T = Ma_1, \quad (4)$$

$$T = ma_2, \quad (5)$$

$$-T = m_3a_3. \quad (6)$$

We have used the observation that the tension in the rope exerts a force T to the left on mass m and therefore by Newton's third law must exert a force T to the right on the beam (and hence on the cart).

We now need a relationship between the accelerations. To do this we look at very small displacements of the three masses. Mass M will fall a distance d_1 that is equal to the distance that mass m moves to the left and the distance the cart moves to the *right*. Therefore,

$$d_1 = d_2 - d_3$$

and

$$a_1 = a_2 - a_3. \quad (7)$$

The easiest way to solve equations (4–7) is to use equation (7) to replace a_1 in equation (4) and use equations (5) and (6) to replace a_2 and a_3. Solving for T we get

$$T = \frac{mMm_3}{Mm + Mm_3 + mm_3}g.$$

Plugging this value for T back into equations (4–6) gives us the values for the accelerations:

$$a_1 = \frac{M(m+m_3)}{Mm + Mm_3 + mm_3}g,$$

$$a_2 = \frac{Mm_3}{Mm + Mm_3 + mm_3}g,$$

$$a_3 = \frac{-Mm}{Mm + Mm_3 + mm_3}g.$$

Notice that in the limit $m_3 \to \infty$ these equations reduce to the answers we obtained in part A.

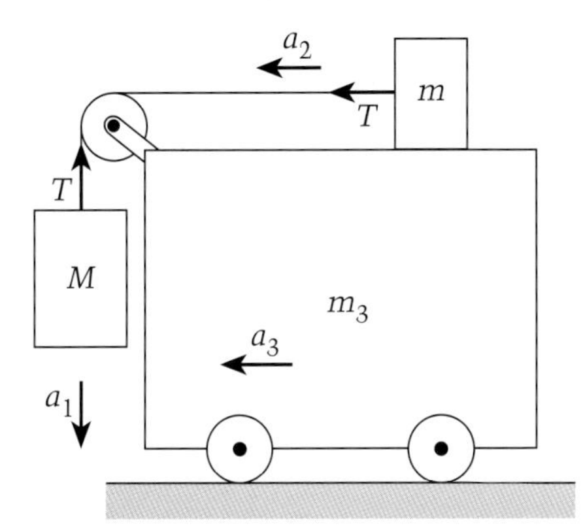

Figure 5

Thrills by design

"Centrifugal power . . . what stillnesses lie at your center resting among motion?"
—Muriel Rukeyser (1913–1980)

AT THE XXIV INTERNATIONAL Physics Olympiad, which was held in Williamsburg, Virginia, during July 1993, students from the 41 participating countries spent a day investigating the physics of some of the amusement part rides at Busch Gardens. Those roller coasters and bumper cars and swings certainly move the awareness of momentum, forces, and acceleration from our brains to our guts.

One ride that appears simple enough to analyze is the rotor. As in many physics explorations, the analysis reveals a hidden effect—a treasure that you may not have previously known. The rotor is a hollow cylinder of radius 2.5 m. Riders stand inside the cylinder with their backs against the wall. As the rotor spins, they feel as if they are being pushed against the wall. When the maximum speed is reached, the floor drops out! As shouts emerge, the riders don't fall. The friction between the wall and the riders keeps them from slipping down.

Let's analyze this ride, keeping our minds ready to discover that extra treasure. From the perspective of a person above the ride, a person is kept from flying through the wall by the push of the wall. This normal force $\mathbf{F}_N$ is directed toward the center of the cylinder. The force of gravity $\mathbf{F}_g$ pulls the person vertically down. The frictional force $\mathbf{F}_f$ is directed vertically up and must be equal to $\mathbf{F}_g$ or the person will slide down.

The horizontal normal force $\mathbf{F}_N$ must supply the required centripetal force to keep the rider moving in a circle:

$$F_N = \frac{mv^2}{R},$$

where m is the mass of the rider, v is the velocity of the rider, and R is the radius of the rotor. The equality of $\mathbf{F}_g$ and $\mathbf{F}_f$ yields the following relationship:

The stage is set for fun and here they are, dozens of spinning rotors enter like twirling dervishes, and as the speed accelerates the riders start to feel the effects. Pushed with their backs to the wall they feel as if they're glued to it, unable to move freely. Some ride upside down, some are being pushed out of the semicircle and land on the floor, some are whirling so fast they take off into space and disappear, and some tops slow down and all the riders slide down in a heap. Mr. Einstein in his space-decorated top is even able to play some rounds on his violin. The scene reminds us that our life, too, is a gigantic ride on a ball, spinning in the universe, glued thanks to gravity to Earth's surface. But unfortunately life is no amusement park, where you come and go whenever you want. You get only one turn and when you're done, you're out.

—T.B.

Tom Bunk

$$mg = \mu F_N,$$

where μ is the coefficient of friction. Substituting for F_N, we can find the required coefficient of friction:

$$\mu = \frac{gR}{v^2}.$$

The coefficient of friction determines the required minimal speed for any rotor! Next time you get to watch or ride on a rotor, take a look at how the ride's designers have increased the coefficient of friction—did they add carpeting to the walls, or did they use rough paint?

Some of you are probably still searching for the surprise discoveries. One minor surprise is that the ride works just as well irrespective of the mass of the rider. The more interesting surprise is that, from the reference frame of the rider, the question "Which way is up?" takes on new meaning. The rider feels gravity pulling one way and a centrifugal force pulling outward. The combination of the two defines a "new gravity" in the rider's balance system. Riders think that they are lying at an angle and are no longer vertical. Next time you're on the rotor, try to be aware of this effect. Estimate the angle of apparent tilt, and check to see if it's consistent with your estimates of $\mathbf{F}_N$ and $\mathbf{F}_g$.

One of the challenges for the engineers working for amusement parks is to develop new, exciting, and safe alternatives to the tried-and-true classics. We thought that *Quantum* readers might enjoy such a design challenge. We'll describe a traditional physics problem—one that was used in the International Physics Olympiad in Budapest, Hungary, in 1976. Your challenge is to understand the physics of the problem and then to incorporate the design into an amusement park ride.

A hollow sphere of radius R = 0.5 m rotates about its vertical diameter with an angular velocity $\omega = 5\ \text{s}^{-1}$. Inside the sphere at the height $R/2$ a small block revolves together with the sphere. (Use $g = 10\ \text{m/s}^2$.)

A. What is the coefficient of friction required for the block to continue to revolve at this height?

B. What is the coefficient of friction required when $\omega = 8\ \text{s}^{-1}$?

C. Investigate the problem of stability (1) for small variations in the position of the block and (2) for small variations in the angular velocity of the sphere.

D. Can the block be replaced by a person as a design for an amusement park ride? Are there any inherent problems with such a ride? Would the public enjoy such a ride? How would you get on and off such a ride?

Solution

The two best solutions came from Stephan Chan of Ontario, Canada, and Chao Ping Iris Yan of Rio de Janeiro, Brazil.

The example provided in our problem explained the physics of the rotor—a hollow cylinder that spins and then "pins" the riders against the wall as the floor drops out. The ride that we hoped readers would design used a rotating hemisphere. The physics of the rotating cylinder and rotating hemisphere are similar in that a centripetal force must be furnished to keep the passenger moving in a circle. In the rotor, the normal force supplied this centripetal force. In the hemisphere, the horizontal components of the normal force and the frictional force must provide the centripetal force. Another difference in the analysis of the two rides is in the measurement of the radius. In the rotor, the radius of circular movement is equivalent to the radius of the cylinder. In our hemisphere, the radius of circular movement is equal to a component of the radius of the hemisphere—$R \sin\alpha$. The final difference is that in the cylinder, we recognized that the frictional force keeps the rider from slipping down. In the hemisphere, the frictional force may keep the rider from sliding down or from sliding up!

We wish to find the minimum coefficient of friction required to keep the rider from sliding down when the angular velocity ω is small

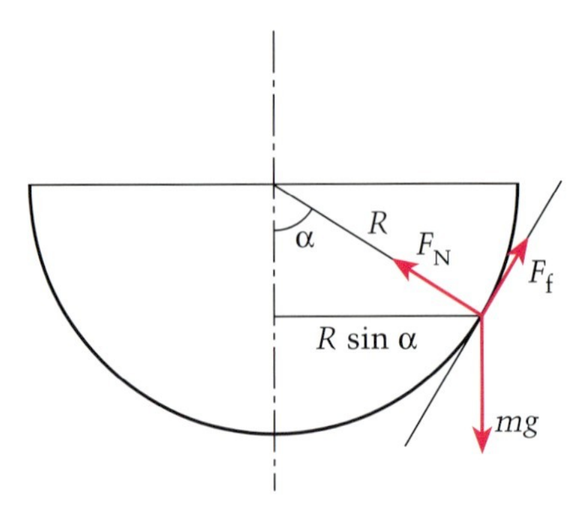

Figure 1

(5 radians per second). As with many physics problems, the first step is a carefully drawn diagram and vector analysis (fig. 1)

The sum of the horizontal components must equal the centripetal force; the sum of the vertical components must equal zero:

$$\sum F_x = F_N \sin\alpha - F_f \cos\alpha = m\omega^2 R \sin\alpha,$$

$$\sum F_y = F_N \cos\alpha + F_f \sin\alpha - mg = 0.$$

Since the frictional force is (less than or) equal to the coefficient of friction μ multiplied by the normal force, we can solve the simultaneous equations for μ:

$$\mu \ge \sin\alpha \frac{1 - \omega^2 R \cos\alpha / g}{\cos\alpha + \omega^2 R \sin^2\alpha / g}.$$

For the values given (ω = 5 rad/s, R = 0.5 m, and $\alpha = 60°$), we get

$$\mu \ge \frac{3\sqrt{3}}{23} = 0.23.$$

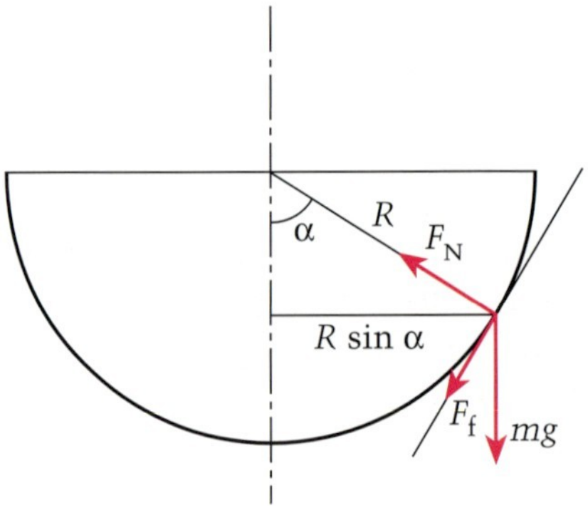

Figure 2

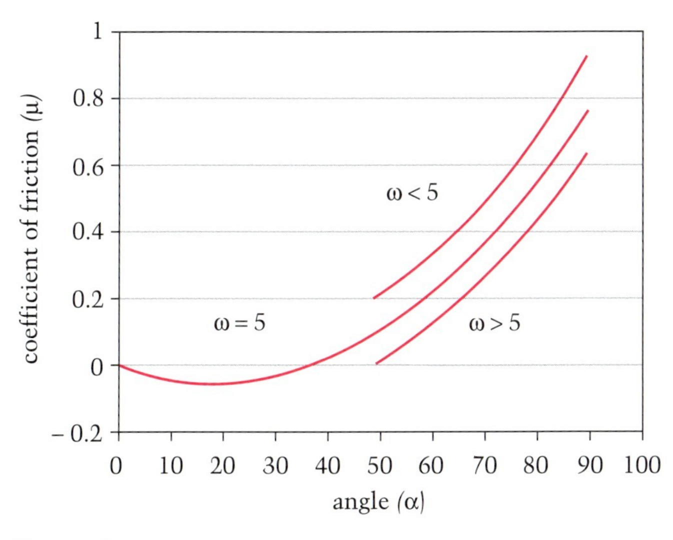

Figure 3

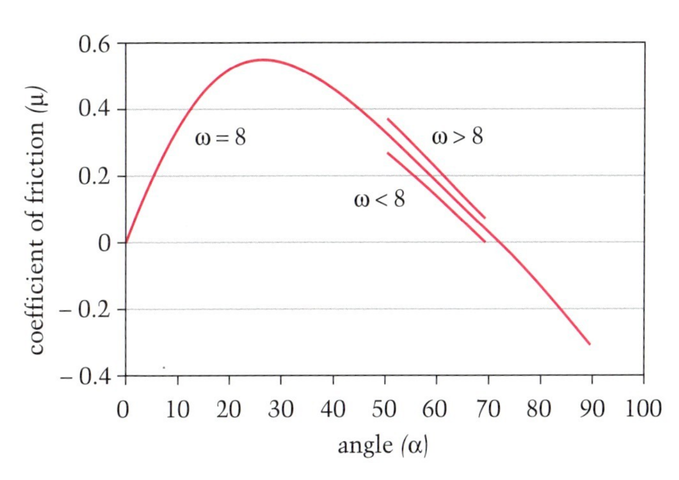

Figure 4

Part B asked for the coefficient of friction required when $\omega = 8$ rad/s. The analysis is similar except that the vector diagram in figure 2 now shows that the frictional force is preventing the object from slipping up the hemisphere. Solving the simultaneous equations again for μ, we get

$$\mu \geq \sin\alpha \frac{\omega^2 R\cos\alpha/g - 1}{\cos\alpha + \omega^2 R\sin^2\alpha/g}.$$

Substituting the values given ($\omega = 8$ rad/s, $R = 0.5$ m, and $\alpha = 60°$) gives us

$$\mu \geq \frac{3\sqrt{3}}{29} = 0.18.$$

In part C of the problem, we wanted to analyze the stability of the mass in the hemisphere ride for small variations of the position of the block and for small variations of the angular velocity of the block.

Using a graphing calculator, a spreadsheet program, or sketching, we can look at a graph with three curves (fig. 3). The main curve shows the relationship between μ and the angle. This μ is the friction required to remain at that angle. The two other curves show that relationship for different values of ω. If the object moves to a higher angle, the minimum friction required to stay at that height is greater. The object does not have that much friction and it slides back down to the original position, where the friction is sufficient to have it remain at that height. If the object moves to a smaller angle, the friction required to remain at that position is less. The object is able to remain at this height. If ω increases, the object will remain where it is (since less friction is required at that new ω); if ω decreases, the object will not have the friction required to maintain its position and will start to slide down to an angle where the friction provided is sufficient for this ω.

At a higher initial ω, the graph reveals a different situation (fig. 4). If the object moves to a higher angle, it stays there; if the object moves to a smaller angle, it will return. If ω increases, the block slides up; if ω decreases, the block will maintain its position.

In part D readers were asked if this hemisphere problem could be a ride for an amusement park and what problems might arise. Chan was able to show that the person would experience an acceleration of approximately 1.4*g* during this ride. He thinks that the person would enter the ride from the bottom, and as the ride spins the rider would slide up against the wall. He doesn't see it as an exciting ride as it stands—he suggests that we increase the speed to make the *g* forces greater.

Yan thought that getting on the ride could be accomplished with a floor at a height equal to $R/2$. The floor would then rotate out of the way during the ride. This would limit the riders to only one side of the hemisphere. Yan suggests that the velocity be increased and decreased during the ride. Yan concludes that the ride may be too scary and people would probably be more secure on the roller coaster.

Electricity in the air

"Go, wondrous creature! mount where science guides,
Go, measure earth, weigh air, and state the tides . . .
Go, teach Eternal Wisdom how to rule—
Then drop into thyself and be a fool!"

—Alexander Pope

THIS PROBLEM IS BASED ON part of one of the theoretical problems given at the XXIV International Physics Olympiad that was held in Williamsburg, Virginia, in July. The problem was written by Anthony French of MIT, who served as the chair of the examinations committee, and is based on an actual application of physics to a real-world situation. The first part of the solution is based on Gauss's law, one of the most fundamental laws of electricity and magnetism.

Carl Friedrich Gauss was the greatest mathematician of his time and along with Archimedes and Newton may have been one of the three greatest mathematicians ever. He developed the method of least squares for fitting curves to data points and used this method to calculate an orbit for Ceres, the largest of the asteroids, after it couldn't be found. He was honored for this work when the name Gaussia was given to the 1001st asteroid. The gauss—a unit of magnetic field strength equal to 10^{-4} tesla—honors his work in magnetism. While still a university student he devised a method of drawing a seventeen-sided regular polygon using only a compass and straightedge. He then went further to show that certain regular polygons (for example, one with seven sides) could not be constructed this way.

Gauss's law tells us that the electric flux through a closed surface is proportional to the electric charge that is enclosed by that surface. To calculate the electric flux, we imagine dividing the surface into many small regions. For each region the contribution to the electric flux is given by the component of the electric field perpendicular to the surface E_n times the surface area A of that region. By convention, the contribution is positive if the electric field is directed out of the enclosed volume and negative if the electric field is directed inward.

Using the well-established 3-finger method, the scientist who seems to be in charge of ruling the electric field activities in front of him is sitting on top of the hill. With his enormous top hat that looks like a pile of brain colored pancakes, the scientist is measuring the electric fields, weighing the air, predicting the tides, and teaching the eternal wisdom in his scientific law school, which is a magnet for school buses. There is one in front of us, hopping 'round the corner and filled with happy ionic electric flux students, whose teacher is sowing electric light bulbs to grow illumination. For the scientist everything seems to be under control, but already Einstein is coming up with some paradigm-shifting theories that could possibly make the pancakes drop into themselves and turn the man in charge into the fool on the hill.

—T.B.

ELECTRIC FIELDS 100 m Alt.
SORRY CLOSED SURFACE
FLUX AIR LINES
GAUSS'S CHARGED CYLINDER
100m
80m
60m
50m
40m
30m
20m
MAGNET SCHOOL
GAUSS'S LAW SCHOOL
ELECTRIC FLUX
BUTTER CANNON
CHARGE IN MOTION
IONS ELECTRIC FLUX TOURS
$4\pi k_e x$?
GAUSS
CHARGE PER UNIT AREA
VISA
ATTENTION ELECTRIC CURRENCY
Tom Bunk
0,0m
$E=mc^2$
RELATIVELY NEW FIELD MAP
VERTICAL ELECTRIC FIELD
V-SPEED SIGNS AHEAD

Because total electric flux is just the sum of all of the individual contributions, we can write Gauss's law in the form

$$\sum E_n A = \frac{q_{enc}}{\varepsilon_0},$$

where $\varepsilon_0 = 8.85 \times 10^{-12}$ C^2/(N · m^2) is the permittivity of free space. For more information about Gauss's law, see the problem entitled "Sources, Sinks, and Gaussian Spheres."

Gauss's law is very useful for finding electric fields in cases of high symmetry. For example, let's find the electric field outside of an infinitely long, straight wire carrying a positive charge per unit length λ. To exploit the symmetry, we choose the gaussian surface to be a cylinder of radius r and length L that is coaxial with the wire. By symmetry, we expect that the electric field will point radially outward from the wire and have the same magnitude at a given distance from the wire. This means that the electric field will be parallel to the ends of the cylinder and will not contribute to the flux. Therefore, the flux is given by the electric field times the area of the curved surface of the cylinder:

$$\sum E_s A = E 2\pi r L.$$

The enclosed charge is equal to the charge per unit length times the length of the cylinder:

$$q_{enc} = \lambda L.$$

Putting these two expressions into Gauss's law, we can solve for the magnitude of the electric field:

$$E = \frac{\lambda}{2\pi\varepsilon_0 r}.$$

Notice that the length of the gaussian cylinder cancels as we expect.

From the standpoint of electrostatics, the surface of the Earth can be considered a good conductor that carries a total charge Q_0 and an average surface charge density σ_0. We can also consider the Earth a perfect sphere with a radius R = 6,400 km to simplify the geometry. Under fair-weather conditions, this surface charge density produces a downward electric field E_0 at the Earth's surface equal to about 150 V/m.

A. Use Gauss's law to calculate the magnitude of the Earth's surface charge density and the total charge carried on the Earth's surface. Is this charge positive or negative?

The magnitude of the downward electric field is observed to decrease with height and is about 100 V/m at a height of 100 m. This occurs because the air above the Earth's surface contains a net charge.

B. Use Gauss's law to calculate the average net charge per cubic meter of the atmosphere between the Earth's surface and an altitude of 100 m. Is this charge positive or negative?

The net charge density you calculate in part B is actually the result of having almost equal numbers of positive and negative singly charged ions (q = 1.6 · 10^{-19} C) per unit volume (n_+ and n_-). Near the Earth's surface, under fair-weather conditions, $n_+ \cong n_- \cong 6 \cdot 10^8$ m^{-3}. These ions move under the action of the vertical electric field and their speed v is proportional to the strength of the electric field:

$$v \cong 1.5 \cdot 10^{-4} \times E,$$

where v is in m/s and E is in V/m.

C. How long would it take for the motion of the atmospheric ions to neutralize half of the Earth's surface charge, if no other processes such as lightning occurred to maintain it?

Solution

We follow the solution provided at the International Physics Olympiad held in the United States in July 1993.

We begin part A by assuming that we have a spherical gaussian surface that is only slightly above the Earth's surface. Therefore, the radius of this surface is R. Because the electric field points radially, the total electric flux through this surface is just the product of the surface area of the sphere A and the electric field E_0. Gauss's law tells us that

$$-E_0 A = \frac{Q_0}{\varepsilon_0},$$

where Q_0 is the total charge enclosed by the surface and the minus sign is included because the electric field is directed into the sphere. Because

$$Q_0 = \sigma_0 A,$$

where σ_0 is the Earth's surface charge density, we can solve for either the charge density or the total charge. Let's find the charge density:

$$\begin{aligned}\sigma_0 &= -\varepsilon_0 E_0 \\ &= \left(-8.85 \cdot 10^{-12} \frac{C^2}{N \cdot m^2}\right)\left(150 \frac{N}{C}\right) \\ &= -1.33 \cdot 10^{-9} \frac{C}{m^2},\end{aligned}$$

where we have replaced the units V/m by N/C. The minus sign tells us that the charge on the Earth is negative, which we also know from the direction of the electric field. We can now find the total charge on the Earth:

$$Q_0 = \sigma_0 A = \sigma_0 4\pi R^2 = -6.85 \cdot 10^5 \text{ C}.$$

Part B required our readers to calculate the average net charge per cubic meter of the atmosphere given the electric field at a height of 100 m. Many students at the International Physics Olympiad solved this part of the problem by considering the gaussian surface to consist of two concentric spheres, one with a radius R and the second with a radius $R + h$ with h = 100 m. However, since $R << h$, the Earth's surface is relatively flat on the scale of the problem. Therefore, it's simpler to consider a cylinder with a cross-sectional area S and a height h sitting just above the Earth's surface, as shown in the figure.

The walls of the cylinder do not contribute to the electric flux, because the electric field is parallel to the walls. Therefore, Gauss's law

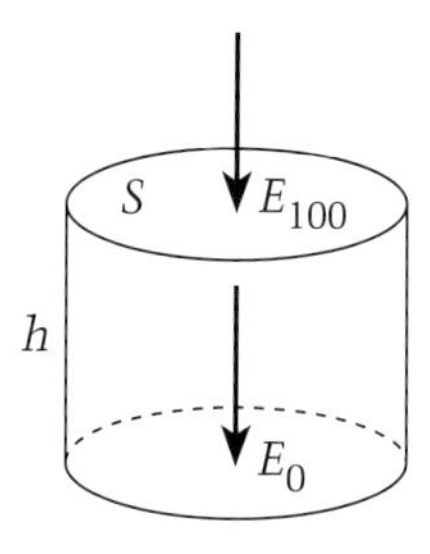

tells us

$$S(E_0 - E_{100}) = \frac{Q_{enc}}{\varepsilon_0} = \frac{\rho Sh}{\varepsilon_0},$$

where ρ is the average charge density inside the cylinder and the contribution of E_{100} to the flux is negative. Using the data in the problem, this yields

$$\rho = \frac{\varepsilon_0}{h}(E_0 - E_{100}) = 4.42 \cdot 10^{-12} \frac{\text{C}}{\text{m}^3}.$$

Notice that the charge density is positive.

To solve part C, we first note that under the influence of the electric field, the positive ions move downward and the negative ions move upward. Therefore, only the positive ions can contribute to the cancellation of the surface charge density. The current per unit area j is given by

$$j = n_+ qv = (1.44 \cdot 10^{-14})E,$$

where we have used the values and relationship given in the statement of the problem. Note that the constant must have units of A/V · m.

Now, j is the rate of change of the surface charge density $(\Delta\sigma/\Delta t)$, and $E = -\sigma/\varepsilon_0$ from part A. Therefore,

$$\frac{\Delta\sigma}{\Delta t} = -1.63 \cdot 10^{-3}\sigma = -\frac{1}{613}\sigma.$$

This is the same relationship that we encounter in radioactive decay. Therefore, its solution is an exponential decrease of σ with time:

$$\sigma(t) = \sigma_0 e^{-t/\tau},$$

with $\tau = 613$ s. Putting $\sigma(t) = \sigma_0/2$ gives

$$t = \tau \ln 2 = 425 \text{ s} \cong 7 \text{ min}.$$

Stop on red, go on green . . .

"When you walk on a path going north, you will only meet people coming from the north. At the crossroads, you'll meet people coming from the east, from the west . . ."

—Nouk Bassomb

WHEN YOU'RE DRIVING down a road and you see a yellow light, don't you wonder when the light will turn red? Maybe there should be an additional light—say, a blue one—that tells you that the yellow light will be changing to red any moment now. But then again, maybe there should be an orange light that tells you that the blue light will be ending soon and that the red light is imminent. But then again . . .

Who needs yellow lights at intersections? Who decides whether the yellow light should be one second, two seconds, or four seconds? Are yellow lights always set to encourage safe driving? Let's analyze what happens when you approach a yellow light.

As you drive down the road at a certain speed, you may see the light turn from green to yellow. You must make a decision to keep going or to step on the brakes and come to a stop. If you're relatively close to the intersection, you know that you can continue at the same speed and make it through while the light is still yellow. If your distance to the intersection is larger, you may decide to stop.

Let's assume that you want to keep going. To calculate your safe distance from the intersection, we simply calculate the distance you must go to get through the intersection while the light is yellow. This may be easier to follow if we use some real numbers as an example. Let's assume that the speed limit is 50 km/h, which is equivalent to 30 mph or 23 m/s. Let's also assume that the yellow light is on for 3.0 s before the light turns red. Therefore, you must travel a distance of 69 m during the time the light is yellow. If the width of the intersection is 15 m, you can safely proceed through the intersection if you are closer than 54 m. We'll call this the "go zone."

If you decide to stop when you see

Crossroads are always a risky challenge, with or without a traffic light. Decisions have to be made quickly, sometimes there is not enough time to take all the possibilities into consideration. We can stop suddenly and have all the cars behind us bumping into each other, or try to speed up and get stuck in the middle of the intersection. We can take a wrong turn and end up off the road. But not only as drivers but as humans in general standing at a crossroads, we sometimes have to make tough decisions not knowing where the turn will lead us, to disaster or success, to glory or tragedy. Not being able to see the future, we may decide not to do anything but to go to sleep and dream about flying cars that can go anywhere anytime. But watch out—there is already a Rolls Royce slipping off the road and on its way to crashing right in the middle of your dreams.

—T.B.

A&O
JUST MARRIED
VIP
Tomas Bunk

the light turn yellow, you must know the distance you will travel as you move your foot from the gas pedal to the brake (the coasting distance) and the distance it takes your car to stop (the braking distance). Once again, let's look at some real numbers and perform a calculation. The car is once again traveling at 23 m/s. If your response time is 1.0 s, the car will travel a distance of 23 m. If the deceleration of the car is 5 m/s², the car will travel an additional 53 m while braking. This distance is calculated according to the following equation:

$$v_f^2 - v_0^2 = 2as,$$

where v_f is the final velocity, v_0 is the initial velocity, a is the acceleration, and s is the distance traveled. The car can be safely stopped if it is at least 76 m from the intersection. We'll call this the "stop zone."

But wait—what happens if you're 65 m from the intersection? If you try to stop, you'll find yourself in the intersection. If you try to continue, you'll find yourself going through a red light. You're in trouble! We'll call this the "dilemma zone."

A safer intersection would not have a dilemma zone. If the yellow light time were 4.0 s, the go zone would be 77 m. The stop zone would still be 76 m. If you are closer than 77 m, you can safely proceed. If you are farther than 76 m, you can safely stop. If you are between 76 and 77 m, you can safely go or stop. This "overlap zone" provides for a safe intersection.

Rather than using data from a single intersection, our problem asks you to do the work of a highway engineer and provide the relevant equations for safe intersections.

A. What is the general equation for the (a) go zone, (b) stop zone, (c) dilemma zone or overlap zone?

Assume a response time t_r, a maximum braking acceleration a, a yellow light time t_y, a speed v_0, an intersection width w, and a car length l.

B. For what speeds will there always be a dilemma zone?

C. Rewrite the equations in part A assuming that the car is going downhill when you see the yellow light.

Solution

An excellent solution was submitted by Ophir Yoktan of Israel. Unfortunately, Yoktan provided no biographical information and so we don't know if Yoktan is a professor or a student. Irrespective of that, the solution presented here closely follows Yoktan's submission.

A. (a) In the "go zone" a person will be able to continue at the traveling speed and get through the intersection within the time that the yellow light is illuminated. This depends on the velocity of the car v_0, the yellow light time t_y, the width of the intersection w, and the length of the car l. (The go zone is quite different for a stretch limo and a compact car.) This gives us

$$d_g < v_0 t_y - w - l.$$

(b) In the "stop zone" a person will be able to stop before the intersection. It depends on the velocity of the car v_0, the acceleration a of the car while braking (a negative number), and the reaction time of the driver t_r:

$$d_s > v_0 t_r - \frac{v_0^2}{2a}.$$

(c) Whether we have a "dilemma zone" or an "overlap zone" depends on the difference between the go zone and the stop zone. If the zone is defined as the go zone minus the stop zone, a negative value will indicate a dilemma zone and a positive value will indicate an overlap zone:

$$\begin{aligned} \text{zone} &= v_0 t_y - w - l - v_0 t_r + \frac{v_0^2}{2a} \\ &= \frac{1}{2a} v_0^2 + (t_y - t_r) v_0 - (w + l). \end{aligned}$$

B. We can find the conditions for which there will always be a dilemma zone by requiring that the zone be negative:

$$0 > \frac{1}{2a} v_0^2 + (t_y - t_r) v_0 - (w + l).$$

See the equation in the box below. A dilemma zone will always exist if the terms within the radical sign are negative. This occurs if the response time is greater than the yellow light time. (In this unrealistic case the yellow light time does not allow any decision making.) This also occurs when the term containing the acceleration is small in comparison to the difference in the yellow light and reaction times.

If the radical is positive, we will always get two positive values for v_0. Calling the smaller root v_1 and the larger root v_2, we see that there will always be a dilemma zone if $v_0 > v_2$ or $v_0 < v_1$ and an overlap zone if $v_1 < v_0 < v_2$. Physically it's easy to understand why a high speed can produce a dilemma zone. Why can a low speed produce a dilemma zone?

Another way of viewing the problem is to realize that people will usually be traveling at a typical speed for this type of road and we wish to set the yellow light time to make it a safe intersection. In this case, we should solve the equation for the yellow light time t_y. The dilemma zone exists for the following values of t_y:

$$t_y < t_r + \frac{(w + l)}{v_0} - \frac{v_0}{2a}.$$

An overlap zone exists when t_y is larger than this value.

This new equation allows us to set the yellow light time at an in-

$$\begin{aligned} v_{1,2} &= \frac{-(t_y - t_r) \pm \sqrt{(t_y - t_r)^2 - 4 \cdot \frac{1}{2a} \cdot (w + l)}}{2 \cdot \frac{1}{2a}} \\ &= a\left[(t_r - t_y) \pm \sqrt{(t_y - t_r)^2 + \frac{2(w + l)}{a}}\right] \end{aligned}$$

tersection, since we can assume that the velocity, the braking acceleration, the response time, and $(w + l)$ are constants.

C. If the car is traveling downhill, there is an acceleration equal to $g \sin \alpha$, where α is the slope. If we let $g' = g \sin \alpha$, then

$$\text{go zone} = v_0 t_y + \frac{1}{2} g' t_y^2 - w - l,$$

$$\text{stop zone} = v_0 t_r + \frac{1}{2} g' t_r^2 - \frac{(v_0 + g' t_r)^2}{2(a + g')}.$$

The overlap zone is, once again, the difference between the go zone and stop zone. The go zone increases as a result of the hill (if the driver allows the car to accelerate down the hill—this is not safe driving!); the stop zone also increases (if the driver loses some braking acceleration as a result of the hill—this is not usually true). We can see from these equations that if the acceleration a is due to a heavier foot on the gas pedal, the go zone does increase, as one might expect. This increase in the go zone is much smaller than we might anticipate, however, and the acceleration would lead to bigger problems if an accident were to occur at the higher speeds.

Fun with liquid nitrogen

"Some say the world will end in fire,
Some say in ice."

—Robert Frost

IRON IS SOLID, MERCURY IS liquid, and nitrogen is a gas. We gain our familiarity with substances at ambient temperature and tend to think of them in that context. Over millennia our technology has found ways to heat iron so that it becomes a liquid and to cool mercury so that it becomes solid. Cooling nitrogen to form a liquid was first achieved in the 1870s. And the world of liquid gases—liquid hydrogen, oxygen, and nitrogen—could not be stranger.

Liquid nitrogen is used to perform lots of interesting experiments. It's also fun. Demonstrations exploiting the extreme cold of liquid nitrogen provide entertainment for children of all ages. Since liquid nitrogen boils at a temperature of 77 K at atmospheric pressure, we keep it cold in a dewar. If we pour some of the liquid nitrogen on the floor, the liquid forms droplets that scoot around the floor like droplets of water on a hot skillet. Of course, the floor *is* like a hot skillet to the very cold liquid nitrogen droplets!

The liquid nitrogen can also be used to superfreeze common materials. In another demonstration we take a rod-shaped piece of rubber sharpened on one end and drop it into the liquid nitrogen. We then remove it with tongs and hammer it into a board. Frozen rubber is as good as a nail—until it thaws. Some things become brittle at these cold temperatures. It's rather spectacular to shatter a frozen banana or a flower with a hammer. It's as if they were made of glass. One class of kindergartners remembered this demonstration five years later.

The expansion and contraction of gases also seem spectacular when liquid nitrogen is used. A blown-up balloon inserted into the liquid nitrogen shrinks down to essentially zero volume, showing that the ideal gas law is not valid for these conditions. The gases in your breath liquify or freeze at these tempera-

Ice or fire? For the skeletons on their balconies to the left and right, it doesn't matter anymore how the world will end, they enjoy the Comedia dell'Arte shtick, where the Harlequin shows how he splits himself into pieces in seven cool steps. Changing many hats on the way, he succeeds in falling apart, never to be put back together again, like Humpty Dumpty, here dressed as the world, who obviously gets pushed off the wall by a jester. While the world in ending by ice at center stage, the end by fire is shown in the orchestra pit, where the mischievous Frank Zappa inflames his hellish musicians to play hot tunes. Blazing comets crash down from the night sky, New York is engulfed in flames, and an iceberg erupts, spitting fire and brimstone, while three cute penguins on an ice floe remind us to repent—why exactly they don't say.

—T.B.

1
2
3
4
5
6
7
LIQUID NITROGEN
REPENT THE END IS NEAR!
TOMAS BUNK
NO SMOKING
FIRE EXIT

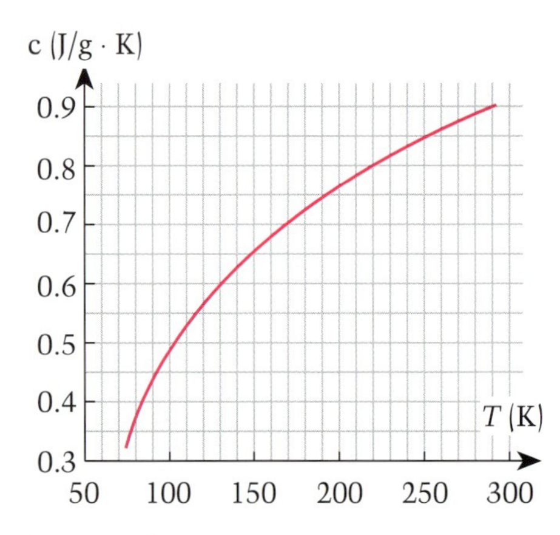

Figure 1
Specific heat of aluminum.

tures. In fact, the balloon makes a good rattle if shaken as it warms up. Usually the balloon expands back to its original volume as it warms, but not always!

In this problem we want to measure the latent heat of vaporization of liquid nitrogen. The latent heat of vaporization is the amount of heat required to convert a unit mass of liquid to vapor at the boiling point of the substance. This is based on one of the two experimental problems given at the XXIV International Physics Olympiad held during the summer of 1993 in Williamsburg, Virginia.

Our method is a variation of the thermal experiment that many of you have performed in your school laboratory. We usually use the known thermal properties of water to measure the specific heat of a block of metal. The specific heat of water c_w is the amount of heat required to raise the temperature of a unit mass of water by 1 degree. We usually assume that c_w is constant with a value of 1 cal/g · C° = 4.186 J/g · K. The heat lost by the metal block is equal to $mc\Delta T$, where ΔT is the change in temperature of the metal. Setting this equal to a similar expression for the heat gained by the water allows us to solve for the specific heat of the metal.

Let's begin this contest problem with an analysis of this common experiment. We then move on to the more challenging Olympiad experiment.

A. Calculate the specific heat of aluminum given the following data: the aluminum block has a mass of 36.2 g and an initial temperature of 100°C. You have an ideal calorimeter (that is, one that loses no heat to the surroundings and does not absorb any heat) containing 100 g of water at an initial temperature of 17°C. After the block is placed in the water, the temperature rises to 23°C.

In the Olympiad experiment we will place a "hot" block of aluminum into liquid nitrogen and determine the latent heat of vaporization of the liquid nitrogen from the amount of liquid that is vaporized. Of course, in a real experiment the calorimeter is not ideal and heat is exchanged with the surroundings whenever there is a temperature difference. In working with liquid nitrogen, there will be a large temperature difference, and the calorimeter will continually allow heat to enter the system. We also cannot assume that the specific heat of the aluminum is a constant. In fact, it varies a lot, as shown in figure 1.

B. Calculate the latent heat of vaporization of liquid nitrogen given the following data: the aluminum has a mass of 19.4 ± 0.1 g and is initially at a room temperature of 20.0° ± 0.1°C. The total mass of the system is monitored as a function of time and gives the following data:

total mass (g)	time (s)
153	0
152	37
151	79
150	121
149	161
148	203
Aluminum block added	
150	332
149	382
148	457
147	489
146	541
145	595

During the Olympiad the students had to measure the changing mass using a triple beam balance. This is why the time was recorded for specific decreases in the mass rather than the mass recorded for specific time intervals.

C. Because every good experimenter gives an uncertainty for each experimental value, estimate the uncertainty in your value for the latent heat.

The actual experimental problem during the Olympiad required students to measure the latent heat of vaporization by two independent methods. The evolution of the experiment began with the decision by Professor Anthony P. French of MIT (chair of the examinations committee) to make use of the ample supplies of liquid nitrogen that the College of William & Mary (the Olympiad host institution) maintains for research. Peter Collings of Swarthmore College accepted the challenge and devised the Olympiad experiment. As with many good ideas, this one was independently created by Gerhard Salinger (National Science Foundation) and published in *The Physics Teacher* in September 1969.

Solution

Part A asked you to calculate the specific heat of aluminum. We begin by writing down an expression for the conservation of energy where the first term is the heat gained by the water (subscript "w") and the second term is the heat lost by the piece of aluminum (subscript "Al"):

$$c_w m_w \Delta T_w + c_{Al} m_{Al} \Delta T_{Al} = 0,$$

where c is the specific heat, m is the mass, and ΔT is the change in temperature. Therefore,

$$\begin{aligned} c_{Al} &= -c\frac{m_w}{m_{Al}}\frac{\Delta T_w}{\Delta T_{Al}} \\ &= \left(-1\frac{\text{cal}}{\text{g}\cdot°\text{C}}\right)\left(\frac{100\text{ g}}{36.2\text{ g}}\right)\left(\frac{6°\text{C}}{-77°\text{C}}\right) \\ &= 0.215\frac{\text{cal}}{\text{g}\cdot°\text{C}}. \end{aligned}$$

Using the conversion factor 1 cal = 4.186 J, we get $c_{Al} = 0.9$ J/g · K at room temperature, in agreement with the graph of the specific heat of alumi-

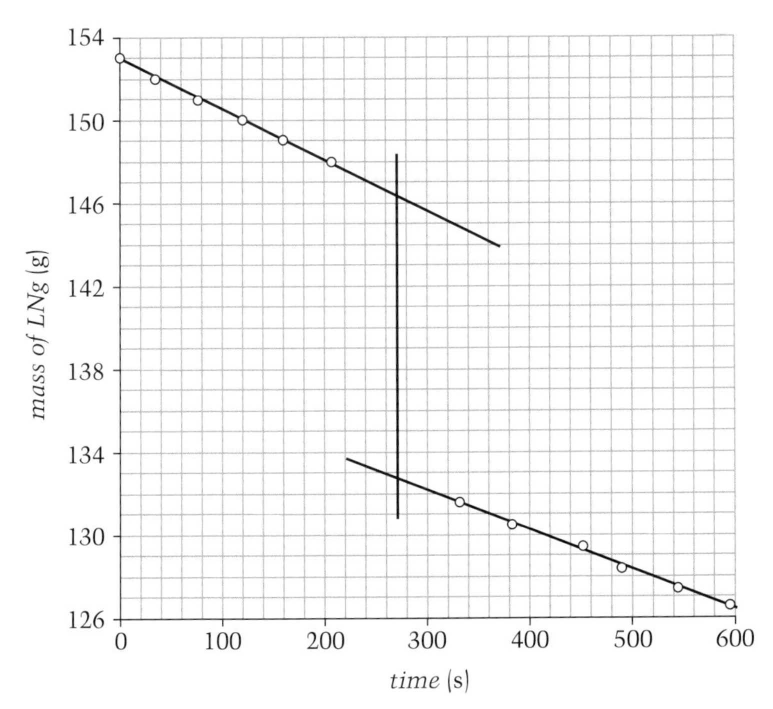

Figure 2

num given in the problem.

Part B used data taken at the XXIV International Physics Olympiad to calculate the latent heat of vaporization of liquid nitrogen. There are two complications involved in this experiment: (1) the change in the specific heat of the aluminum as a function of temperature and (2) the loss of heat to the surroundings, since the liquid nitrogen boils at 77 K.

The specific heat of aluminum as a function of temperature is shown in the graph given in the problem. Because the amount of heat q required to change the temperature of 1 g of aluminum by ΔT is just $c\Delta T$, this heat can be obtained graphically by computing the area under the curve between the two temperatures. The easiest way of doing this is to count boxes (estimating fractional boxes) under the curve between T = 77 K and T = 293 K. We obtain 300 ± 6 boxes with the area of each box being (0.05 J/g · K)(10 K) = 0.5 J/g. (Don't forget the 130 boxes below the x-axis.) This gives us a total q = 150 ± 3 J/g. Therefore, the total heat given up by the aluminum is

$$Q = m_{Al}q = (19.4 \text{ g})(150 \text{ J/g}) = 2{,}910 \text{ J}.$$

We now need to obtain the mass of liquid nitrogen that was evaporated with this heat. To do this we plot the mass of the liquid nitrogen in the calorimeter as a function of time as shown in figure 2. (Don't forget to subtract the mass of the aluminum after it is put into the liquid nitrogen.) This graph shows that the loss of mass due to heat from the surroundings is quite important and occurs at a different rate before and after the aluminum is put into the liquid nitrogen. We can obtain a very good estimate of the mass of liquid nitrogen vaporized by looking at the difference in the two lines at the middle of the time interval—that is, around 270 s. This yields m_N = 14.4 ± 0.3 g. We can now calculate the latent heat of vaporization:

$$L = \frac{Q}{m_N} = \frac{2{,}910 \text{ J}}{14.4 \text{ g}} = 202 \text{ J/g}.$$

In part C you were asked to calculate the error in your value for the latent heat. You may remember the simple rule that when experimental values are multiplied or divided, the percentage errors in the measured quantities add. This gives an overestimate and can be refined by adding the errors in quadrature—that is, add the squares of the errors and then take the square root.

The error in q is 2.0%, while the error in m_{Al} is 0.5%. Adding these in quadrature yields a 2.1% error in Q, or Q = 2,910 ± 60 J. The error in m_N is 2.1%. Adding this in quadrature to that in Q yields an error in L of 3%. Therefore, our experimental value is

$$L = 202 \pm 6 \text{ J/g}.$$

Laser levitation

"Suspend here and everywhere, eternal float of solution!"
—*Walt Whitman,* Leaves of Grass

HOW CAN SOMEONE LEVItate an object? Magicians do it all the time. Can physicists do it as well? The easiest technique is to attach a string to the object and secure the string to the ceiling. The weight of the object is balanced by the tension in the string. If the suspended object is a magnet, then a second magnet can keep it in place. A third technique is to shoot pellets at the object so that the force of the pellets balances the weight of the object.

Let's assume that the object we wish to suspend is a rectangular box oriented so that its bottom is horizontal. If we shoot pellets vertically upward at the box, the pellets just provide an average force on the box that is equal to its weight. If the pellets rebound from the box downward with the same speed, then the momentum change of each pellet is given by

$$\Delta p_{\text{pellet}} = 2mv_0,$$

where v_0 is the initial speed of the pellets and m is the mass of each pellet. The impulse-momentum theorem and Newton's third law tell us that the beam of pellets exerts a force on the box equal to

$$F_{\text{box}} = R\Delta p_{\text{pellet}},$$

where R is the number of pellets hitting the box each second.

We can get a feeling for the problem by solving it with some appropriate values. If the pellet gun shoots 5 pellets per second, and each of these 2-g pellets hits the box with a speed of 50 m/s and rebounds with the same speed, what is the heaviest box that can remain suspended? Let's work it through:

$$p_{\text{pellet}} = mv = (2 \cdot 10^{-3}\ \text{kg})(50\ \text{m/s}) = 0.1\ \text{kg m/s},$$

$$\Delta p_{\text{pellet}} = 0.2\ \text{kg m/s},$$

$$F = (5\ \text{pellets/s})(0.2\ \text{kg m/s}) = 1\ \text{N}.$$

Therefore, a 0.1-kg box can be suspended with these high-speed pellets.

Most people who are high-minded like to lift themselves up and float above the daily burden of gravity. There are many ways to elevate oneself—the usual one is by pulling oneself by one's own hair or by somebody else's hair, or by renting a magician. If you are a knight, a magnet will help, or maybe you are lucky and are light-headed. Some meditate to rise above the profane, some acquire a propeller hat, some heave themselves up with sheer muscle power, and some use a pulley and a sandbag, or get struck by Cupid's arrows. On top of Rapunzel's tower a lady's pizza party is taking place, automatically causing many men to try to get up there to lift their spirits by enjoying the high life on the lofty top and ignoring gravity.

—T.B.

SAL'S PIZZA
ELEVATOR
Tomas Bunk

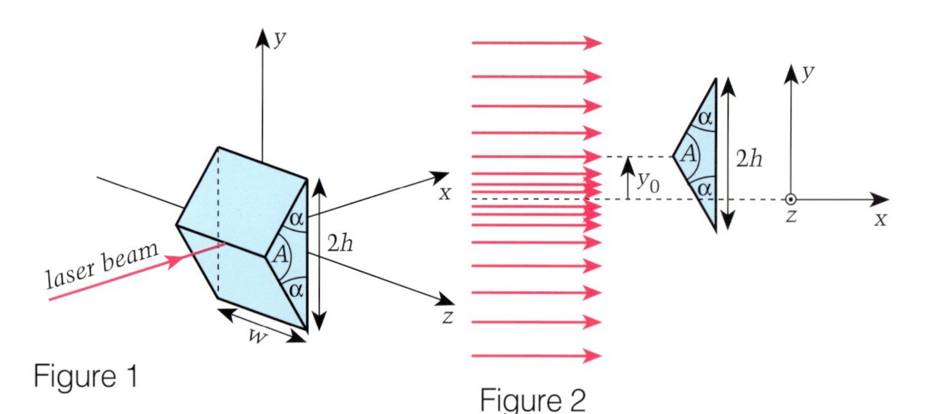

Figure 1

Figure 2

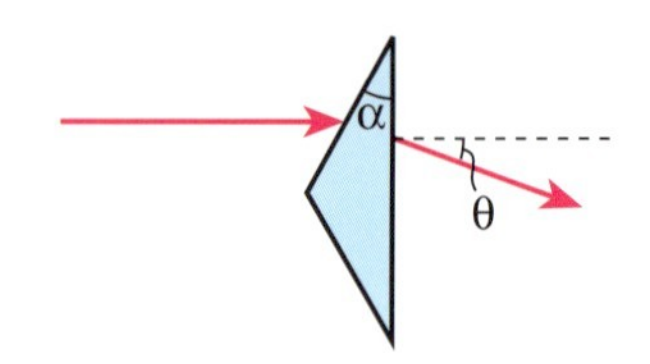

Figure 3

A. The problem becomes more challenging to solve if the pellets hit the box at an angle. Assume that the pellets are identical to those in the example, but that they hit the box at an angle of 53° from the vertical. Once again, the pellets rebound at the same speed (50 m/s) and at the same angle. (Assume that the pellets hit at random orientations about the vertical so that there is no horizontal component of the net force.) What is the heaviest box that can remain suspended?

What do we do if the object to be suspended is so small that its weight is of the order of 10^{-10} newtons? If the object is transparent, it can be levitated by a laser beam! How to do this was one of three theoretical problems that were given to students who participated in the XXIV International Physics Olympiad, which was hosted in the United States in July 1993. This theoretical problem was created by Charles Holbrow of Colgate University. We adapted it for *Quantum* readers.[1]

By means of refraction a strong laser beam can exert appreciable forces on small transparent objects. To see that this is so, consider a small glass triangular prism with an apex angle $A = \pi - 2\alpha$, a base of length $2h$, and a width w. The prism has an index of refraction n and a mass density ρ.

Assume that the prism is placed in a laser beam aimed horizontally in the x-direction. (Throughout this problem assume that the prism does not rotate—that is, its apex always points opposite to the direction of the laser beam, its triangular faces are parallel to the xy-plane, and its base is parallel to the yz-plane, as shown in figure 1.) Take the index of refraction of the surrounding air to be $n_{air} = 1$. Assume that the faces of the prism are coated with an antireflective coating so that no reflection occurs. The momentum of a photon is given by $p = E/c$.

The laser beam has an intensity that is uniform across its width in the z-direction but falls off linearly with the vertical distance y from the x-axis such that it has a maximum value I_0 at $y = 0$ and falls to zero at $y = \pm 4h$ (fig. 2).

B. Write equations from which the angle θ (see figure 3) may be determined in terms of α and n for the case when the laser beam strikes the upper face of the prism.

C. Express, in terms of I_0, θ, h, w, and y_0 the x- and y-components of the net force exerted on the prism by the laser light when the apex of the prism is displaced a distance y_0 from the x-axis, where $h \le y_0 \le 3h$. If we want the prism to be suspended, should the prism be placed above or below the axis of the laser beam?

D. Plot graphs of the values of the horizontal and vertical components of force as functions of the vertical displacement y_0.

E. Suppose that the laser beam is 1 mm wide in the z-direction and 80 μm thick (in the y-direction). The prism has the following characteristics: $\alpha = 30°$, $h = 10$ μm, $n = 1.5$, $w = 1$ mm, and $\rho = 2.5$ g/cm^3. How many watts of power would be required to balance this prism against the pull of gravity when the apex of the prism is at a distance $y_0 = 2h = 20$ μm?

This problem is certainly difficult enough. Olympiad students from 42 countries took the problem one step further and solved parts C, D, and E for prism positions where $y_0 < h$! And some of them correctly completed this analysis within the allocated time of 100 minutes!

Solution

An excellent solution was submitted by Scott Wiley of Weslaco, Texas.

A. The preliminary problem involves suspending a box with bullets striking the bottom of a box at an angle θ. Because momentum is a vector quantity, we can break the initial and final momentum of the bullets into vertical and horizontal components. Ignoring the horizontal components as instructed, we can calculate the change in the vertical component of the momentum for each bullet:

$$\Delta p = -mv_0 \cos\theta - mv_0 \cos\theta$$
$$= -2mv_0 \cos\theta,$$

where the minus sign indicates that the change is in the downward direction. Therefore, the force F_{box} exerted on the box is

$$F_{box} = R\Delta p$$
$$= 2Rmv_0 \cos\theta$$
$$= 0.6 \text{ N},$$

based on the data given in the problem.

B. Let's now move on to the laser beam. From the geometry of figure 4 we see that the angle of incidence for the incoming light beam is α. Using Snell's law and setting the index of refraction of air equal to 1, we have

$$\sin\alpha = n \sin\phi,$$

[1]The entire XXIV International Physics Olympiad Examination has been published in *Physics Today* (November 1993) in an article by Anthony P. French, chair of the examination committee.

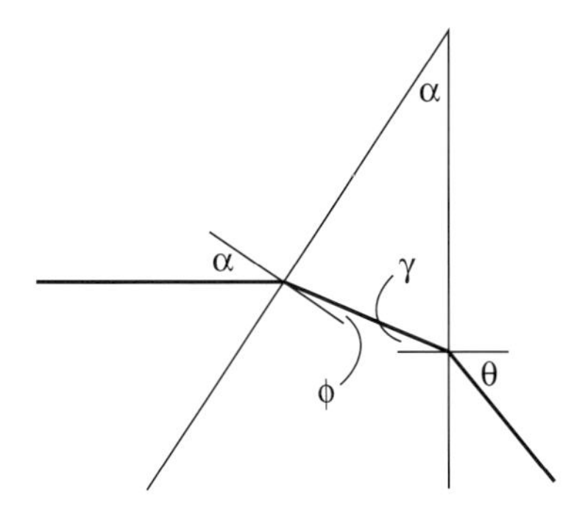

Figure 4

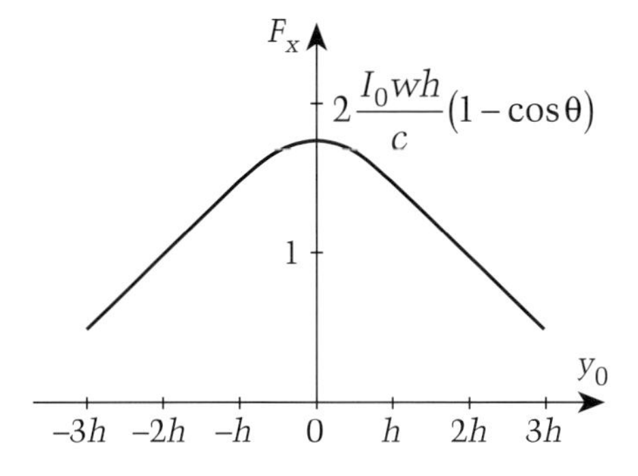

Figure 5

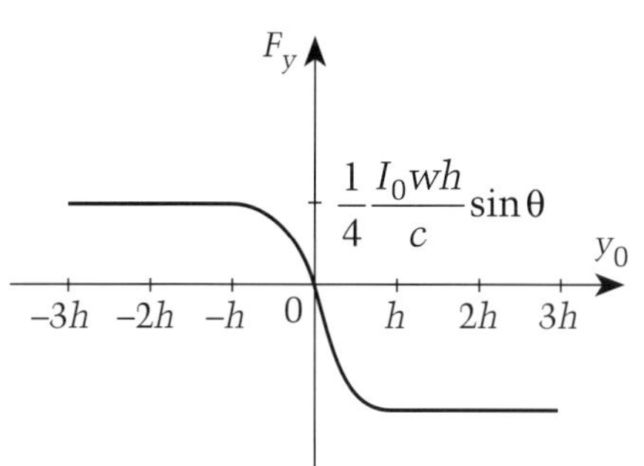

Figure 6

where ϕ is the angle of refraction at the first surface. A little more geometry shows that

$$\phi + \gamma = \alpha,$$

where γ is the angle of incidence at the second surface. Finally, using Snell's law at the second surface yields

$$n \sin \gamma = \sin \theta,$$

with θ being the angle of refraction back into the air.

Solving these equations, we find

$$\theta = \arcsin [n \sin (\alpha - \phi)],$$

with

$$\phi = \arcsin \frac{\sin\alpha}{n}.$$

C. By conservation of linear momentum, the total momentum of the system must remain unchanged. Since the wavelength of the laser light is the same before and after it enters the prism, the magnitude of the momentum of each photon is the same and equal to E/c, where E is the energy of each photon. However, the angle of the beam has changed. Therefore, the horizontal and vertical components of the momentum change of the prism are

$$\Delta p_{px} = p - p\cos\theta = \frac{E}{c}(1-\cos\theta),$$

$$\Delta p_{py} = p\sin\theta = \frac{E}{c}\sin\theta.$$

Therefore, using the technique from part A, we find that the components of the force on the prism are

$$F_x = \frac{N}{t}\Delta p_{px} = \frac{NE}{ct}(1-\cos\theta),$$

$$F_y = \frac{N}{t}\Delta p_{py} = \frac{NE}{ct}\sin\theta.$$

Knowing that the power P in terms of the intensity and the number of photons N is

$$P = IA = \frac{NE}{t},$$

we have

$$F_x = \frac{IA}{c}(1-\cos\theta),$$

$$F_y = \frac{IA}{c}\sin\theta.$$

However, the intensity is not uniform over the prism. Because it falls off linearly with distance, we can avoid doing an integral and just use the average value for each face. The average intensities for the upper and lower faces for the case $h \le y_0 \le 3h$ are

$$\overline{I_u} = \frac{I(y_0) + I(y_0+h)}{2} = I_0\left(\frac{7h-2y_0}{8h}\right),$$

$$\overline{I_l} = \frac{I(y_0) + I(y_0-h)}{2} = I_0\left(\frac{9h-2y_0}{8h}\right).$$

As shown in part B, the top face provides an upward momentum to the prism. Conversely, the lower face provides a downward momentum. Therefore, the net vertical force is

$$F_y = \left(\overline{I_u} - \overline{I_l}\right)\frac{wh}{c}\sin\theta = -\frac{I_0wh}{4c}\sin\theta.$$

The minus sign indicates that the prism should be placed below the axis of the beam for it to be suspended.

Because the horizontal components act in the same direction, we have

$$F_x = \left(\overline{I_u} + \overline{I_l}\right)\frac{wh}{c}(1-\cos\theta) = -\frac{I_0w}{2c}(4h-y_0)(1-\cos\theta).$$

D. Similar calculations for the range $0 \le y_0 \le h$ yield the graphs shown in figures 5 and 6.

E. Using the dimensions and density given in the problem, the weight of the prism is $W = 1.42 \cdot 10^{-9}$ N. To levitate the prism, the upward force on the prism must equal its weight. This requires $I_0 = 6.19 \cdot 10^8$ W/m^2. The average intensity of the beam is $I_0/2$, so the power of the beam must be

$$P = IA = 24.8 \text{ W}.$$

Mirror full of water

"The empty mirror. If you could really understand that, there would be nothing left to look for."
—van de Wetering

"IT'S DONE WITH MIRrors." Whether we attend magic shows and see the magician push sharp swords through a box containing the "lovely assistant" or ride the "Haunted Mansion" and see the ghost flying through the room at Disneyland, we are often surprised and pleased by clever manipulations of images.

In this problem, we'll look at the image produced by a concave mirror filled with water. Because our confidence in a physics solution increases if different approaches to the problem yield the same result and there are many ways of obtaining the position of the image, we will want to discover as many of them as possible. Perhaps you will come up with a solution that is fundamentally different from the ones we expect.

Texts on geometrical optics often begin by showing that the reflection of light from plane mirrors follows the principle that the angle of incidence is equal to the angle of reflection. If the mirror is curved, this behavior still holds, but the geometry of the parabolic mirror is such that all parallel rays come to a focus for a concave mirror, or appear to diverge from the focus in the case of a convex mirror. For a spherical mirror, the spherical surface approximates the parabolic curve and parallel rays near the axis also come together at (or diverge from) the focus.

The relationship between the image and object is given by the mirror formula

$$\frac{1}{s}+\frac{1}{s'}=\frac{1}{f},$$

where s and s' are the distances of the object and image from the surface of the mirror and f is the focal length of the mirror. The focal length is often stamped on the mirror and is equal to one half of the radius of the spherical surface from

When we add water to a concave mirror, we get a funny brew of unreal imagery that makes us doubt our own eyes. This image has two magic parts, one black and the other yellowish. In the black imagery we meet numerous characters that don't even have a mirror image, because they live in a dark gray zone of ghostly nonexistence. On the yellowish plane we find situations mirroring the lighter side of flat mirrors, like the danger of falling in love with one's own mirror image or the consequences of not liking oneself or the surprise of seeing somebody else much less attractive in the mirror. One old person rises from the grave to jump through a mirror and become a newborn baby on the other side. After all, mirrors only reflect light, and this is the lighter side, where everything is possible.

—T.B.

LENS
IMAGE
WATER
MIRROR
LOOK
OBJECT
FOPT
KHNP
STPX
FLTZY
KLPBH
RIP
TOMAS BUNK 94

which the mirror is made. The focal length can be measured by shining a beam of parallel light onto a concave mirror and measuring the distance from the surface of the mirror to the point where the beam is brought to a focus. For a convex mirror the light appears to diverge from a focal point located behind the mirror. Both of these points can be determined by drawing several rays parallel to the axis of the mirror, using the law of reflection at the surface, and locating where the rays cross.

To make effective use of the mirror formula we must remind ourselves of a number of conventions. The distance s is positive if the object is located in front of the mirror. This will always be the case for real objects, but the "object" could be an image produced by another optical device. In this case, the object could be located behind the mirror and s would be negative. If the image is located in front of the mirror, the image distance s' is positive; if the image is behind the mirror, s' is negative. Finally, f is positive for a concave mirror and negative for a convex mirror.

As an example, consider an object located a distance $3f$ in front of a concave mirror:

$$\frac{1}{s'} = \frac{1}{f} - \frac{1}{s} = \frac{1}{f} - \frac{1}{3f} = \frac{2}{3f}.$$

Therefore, the image is located a distance $3f/2$ in front of the mirror. This can also be shown with a diagram that traces the rays. Convince yourself that the image would be $3f/4$ behind the mirror if we use a convex mirror instead of the concave mirror.

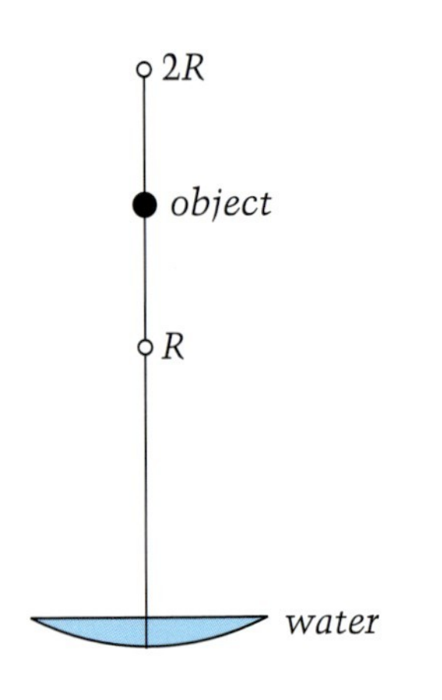

Figure 1

The mirror formula also works for lenses if we adopt the following conventions: s is positive if the object is located in front of the lens, negative if the object is located behind the lens; s' is positive if the image is located behind the lens, negative if the image is located in front of the lens. Converging lenses (thicker in the center than at the edges) have positive f, while diverging lenses (thinner at the center) have negative f.

Another useful relationship is the lens maker's formula. For the special case when one of the surfaces is planar, it tells us that

$$\frac{1}{f} = \frac{n-1}{R},$$

where n is the index of refraction of the lens material and R is the radius of the curved surface.

Now that we have completed this very brief review, let's take a look at our problem. A concave mirror of radius R resting face up on a table top has been filled with a small amount of water (index of refraction $n = 4/3$) as shown in figure 1. A small object is located a distance $d = 3R/2$ from the mirror along the optic axis. Where is the image located? In the spirit of the "thin lens approximation" often used in such problems, we will neglect the thickness of the water.

A. Let's begin by using a technique used by eye doctors. Often the doctor will place a lens in front of your glasses to show you how the new lenses will work. This works because the effective focal length f' of two (or more) lenses (or mirrors) in close proximity is given by

$$\frac{1}{f'} = \sum \frac{1}{f_i}$$

—that is, the focal lengths add as reciprocals. Therefore, the mirror–water combination can be replaced by a mirror with an effective focal length and you can use the mirror formula given above. Does the water lens appear in the sum once or twice? Use the other methods to check yourself.

B. You can also obtain the effective focal length by tracing a ray parallel to the optic axis as it enters the water and bends according to Snell's law, reflects from the mirror surface, and exits the water again. Don't forget to make suitable approximations.

C. Our third method makes use of the observation that images formed by one optical element act as objects for subsequent optical elements. Begin by finding the location of the image formed by the air–water interface. Use this image as the object for the mirror (without the water) and find the new image location. Then find the image of this image formed by the water–air interface when the light exits the water. This is the final image produced by the combination.

D. The trickiest method treats the combination as a water lens, a mirror, and a water lens in combination. Find the location of the image produced by each element and then use it as the object for the next element. This is tricky because it's very easy to make mistakes with the sign conventions.

Solution

We model our solution after the very complete solution submitted by Prof. Anthony F. Behof of DePaul University in Chicago, Illinois.

We begin by using the lens maker's formula to find the focal length f_w of the converging lens formed by the water:

$$\frac{1}{f_w} = \frac{n-1}{R} = \frac{4/3-1}{R} = \frac{1}{3R}.$$

Therefore, $f_w = 3R$.

Method A. We use the idea that the effective focal length f' of the combination of optical elements is the sum of the reciprocals of the individual elements. When we do this, we must use the focal length of the water lens twice because the light

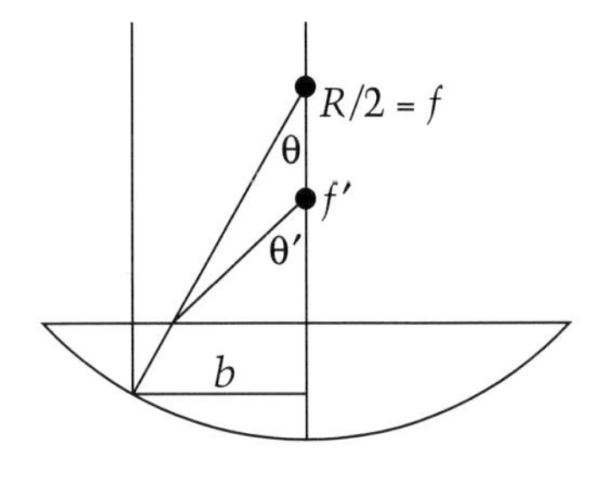

Figure 2

passes through the lens, strikes the mirror, and then passes through the lens again. The resulting formula is

$$\frac{1}{f'} = \frac{1}{f_w} + \frac{1}{f_m} + \frac{1}{f_w}$$

$$= \frac{1}{3R} + \frac{2}{R} + \frac{1}{3R} = \frac{8}{3R}.$$

This effective focal length of $3R/8$ can be used in the mirror formula to find the image location:

$$\frac{1}{d'} = \frac{1}{f'} - \frac{1}{d} = \frac{8}{3R} - \frac{2}{3R} = \frac{2}{R}.$$

This tells us that the image is located at $d' = R/2$.

Thomas A. Davidson of Amarillo, Texas, points out that a mirror behaves the same whether it is immersed in air, water, or a vacuum. Introducing an air/water interface in front of the mirror effectively shortens the focal length of the mirror by the ratio of the indices of refraction. Thus, the effective focal length is $R/2n$, in agreement with the answer we obtained above.

Method B. An alternate method of finding the effective focal length of the water–mirror combination is to look at a ray parallel to the optic axis as shown in figure 2, where the angles and the thickness of the water have been exaggerated for clarity. Without the water this ray would pass through the focal point f of the mirror. However, because of the refraction at the surface, the ray intersects the optic axis at the effective focal point f'. Snell's law tells us that

$$\frac{4}{3}\sin\theta = \sin\theta'.$$

Ignoring the thickness of the water, we also know that

$$\tan\theta \cong \frac{b}{f}$$

and

$$\tan\theta' \cong \frac{b}{f'}.$$

Because the angles are actually quite small, we use the small angle approximation that $\tan\theta \cong \sin\theta$ and $\tan\theta' \cong \sin\theta'$. Therefore, we have

$$\frac{4}{3}\frac{b}{f} \cong \frac{b}{f'},$$

or

$$f' \cong \frac{3}{4}f = \frac{3}{8}R.$$

Method C. The top surface of the water forms a virtual image at a distance

$$nd = \frac{4}{3}\frac{3R}{2} = 2R$$

above the surface of the water. This image acts like an object for the mirror without the water. Using the mirror formula with the focal length $f_m = R/2$ and $d = 2R$, we obtain

$$\frac{1}{d'} = \frac{1}{f} - \frac{1}{d} = \frac{2}{R} - \frac{1}{2R} = \frac{3}{2R},$$

or $d' = 2R/3$.

The light leaving the water gets bent once again to form an image at

$$\frac{d'}{n} = \frac{3}{4}\frac{2R}{3} = \frac{1}{2}R.$$

Method D. Finally we can treat the system as a combination of a water lens, a mirror, and a water lens by finding the image produced by each one and using the image as the object (sometimes imaginary) for the next one. Applying the lens formula to the water lens yields

$$\frac{1}{d_1'} = \frac{1}{f_w} - \frac{1}{d_1} = \frac{1}{3R} - \frac{2}{3R} = -\frac{1}{3R}.$$

Therefore, the image is virtual and located a distance $3R$ above the surface of the water.

This image acts as an object for the mirror. Ignoring the thickness of the water, the object distance for the mirror is $d_2 = 3R$. Using the mirror formula, we obtain

$$\frac{1}{d_2'} = \frac{1}{f_m} - \frac{1}{d_2} = \frac{2}{R} - \frac{1}{3R} = \frac{5}{3R},$$

and the image is located a distance $3R/5$ above the mirror.

Since this image is located on the "wrong" side of the water lens, the new object distance $d_3 = -3R/5$. Thus,

$$\frac{1}{d_3'} = \frac{1}{f_w} - \frac{1}{d_3} = \frac{1}{3R} - \frac{-5}{3R} = \frac{2}{R},$$

and the final image is located a distance $R/2$ above the surface as before.

Prof. Behof goes on to provide a fifth solution using matrix optics and a way of verifying the result experimentally. To see how to measure the effective focal length, let's first consider the case with no water and set the object and image distances equal to each other—that is, $d = d'$. We then find the well-known result that $d = 2f = R$. If we now add water, we find that $d = R/n = 3R/4$.

Now hold a lighted target above a mirror that has been filled with a few millimeters of water. Adjust the height of the target above the water until the target's image is in focus on a screen held at the same height. This height d is the effective radius of curvature of the combination and the effective focal length is $d/2$. ◘

Rising star

"What is the sound of one hand clapping?"
—Zen koan

"Can two sounds make silence?"
—Physics challenge

WE SUPPOSE THAT YOUNG people are first introduced to waves while attending or watching sports events. These stadium waves can provide some useful insights into the most counterintuitive property of waves: the wave moves, but the medium does not. In a stadium wave, a group of spectators at one end of the stadium stands and then sits. This triggers the adjoining section of fans to stand and sit, followed by the next section, and so on. While the wave of people standing and sitting moves around the stadium, no person moves in that direction—that is, the people remain at their seats. Leonardo da Vinci noticed this wave property in water and remarked that the wave flees the place of creation while the water does not.

An interesting wave phenomenon that is not easily demonstrated in a stadium wave is interference. What happens when two waves meet? A first step in our understanding will be to look at two pulses passing each other on a spring. What one notices is that as the peaks of the pulses meet, a momentary superpeak is created (fig. 1a). What is perhaps more surprising is that when a peak pulse meets a valley pulse, there may be a point on the string that doesn't move. For this point, it's as if no pulse passed by (fig. 1b).

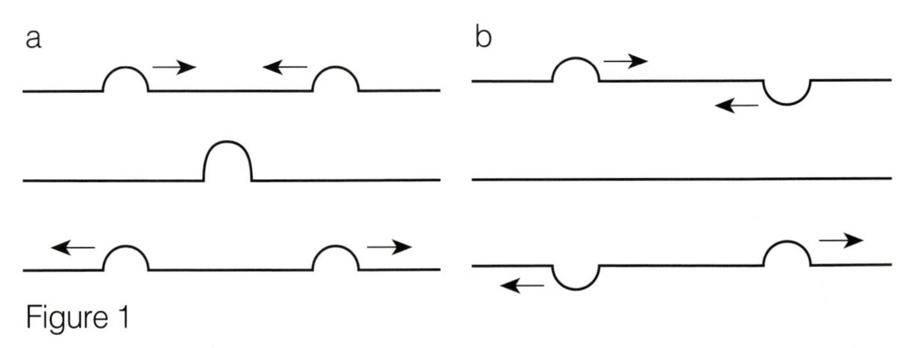

Figure 1

Every person creates waves just by being alive. Some great thinkers make huge waves that affect history for thousands of years, and some people's waves go straight out the window and disappear. Even little ducks can make waves just by wiggling their tails, others like to ride the waves, and some generate ripples in the sunset sky with a pebble. Destructive people cause interference with their dreadful motorboats, while one-handed Zen monks enjoy producing good vibrations by clapping with whatever body part they can use. Below them, all the people of the world create a wave of peace and harmony. The scientist is trying to find out if two sounds coming together can extinguish each other and produce silence, and the policeman proves that with two flashlights he can create darkness at noon.

—T.B.

SOUNDS OF SILENCE IN STEREO
DARKNESS AT NOON
INTERFERENCE
REFLECTION!
tom bunk

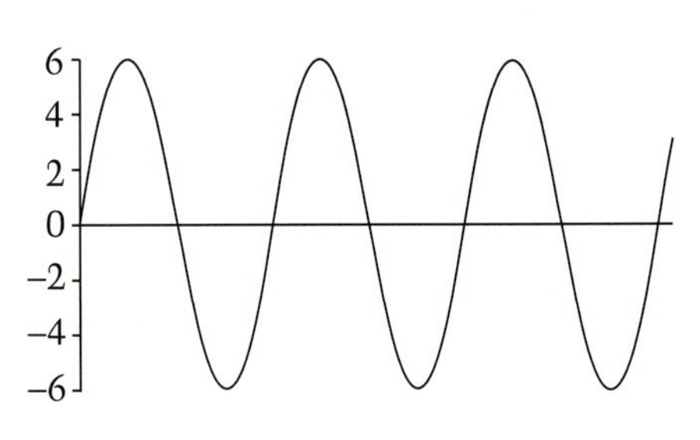
Figure 2

A periodic wave is a continuous series of pulses (fig. 2). This representation can be assigned to any wave phenomenon. A sound wave, which propagates by disturbing the air in compressions and rarefactions, can use figure 2 as a graphical representation, where the peaks are compressions and the valleys are the rarefactions. A series of sketches in which two waves pass each other will reveal that the sum of the waves produces points on the string that undergo large displacements and other points that undergo *no* displacement whatsoever, as shown in figure 3. The points of maximum disturbance are called *antinodes*, while the points of no disturbance are referred to as *nodes*.

The interference of sound waves can create these nodal points, and one of the jobs of the acoustical engineer is to ensure that a new concert hall does not have places (due to reflections) where aspects of the music cannot be heard. Acoustical engineering is both an art and a science. It is part good fortune and part mystery why Carnegie Hall and La Scala have such exceptionally fine acoustics.

And so, in answer to the physics challenge posed above, two sounds *can* create silence. Two light sources can also create darkness, as in Young's double-slit experiment or a Michelson interferometer. And two beams of electrons can produce locations where no electrons will reach in what is arguably the most profound discovery of the 20th century. The Zen koan about one hand clapping must remain a mystery. We're not sure what insights physics can offer to this puzzle.

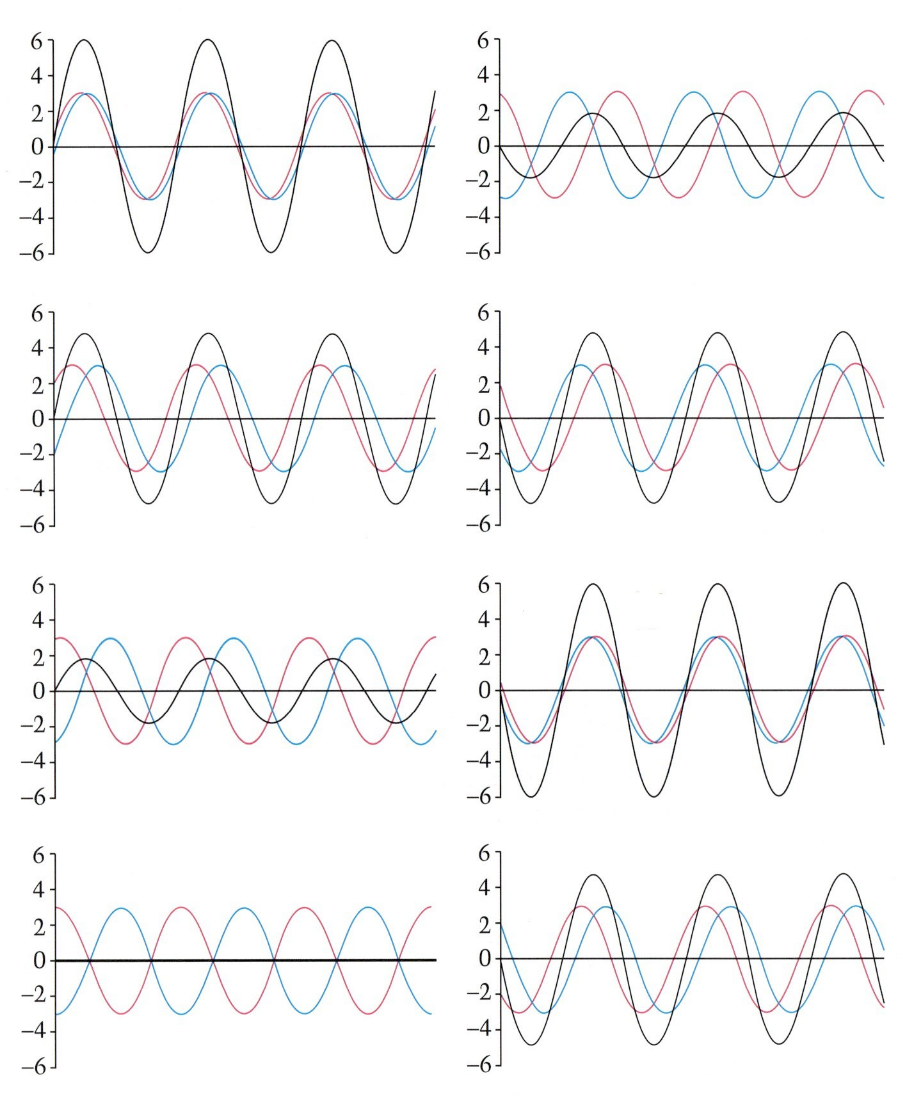
Figure 3

Our problem is from the XII International Physics Olympiad, held in Bulgaria in 1981. The receiver of a radio observatory is placed on an island, near the shore, at a height of 2 m above sea level. It detects only the horizontal components of the electric field. When a radiostar radiating waves with a wavelength of 21 cm rises above the horizon, the receiver records maxima and minima.

A. Determine the altitude of the star when maximum and minimum are observed.

B. Does the intensity decrease or increase after the star first appears above the horizon?

C. Investigate the ratio of the intensity of the successive maxima and minima.

(Note: The ratio of the amplitudes of the incident and reflected waves is $(n - \sin\theta)/(n + \sin\theta)$, where θ is the angle of the incident wave measured from the horizontal and n is the index of refraction. For radio waves and water, $n = 9$.)

Solution

In this problem the radio receiver records maxima and minima. This is our clue that some interference effect is occurring. The insight that solves the problem comes from the description of how the receiver is placed on an island near the shore. The interference is probably a result of the electromagnetic wave arriving directly from the star and a second

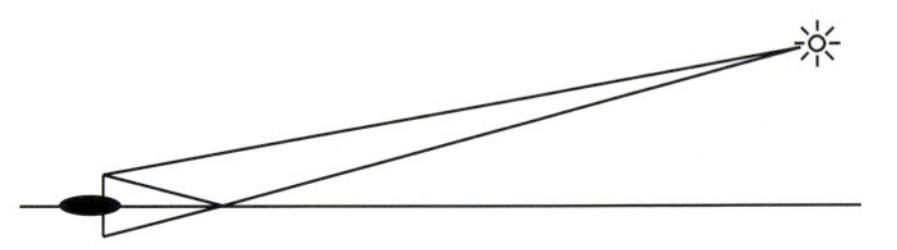
Figure 4

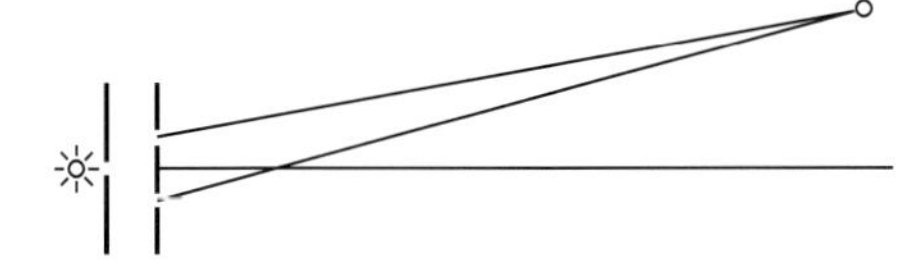

Figure 5

electromagnetic wave arriving after reflecting from the surface of the water. Figure 4 shows these two rays.

The problem can be solved by recognizing that the reflected wave has traveled a longer path and has undergone a phase change at the surface equal to $\lambda/2$. The total path difference must be equal to an integral number of wavelengths for constructive interference (producing maxima) and an odd half-integral number of wavelengths for destructive interference (producing minima.)

The geometry, though not difficult, is unfamiliar. By drawing reflections of the reflected ray and the radio receiver, we are reminded of Young's double slit experiment, in which light emerging from a pair of slits forms an interference pattern on a distant screen. Figure 5 shows Young's double slit geometry for comparison. Young's double slit geometry is analyzed in all physics texts covering optics. By drawing a line perpendicular to the line-of-sight to the star as shown in figure 6, we see that the path difference δ is given by

$$\delta \cong 2h \sin\theta,$$

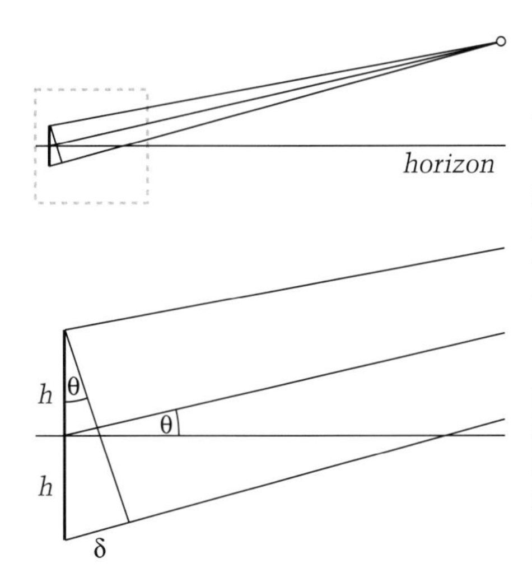

Figure 6

where θ is the altitude of the star, h is the height of the radio tower, and we have assumed that for small θ the triangle is approximately a right triangle.

In our radio receiver problem, the conditions for maxima and minima (remembering the phase shift upon reflection from the water) are

$$\delta_{\max} = \left(k + \frac{1}{2}\right)\lambda$$

and

$$\delta_{\min} = k\lambda,$$

where k is an integer 0, 1, 2, 3, … .

Part A of the problem asked for the altitude of the star when maxima and minima are observed:

$$\sin\theta_{\max} = \frac{\delta_{\max}}{2h} = \frac{\left(k + \frac{1}{2}\right)\lambda}{2h},$$
$$\sin\theta_{\min} = \frac{\delta_{\min}}{2h} = \frac{k\lambda}{2h}.$$

Part B asked whether the intensity increases or decreases as the star rises over the horizon. The star just peeks over the horizon when $\theta = 0$. Since this corresponds to a minimum, the intensity of the star will increase until it reaches its first maximum.

Part C asked for an investigation of the intensities of successive minima and maxima given that the ratio of the incident and reflected rays at the water's surface is

$$\frac{n - \sin\theta}{n + \sin\theta},$$

where $n = 9$.

The maximum amplitude can be found by adding the incident and reflected electric fields. Assuming that the incident electric field is E, the amplitude would be

k	$\theta_{\min}$	$\theta_{\max}$	$intensity_{\min}$	$intensity_{\max}$
0	0	1.50°	0	3.9768
1	3.01°	4.52°	0.000135	3.9309
2	6.03°	7.54°	0.000532	3.8858
3	9.06°	10.59°	0.00118	3.8415

Figure 7

$$amplitide_{\max} = E + E\frac{n - \sin\theta}{n + \sin\theta}$$
$$= E + E\left(\frac{n - \frac{\lambda\left(k + \frac{1}{2}\right)}{2h}}{n + \frac{\lambda\left(k + \frac{1}{2}\right)}{2h}}\right)$$
$$= \left(\frac{4nh}{2nh + \lambda\left(k + \frac{1}{2}\right)}\right)E.$$

Since intensity is proportional to the square of the amplitude,

$$intensity_{\max} \sim \left[\left(\frac{4nh}{2nh + \lambda\left(k + \frac{1}{2}\right)}\right)E\right]^2.$$

Similarly, the minimum amplitude can be found by subtracting the incident and reflected electric fields. The corresponding intensity is:

$$intensity_{\min} \sim \left[\left(\frac{2k\lambda}{2nh + \lambda k}\right)E\right]^2.$$

Relative numerical values can now be computed. They are summarized in the spreadsheet chart in figure 7. ◘

Superconducting magnet

"Know then that this is the law: the north pole of one lodestone attracts the south pole of another."
—Petrus Peregrinus (13th century A.D.)

FOR THE XXV INTERNATIONAL Physics Olympiad held in July 1994 in Beijing, the Chinese hosts prepared problems that are an interesting mix of the modern and the traditional. We have used one of the theoretical problems as the basis for this month's contest problem.

For those of us who grew up with conventional electromagnets, it is very strange to see an electromagnet that is not connected to an external power source. But that is what happens with a superconducting magnet. After a current has been established in the magnet, the magnet can be disconnected from the external power and it will continue to produce a steady magnetic field for a very long time.

In a conventional electromagnet, a large current passes through a solenoid (a coil of wire) producing a magnetic field inside the solenoid. But the current produces a lot of heat due to the resistance of the wire. The production of this heat means that a source of energy is required to maintain a steady state.

In a superconducting magnet, the coil is immersed in liquid helium at a temperature of 4.2 K. At this temperature, the wire becomes superconducting—that is, its electrical resistance drops to zero. Therefore, no heat is produced and the need for an external power supply is eliminated.

The current in the magnet is controlled with a specially designed superconducting switch wired in parallel with the coil as shown in figure 1 (on page 80). The superconducting switch is usually a small length of superconducting wire wrapped with a heater wire and thermally insulated from the liquid helium bath. When the wire is heated, the wire reverts to the normal state and its resistance suddenly changes from $r = 0$ to $r = r_n$.

This very modern device can be analyzed using the very traditional physics that we learn in an introduc-

To make physics more accessible, the management has arranged a complete switchboard system on the stage. In the center we present the superconducting magnet star taking a superhot shower in a tub of liquid helium at a temperature of 4.2 K, attracting the madly excited attention of the backstage crew. In the same tub and attached to it is a superconducting switch, which is connected to a variable resistor that reduces the current to zero by pulling the plug from the annoyed superpower source. In control of the current, naturally, is the divine finger, operating the external main power switch while the statue of the existential thinker with a bulb in its hand is being enlightened thanks to an operator in front of the stage. There also we find the students who are being instructed to get all this information into their heads.

—T.B.

ELECTROMAGNETIC FIELDS FOREVER
LET ME TAKE YOU DOWN..
K
I
E = mc²
E
LIQUID HELIUM HOT
OHM
V = IR
I = 1A
V_c = V_s
I = I_c + I_s
PROBLEMS
Tom Bunk

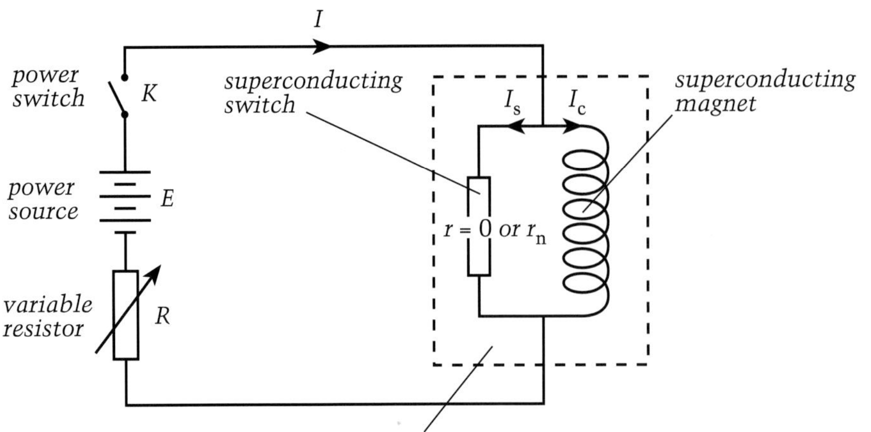

Figure 1

tion to circuits. We start out with Ohm's law, $V = IR$. We then add Kirchhoff's two laws. Kirchhoff's voltage rule tells us that the voltage drop across the superconducting coil must be the same as the voltage drop across the superconducting switch, $V_c = V_s$. This is just a statement of the conservation of energy. Kirchhoff's current rule tells us that the current flowing into a junction must equal the current flowing out of the junction—a statement of the conservation of charge. Using the directions indicated in figure 1, we have $I = I_c + I_s$.

The remaining physics that we require involves the coil. It is interesting because we have a pure inductor—the wire in the coil is superconducting and its resistance is zero. The voltage drop across an inductor depends on the geometry and size of the coil (contained in the inductance L) and the change in the current. Notice that it is only the change in the current that matters, not the actual value of the current. Thus,

$$V_c = -L\frac{\Delta I_c}{\Delta t}.$$

As a consequence of this, when a solenoid is wired in series with a resistor and a battery, the current cannot instantaneously reach the final value of V/R. The voltage drop produced across the inductor as the current increases means that the current must climb to this final value exponentially. That is,

$$I = \frac{V}{R}\left(1 - e^{-t/\tau}\right),$$

where $\tau = L/R$ is the time constant characteristic of this circuit.

Let's now use these basic ideas about circuits to see how the superconducting switch can be used to control the operation of the superconducting magnet. Let's assume that $r_n = 5\ \Omega$ and $L = 10$ H. We start out with switch K closed.

A. Assume that from $t = 0$ until $t = 3$ min, we have $I = 1$ A, $I_c = I_s = 0.5$ A, and $r = 0$. We now use the variable resistor R to reduce the current I linearly to zero from $t = 3$ min to $t = 6$ min while keeping $r = 0$. Plot graphs of I_c and I_s as functions of time and explain why they behave this way.

B. Assume that from $t = 0$ until $t = 3$ min, we have $I = I_c = 0.5$ A, $I_s = 0$, and $r = 0$. At $t = 3$ min we use the heater to suddenly put the superconducting switch in the normal state with $r = r_n$. At $t = 6$ min we cool the switch so that it suddenly returns to the superconducting state with $r = 0$. Plot graphs of I, I_c, and I_s as functions of time and explain why they behave this way.

C. Let's now change the initial conditions by putting the initial current through the switch instead of the coil. Assume that the external resistor has a constant value $R = 5\ \Omega$ and that from $t = 0$ until $t = 3$ min we have $I = I_s = 0.5$ A, $I_c = 0$, and $r = 0$. At $t = 3$ min the switch changes to the normal state with $r = r_n$. At $t = 6$ min the switch returns to the superconducting state with $r = 0$. Plot graphs of I, I_c, and I_s as functions of time and explain why they behave this way.

D. When the switch is in the superconducting state, the magnet may be operated in the "persistent mode." In the persistent mode switch K is open and a current circulates through the coil and the superconducting switch indefinitely. Suppose that the magnet is operating in the persistent mode (that is, $I = 0$, $I_c = I_0$, and $I_s = -I_0$) with $I_0 = 20$ A from $t = 0$ to $t = 3$ min. We now want to shut the magnet down by reducing I_0 to zero. However, we will destroy the switch if the current through the switch in the normal state exceeds 0.5 A. What steps can you use to shut the magnet down? Be sure to plot I, I_c, I_s, and r as functions of time to illustrate your method.

Solution

A. Because the resistance r of the superconducting switch is zero, the voltage drop across the switch must be zero—that is, $V_s = 0$. Because the coil is in parallel with this switch, the voltage drop across the coil must

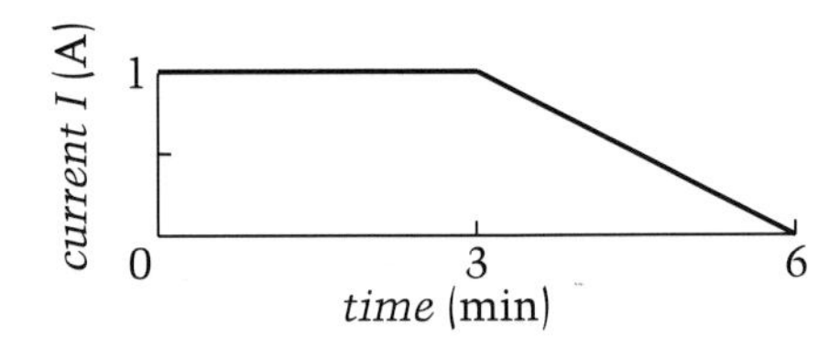

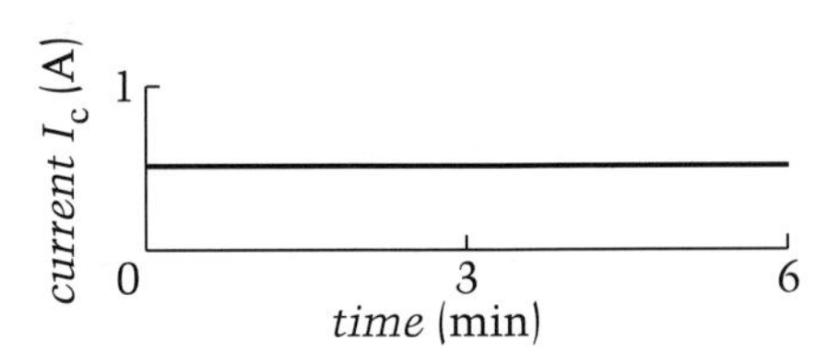

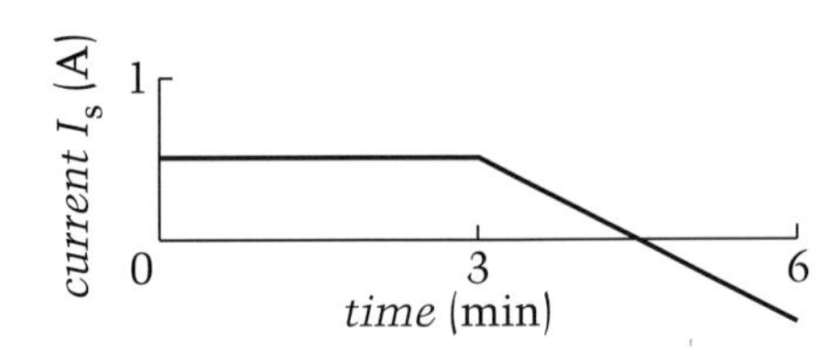

Figure 2

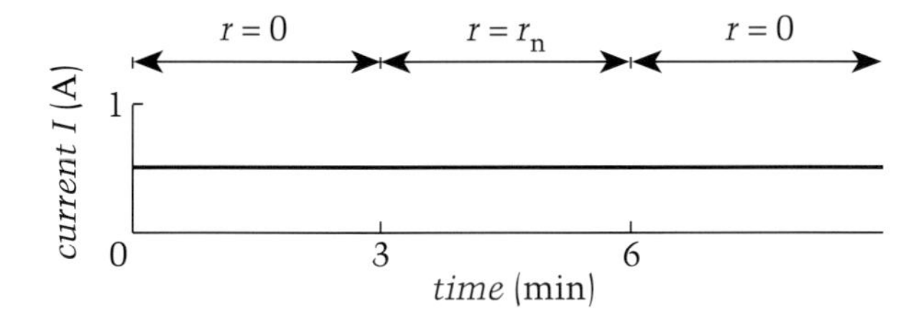

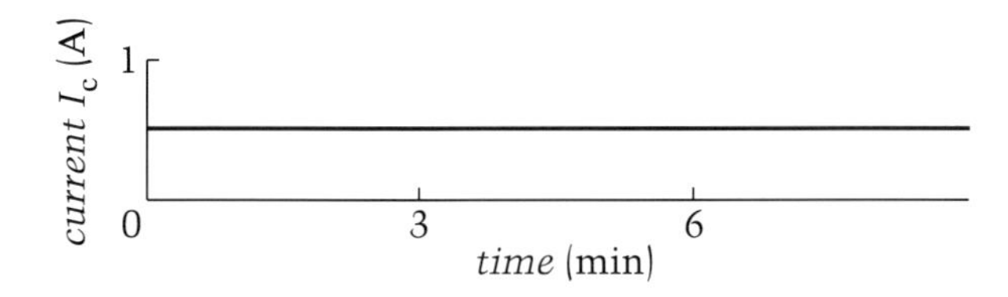

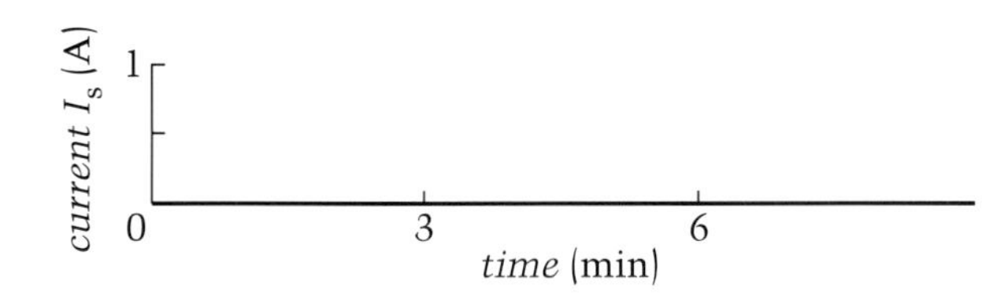

Figure 3

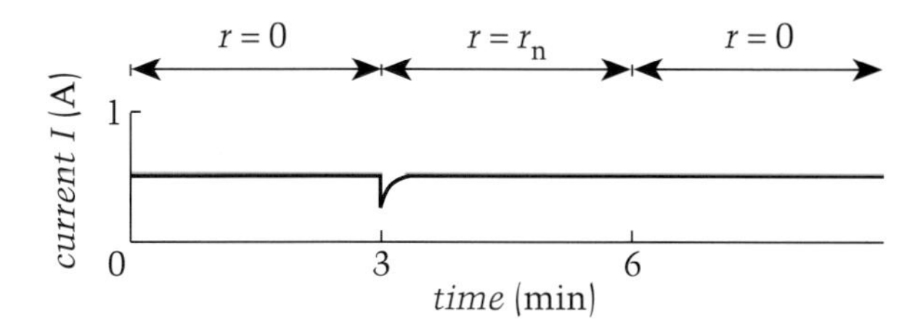

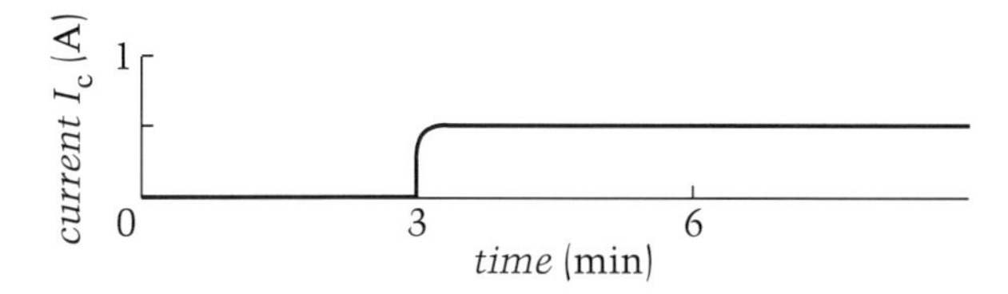

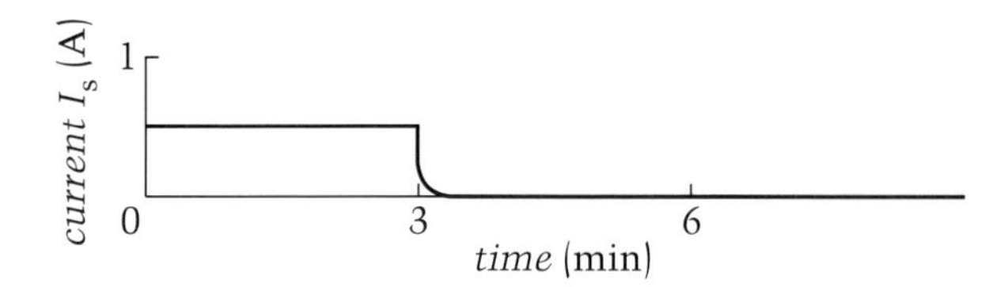

Figure 4

also be zero—that is, $V_c = 0$. Therefore,

$$V_c = 0 = -L\frac{\Delta I_c}{\Delta t}.$$

This means that the current I_c through the coil cannot change (see figure 2). Conservation of charge (or equivalently Kirchhoff's current rule) requires that

$$I = I_c + I_s.$$

Therefore, any change in the total current I must result in a change in the current I_s through the superconducting switch, and I_s must drop linearly to $-\frac{1}{2}$ A. At t = 6 min, the current is flowing clockwise around the loop containing the coil and the superconducting switch.

B. Because the current through the superconducting switch I_s is zero, the voltage across the switch V_s remains zero even when the switch changes to the normal state and returns to the superconducting state. This means that the voltage across the coil V_c must remain zero, and the current I_c through the coil cannot change. Therefore, the total current I will also not change (see figure 3).

C. At t = 3 min, the resistance of the superconducting switch suddenly jumps from 0 to $r_n = 5\ \Omega$. Because the current I_c through the coil cannot change instantaneously due to its inductance, the total current I (and thus the current I_s through the superconducting switch) must drop from E/R to $E/(R + r_n)$. With $R = r_n$, both currents will drop to one half their original values, as shown in figure 4.

The currents now approach their steady-state values exponentially with a time constant $\tau = L/R_t$, where R_t is the total resistance connected to the inductance. Since the two resistances are in parallel, we have

$$\tau = \frac{L}{R_t} = \frac{10\text{ H}}{2.5\ \Omega} = 4\text{ s}.$$

At steady state, there cannot be any current through the superconducting switch. Otherwise, there would be a voltage drop across the coil, necessitating a changing current through the coil in violation of the steady-state condition. Therefore, the current from the power source returns to its original value and all of this current must pass through the coil.

As in part B, at t = 6 min there is no current through the superconducting switch and, therefore, there are no changes in the currents when the superconducting switch returns to zero resistance.

D. Begin by closing the power switch K and increasing the total current I to 20 A. Note that this is the value of the circulating current. Because the superconducting switch has $r = 0$, I_c cannot change and I_s must increase by 20 A. In other words, I_s changes from −20 A to zero.

Because there is no current through the superconducting switch, we can now change it to the normal state $r = r_n$. We now gradually reduce the total current I to zero while keeping $I_s < 0.5$ A. Because

$$V_s < (0.5\text{ A})(5\ \Omega) = 2.5\text{ V},$$

the current through the inductor must obey

$$-\frac{\Delta I_c}{\Delta t} = \frac{V_s}{L} < \frac{2.5\text{ V}}{10\text{ H}} = 0.25\text{ A/s}.$$

Therefore, the current must be

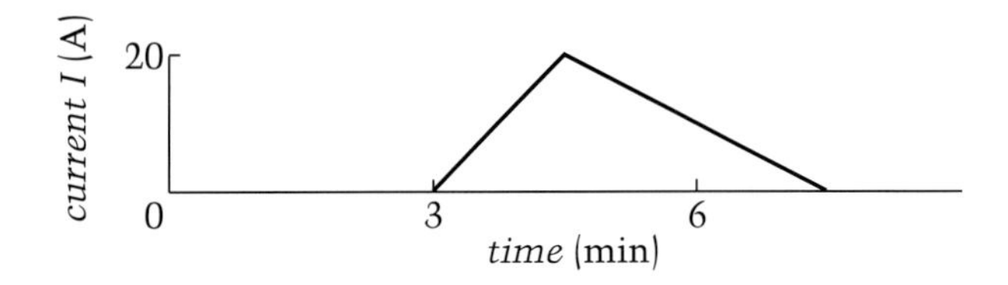

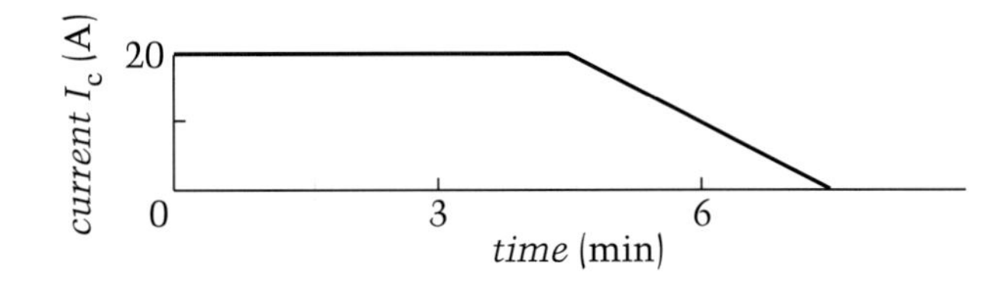

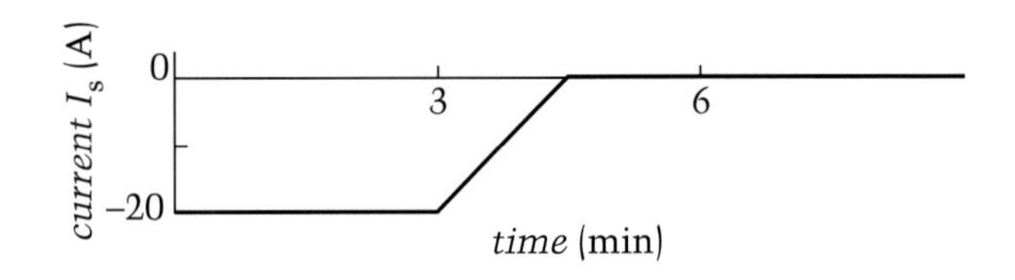

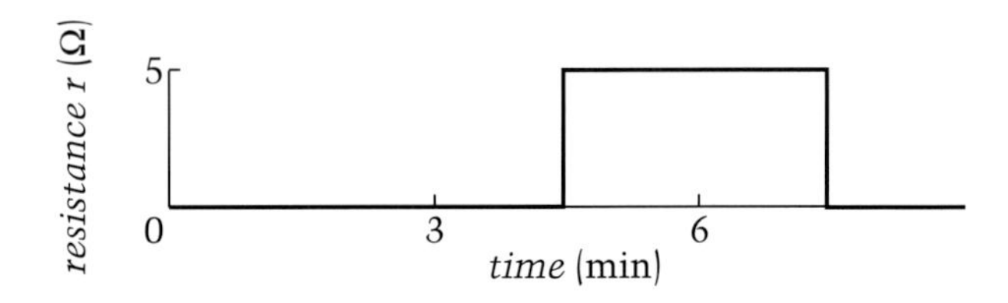

Figure 5

dropped to zero over a minimum of 80 s. These conditions are satisfied in figure 5 with $\Delta I/\Delta t \cong 0.1$ A/s.

As a final step, we can return the superconducting switch to the superconducting state and open the power switch *K*. ◘

Cloud formulations

"Eternal Clouds, let us appear; let us arise from the roaring depths of Ocean, our father; let us fly towards the lofty mountains [and] spread our damp wing over their forest-laden summits . . ."

—Aristophanes

HOW CAN YOU TAKE TEN thousand gallons of water and suspend them in mid-air? Build a cloud! It seems almost counterintuitive that wet air should be less dense than dry air and float in the sky. But the beauty of the cirrus and cumulus attest to this as we gaze at the myriad shapes and forms above us. Cloud formation reveals to us properties of the environment as well as properties of gases. As we notice that the western slopes of the Rockies are moist while the eastern slopes are deserts, we deduce that the air currents must be coming from the Pacific Ocean.

Last year, I went into a variety store to buy a mylar balloon. When asked what I wanted on the balloon, I wrote down on a piece of paper: $PV = nRT$. The employee was somewhat surprised and asked what the expression meant. After she acknowledged that she had studied high school chemistry, I hoped that she could now learn the ideal gas law within the context of her job. I asked her if anybody had ever bought a balloon during the winter and returned a few minutes later to complain that the balloon had a leak. She said that this had indeed happened, but that she would explain to the consumer that the balloon would re-inflate as soon as they got it home. In fact, as she was explaining this to the doubting customer, the balloon would inflate before their eyes. The mysterious equation on my balloon could explain this phenomenon.

The ideal gas law, $PV = nRT$, describes the relationship between the macroscopic properties of an enclosed gas. In the equation, P is the pressure, V is the volume, T is the temperature of the gas in kelvins, n is the number of moles of gas, and R is the gas constant. In the mylar balloon example, the pressure of the balloon is a constant—the pressure of the atmosphere pushing on the balloon. The balloon is fully inflated

A Tower of Babel-shaped rock rises high above the sea with a winding road leading to the top. One side is warm and summery, the other cold and icy. The balloon on the warm side, fully inflated, joyfully floats up into the sky while the same balloon on the frosty side decreases in volume and sadly drops down with its three scholarly passengers reacting accordingly. Up the winding road we follow a procession of evolutionary development, from the sea creature leaving the ocean and pushing a balloon full of promise all the way up to the modern humans, who are leaving the Earth to ride the lofty skies toward heaven. Some, though, end up on the cool side, and with their balloons they plummet into the icy waters of the dead. They should have learned their physics.

—T.B.

$E=mc^2$
TOM BUNK

inside the store while the temperature of the gas is equal to the store's room temperature. As the new balloon owner steps outside into the cold winter air, the temperature of the helium gas inside the balloon decreases. Since the pressure remains the same, the decrease in temperature is matched by a corresponding decrease in volume and the mylar balloon appears to be only partially inflated. Stepping back into the store, the balloon will magically become fully inflated again as the gas warms up.

A second illustration of the gas law occurs when a bicycle tire is inflated. In this example, the volume of the tire remains constant. As more and more gas is pumped into the tire, the pressure increases and there is a corresponding increase in the temperature of the tire. Feel the tire and it will be warm. Automobile tires in the winter will be slightly underinflated when you begin your journey but will be just right when the tires warm up from the friction with the road and the flexing of the side walls.

The ideal gas law can also help our readers understand how a pressure cooker works, how our lungs inhale through the movement of the diaphragm, and how a hot air balloon rises and falls in the atmosphere. Physicists and engineers often summarize the behavior of a gas on a P–V diagram, where the pressure is plotted on the y-axis and the volume is on the x-axis. For example, processes with constant temperatures are hyperbolas, since $PV = nRT$ and nRT is a constant.

Four processes are of special interest. The first three are changes that occur at constant temperature, constant volume, and constant pressure. In the fourth process no heat is transferred into or out of the system. This adiabatic process occurs when the change occurs very quickly—for instance, when sound waves move through the room. The changes in pressure occur so quickly, any heat transfer can be neglected. An adiabatic process also occurs when the system is thermally isolated from its surroundings. In this case the process can be very slow. As an example, a gas confined to an insulated container can expand adiabatically if weight on the piston is slowly removed.

When gases expand adiabatically, we expect that the pressure, the volume, and the temperature will all change. Fortunately, there is a relationship between the pressure and the volume during an adiabatic process: PV^γ = constant, where γ is the ratio of the specific heats for the gas and is equal to 1.4 for diatomic gases.

This brief introduction provides the background for this month's contest problem concerning cloud formation on the side of a mountain. The problem is adapted from the XVIII International Physics Olympiad which was held in Jena, East Germany, in 1987 (a few years before the German unification).

Moist air is streaming adiabatically across a mountain range as indicated in the figure. Equal atmospheric pressures of 100 kPa are measured at meteorological stations M_0 and M_3, and a pressure of 70 kPa at station M_2. The temperature of the air at M_0 is 20°C.

As the air ascends, cloud formation begins at location M_1, where the pressure is measured to be 84.5 kPa.

A quantity of moist air, with a mass of 2,000 kg over each square meter, ascends the mountain. This moist air reaches the mountain ridge (M_2) after 1,500 s. During this time, 2.45 g of water per kilogram of air precipitates as rain.

A. Determine the temperature T_1 at M_1, where the cloud forms.

B. Assuming that the atmospheric density decreases linearly with height, what is the height h_1 of station M_1?

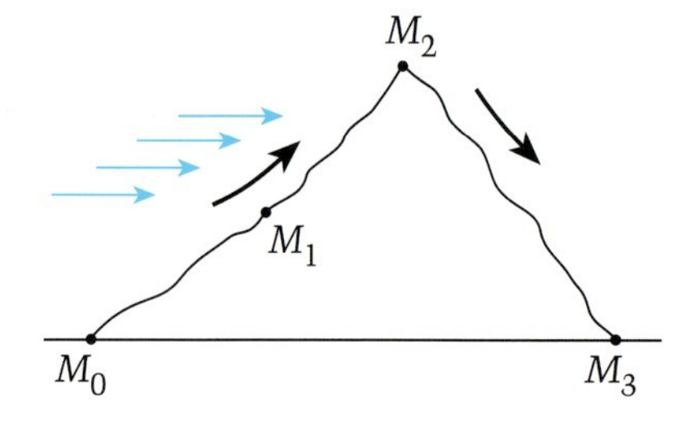

C. What temperature T_2 is measured at the ridge of the mountain?

D. Determine the height of the water column precipitated by the air stream in 3 hours, assuming a homogeneous rainfall between points M_1 and M_2.

E. What temperature T_3 is measured on the back side of the mountain range at station M_3? Compare the atmospheric conditions at M_0 and M_3.

Hints: Assume that the atmosphere is an ideal gas. Influences of the water vapor on the atmospheric density are to be neglected.

The atmospheric density for P_0 and T_0 at station M_0 is ρ_0 = 1.189 kg/m³. The specific heat of vaporization of the water within the volume of the cloud is L_v = 2,500 kJ/kg.

Solution

The clouds certainly cleared for Alex Lee, a student at Choate Rosemary Hall, as he solved this problem. We will follow along with Alex as the problem unfolds.

A. The first part of the problem asked readers to determine the temperature T_1 at M_1, where the cloud forms.

Since the air is streaming adiabatically, we have two equations that hold:

$$PV^\gamma = \text{constant},$$

$$\frac{PV}{T} = \text{constant}.$$

We can combine these two equations to get

$$\frac{T_1}{T_0} = \left(\frac{P_1}{P_0}\right)^{1-\frac{1}{\gamma}}.$$

Therefore,

$$T_1 = 279.4 \text{ K} = 6.4°\text{C}.$$

B. Consider the pressure difference between M_0 and M_1. This must be caused by the extra chunk of air below M_1 and above M_0. Consider an imaginary cylinder with base area A and height h_1

from M_0. Then we have

$$-P_1A + P_0A = mg,$$

where m is the mass of air within the cylinder. Since the density of air varies linearly, the mass is

$$m = Ah_1 \frac{\rho_0 + \rho_1}{2},$$

where ρ is the density of air. The density is calculated from the equation of state:

$$\frac{P_0}{\rho_0 T_0} = \frac{P_1}{\rho_1 T_1},$$

$$\rho_1 = 1.054 \text{ kg/m}^3.$$

Solving for h_1 we get

$$h_1 = \frac{P_0 - P_1}{g\left(\frac{\rho_0 + \rho_1}{2}\right)},$$

$$h_1 = 14.08 \text{ m}.$$

C. At M_2 the air again streams adiabatically up the slope. In this movement, we must also take into consideration that there is an additional temperature change due to the condensation of water:

$$T_2 = T_i + \Delta T,$$

where T_i is the temperature from the adiabatic process. From this we get

$$\frac{T_i}{T_2} = \left(\frac{P_2}{P_1}\right)^{1-\frac{1}{\gamma}},$$

$$T_i = 264.8 \text{ K}.$$

As for ΔT, we know that the latent heat is

$$Q = mL_V = C_P \Delta T,$$

$$\Delta T = \frac{mL_V}{C_P}$$

$$= \frac{(2.45 \text{ g})(2500 \text{ kJ/kg})}{1000 \text{ H/kg} \cdot \text{K}}$$

$$= 6.1 \text{ K},$$

$$T_2 = 264.8 \text{ K} + 6.1 \text{ K} = 270.9 \text{ K}.$$

D. The precipitation separated from the ascending column of air per square meter per second is

$$\frac{\left(2000 \frac{\text{kg}}{\text{m}^2}\right)\left(2.45 \frac{\text{g}}{\text{kg}}\right)\left(10^{-3} \frac{\text{kg}}{\text{g}}\right)}{1500 \text{ s}}$$

$$= 3.3 \cdot 10^{-3} \frac{1}{\text{m}^2 \cdot \text{s}},$$

$$\left(3.3 \cdot 10^{-3} \frac{1}{\text{m}^2 \cdot \text{s}}\right)(3 \text{ hr})\left(\frac{3600 \text{ s}}{1 \text{ hr}}\right)$$

$$= 35.3 \text{ kg/m}^2.$$

Since 1 kg/m^2 results in a precipitation level of 1 mm, the depth of the precipitated water is 35.3 mm.

E. Whatever air that has passed over the mountain will probably descend adiabatically:

$$T_3 = T_2 \left(\frac{P_3}{P_2}\right)^{1-\frac{1}{\gamma}},$$

$$T_3 = 300.0 \text{ K} = 26.9°\text{C}.$$

If there were no condensation and rain from the air, T_3 should be equal to T_0. Because of the rainfall, the air at M_3 is colder and less moist than at M_0.

Weighing an astronaut

"The earth before us is a handful of soil,
but it sustains mountains without feeling their weight
and contains the rivers and seas without their leaking away."
—Confucius

ONE DAY SOME OF US WILL be living for extended periods of time in a space colony. What health problems might arise? Will we lose weight, or will our bones weaken in "zero gravity"? Medical questions were very important during the Skylab mission from May 1973 until February 1974. At the most basic level the scientists wanted to know if the astronauts would lose weight during prolonged stays in space. Let's begin by taking a closer look at the concept of weighing an astronaut.

When you want to know your weight in the morning, you simply step on a bathroom scale and read your weight. But how does the scale "know" your weight? You are actually measuring the amount of stretch or compression of a spring inside the scale. If we assume that the spring is ideal (that is, it obeys Hooke's Law), we know that $F = -kx$, where F is the force of the spring, k is the spring constant (a measure of the stiffness of the spring), and x is the extension of the spring. In this case, the applied force is just the force of gravity acting on you. Since the force of gravity is given by mg, if you have a mass of 60 kg and the local acceleration due to gravity is 9.80 m/s^2, the reading will be 588 N, or an equivalent value in pounds, stones, or kilograms. It seems that only physics teachers' scales are calibrated in newtons!

How much would you weigh if you had been a resident of Skylab in 1973? The orbit of Skylab had an altitude of 386 km above the Earth's surface, or a radius of 6,378 km + 386 km = 6,764 km. Since gravity is an inverse-square force, you can calculate the force of gravity on you in Skylab:

$$F = F_E\left(\frac{R}{R_E}\right)^2$$
$$= (588\text{ N})\left(\frac{6{,}378\text{ km}}{6{,}764\text{ km}}\right)^2$$
$$= 523\text{ N},$$

There are many ways to lose weight, the most painful one being to stop eating too much chocolate cake. The more exciting one is to go into space (doesn't have to be further than the moon), where you can eat as much as you want without gaining even an ounce, according to the scale, and the scale never lies. You can also stay on Earth and take a real fast, almost free-fall ride standing on a scale from high up shooting downwards. The scale will take off a couple of pounds during the ride, but the ride takes you straight to hell because you are cheating reality, especially if the ride ends right in front of the tree of knowledge. There we find Newton sitting and reflecting, while in the treetop appear two divine hands, one dropping an inspirational apple on Newton's brain, the other expelling him from the amusement park, showing that as a regular mortal you can't win.

—T.B.

DOWN
BURP!
CHOCOLATE
NASA
HERS
NASA
BMMD
LOSE WEIGHT!
GO INTO SPACE!
NEXT
LAUNCHING
Tree
OF
KNOW-
Ledge
I. NEWTON
UP
GRAVITY
BY
ISAAC
Tom Bunk

where the subscript E refers to Earth values. Therefore, you expect a scale reading of 523 N.

However, when you step on the bathroom scale in Skylab, it reads zero! In the vernacular of space, you are weightless. What does it even mean to "step on a scale" in a zero-gravity environment? We get into trouble because we have used weight to mean two different things—the force of gravity and reading on the scale. This does not cause complications if you are in an inertial system—that is, a system that is not accelerating. We assume that your bathroom on Earth is an inertial reference system. (It is not really an inertial system because the Earth is rotating on its axis and revolving about the Sun, but it is approximately an inertial system.) When you stand on the bathroom scale, you have no acceleration and Newton's Second Law tells us that there is no net force acting on you. Therefore, the downward force of gravity must be equal in size to the upward force of the scale acting on you, and it does not matter which force we measure.

Problems arise when your reference system is accelerating. Because Skylab is in orbit about Earth, it has a centripetal acceleration of v^2/R. And so do you—you are also in orbit. The force of gravity provides the centripetal acceleration needed to keep you in orbit and there is no need for the scales to hold you up (or anything else, for that matter).

We can create a similar situation on Earth. Some amusement park rides take you to the top of a structure and drop you over the edge. If you were to stand on a bathroom scale during this free fall, it would read zero. Both you and the scale are in free fall and the scale does not need to support you. For this reason, many physics teachers carefully distinguish between the force of gravity and the weight. Weight is the reading on the bathroom scale, or the support force needed to keep you at rest in the noninertial reference system. Other teachers use the term "apparent weight" to refer to the scale reading and use weight to refer to the force of gravity.

Let's return to the elevator and cause it to accelerate *upward* at 9.80 m/s^2. What will you weigh? The net upward force on you must be mg to give you an upward acceleration of g. Because the force of gravity is mg downward, the spring in the scale must push upward with a force of $2mg$. Therefore, the scale would read $2 \cdot 588$ N = 1,176 N.

We could have avoided our weight problem by realizing that the NASA scientists were really concerned with the mass of the astronaut. But how do we measure the mass of an astronaut in space? We obviously cannot ask the astronaut to stand on a scale. An easy way to do this is to use Newton's Second Law, apply a known force to the astronaut, and measure the resulting acceleration. NASA accomplished this by designing a Body Mass Measuring Device (BMMD for short). It is basically a chair mounted on a pair of leaf springs. Whenever a mass m is attached to a spring with a spring constant k and displaced a small distance from the equilibrium position, it executes simple harmonic motion with a period T given by

$$T = 2\pi\sqrt{\frac{m}{k}}.$$

In practice, an astronaut sits in the BMMD and measures the period of oscillation to obtain the mass. This brings us to our problems.

A. The BMMD was calibrated by loading the chair with a known mass and measuring the corresponding period of oscillation. Graph the following data supplied by NASA to find the combined spring constant of the leaf springs and the mass of the empty chair for the BMMD:

Mass (kg)	Period (s)
0.00	0.90149
14.06	1.24979
23.93	1.44379
33.80	1.61464
45.02	1.78780
56.08	1.94442
67.05	2.08832

B. One of the crew members in the second group of three astronauts to live in Skylab was electrical engineer Dr. Owen K. Garriott. He measured periods of 2.012 s and 1.981 s while sitting in the BMMD at the beginning and end of a 58-day interval. How much weight (read mass) did he gain or lose during this time?

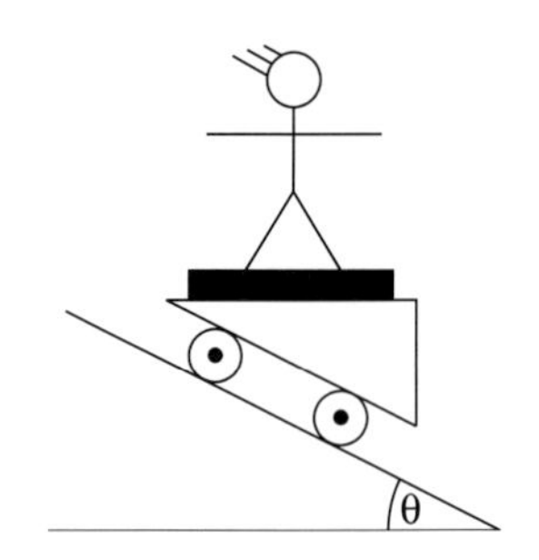

Figure 1

C. Assume that you ride a skateboard down an inclined plane that makes an angle θ with the horizontal. A scale has been mounted on the skateboard in a horizontal position as shown in figure 1. There is no friction between the skateboard and the inclined plane and you have a mass of 60 kg. What does the scale read? (A version of this question appeared on the preliminary examination used to select members of the 1995 US Physics Team that competed in the International Physics Olympiad in Canberra, Australia.)

Solution

This problem was given as a class assignment by Art Hovey in Woodbridge, Connecticut, and by your author (LDK) to sophomores at Montana State University. The best solutions at ARHS were submitted by Kurt Rohlooff and Lori Sonderegger, and the best one at MSU was written by Dave Peters.

A. The equation for the period T of the simple harmonic motion of a mass m on a spring is given by

$$T = 2\pi\sqrt{\frac{m}{k}},$$

where k is the spring constant. In our case, the total mass m is the sum of the mass of the chair m_c and the mass of the astronaut m_a. Squaring both sides and solving for T, we obtain

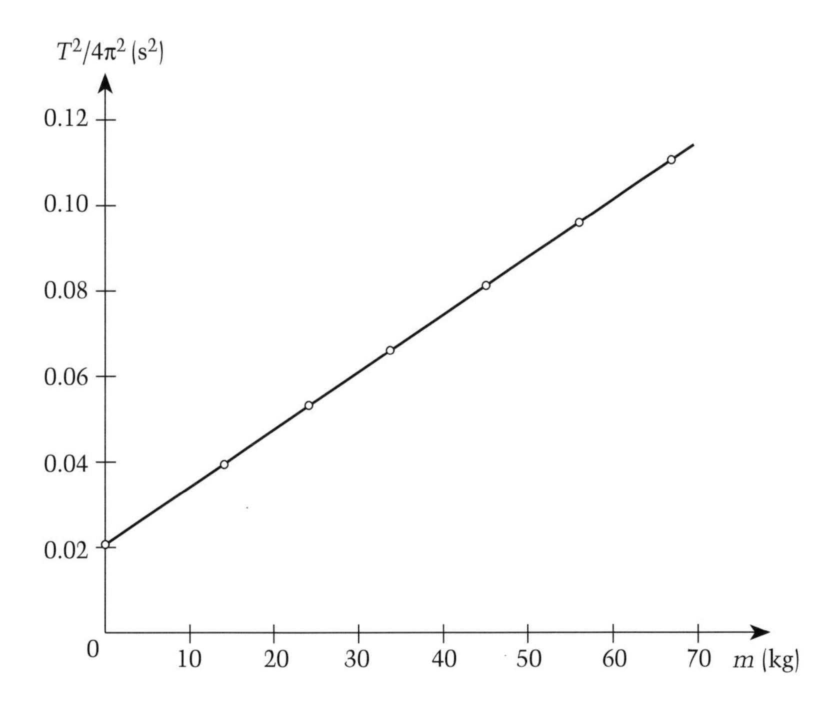

Figure 2

$$\frac{T^2}{4\pi^2} = \frac{1}{k} m_a + \frac{m_c}{k}.$$

This is of the form $y = ax + b$. Therefore, if we plot a graph of $T^2/4\pi^2$ versus the mass used in the calibration of the BMMD, we should obtain a straight line with a slope of $1/k$ and a y-intercept of m_c/k (fig. 2). This allows us to obtain the numerical values $k = 748$ N/m and $m_c = 15.4$ kg.

B. We can now use these values and the numerical data for astronaut Garriott to discover that he lost 2.3 kg of mass during 58 days in space.

C. The third part of this problem asked you to calculate the reading on the scale as a person rides down an incline as shown in figure 1. We begin by calculating the acceleration of the entire system (skateboard, scale, and person) down the incline. Because the component of the force of gravity down the incline is $m_s g \sin\theta$, the acceleration of the system down the incline is $g \sin\theta$. This is also the acceleration of the person, which allows us to use Newton's second law to find the forces acting on the person. Assuming that the scale reads the force applied normal to its surface, let's look at the vertical forces acting on the person. The force of gravity mg acts downward and the force of the scale F_s acts upward. Therefore, the difference in these two forces must yield the mass times the vertical component of the acceleration:

$$mg - F_s = (mg \sin\theta) \sin\theta,$$

or

$$\begin{aligned} F_s &= mg\,(1 - \sin^2\theta) \\ &= mg \cos^2\theta \\ &= (588\ \text{N}) \cos^2\theta. \end{aligned}$$

This is the standard solution given by the top students on the preliminary exam to select the 1995 US Physics Team that competes in the International Physics Olympiad and the one expected in textbooks. However, there is a problem—the person will have a very difficult time standing vertically on the scale. This difficulty was pointed out by Dr. Albert A. Bartlett, professor emeritus at the University of Colorado and a past president of the American Association of Physics Teachers. ◘

The first photon

"Have other eyes, new light! And look!
This is my glory, unveiled to mortal sight."
—Bhagavad Gita

AN ENVIRONMENTALLY rich village inhabited by curious people had been fully explored. All of the interesting corners and crevices of this remarkable land had revealed their secrets. There were no mysteries left. Oh, of course, some of the villagers remarked that a finer microscope may reveal a little more detail. But most were pleased with the comfort they felt in the familiar surroundings.

Such was the state of physics at the turn of the 20th century. The great syntheses of Newton and Maxwell remarkably explained so much about forces and motion, electricity, magnetism, and optics. Albert Michelson, America's first physics Nobel laureate, remarked that physics was complete and the following years would be devoted to simply increasing the precision of the experiments. Yet there existed a problem or two that appeared to be stumbling blocks. One was the photoelectric effect—the ability of light to free electrons from a metallic surface.

Maxwell and Hertz provided the theoretical and experimental evidence to convince the entire physics community that light was an electromagnetic wave. Almost a century earlier, Thomas Young had argued that light was a wave phenomenon and even measured the wavelength of this light. Yet a wave picture of light presented numerous difficulties in explaining the photoelectric effect. If the light is very dim, it should take hours to free an electron. Contrary to this prediction, the electrons are freed almost instantly. If the light is very intense, the expectation was that many electrons would be freed. However, intense, bright red light does not free even a single electron. Finally, when the light is able to free electrons, the kinetic energies of the electrons do not depend on the energy reaching the surface of the metal in a given time.

What a depressing life to be a cute and innocent electron and be a part of a chain gang hammering away on a metallic surface! But there is light on the horizon—plain light will free the electrons from their bondage. Red light won't do the trick, though, as one of the unfortunate captives standing in a beam of red light learns while waiting in vain for the photoelectric effect to occur. Under the dark city of enslaved photons we see the underground station where people from previous centuries happily ride the electromagnetic wave train. That was before Einstein introduced the more comfortable electromagnetic particle ride applied to moving bodies. High in the night sky, sitting in the same boat, are Planck, Newton, Maxwell, and Einstein, who is working the long bow for the entire think quartet.

—T.B.

E=h<nu>
FREE AT LAST!!
POTAT
CHIPS
10¢
Tomas Bunk

In what now seems like physics folklore, the young patent clerk Albert Einstein stepped onto the scene to propose that light behaves like a particle (known as a photon) and that each photon has an energy that depends on its frequency. More precisely, Einstein attributed an energy to each photon of light according to the equation

$$E = h\nu,$$

where h is Planck's constant ($6.63 \cdot 10^{-34}$ J · s) and ν is the frequency of the light.

The electrons are bound to the metal with a certain energy defined as the work function ϕ. When a photon strikes an electron in the metal, the electron acquires all the photon's energy and the photon disappears. The maximum kinetic energy a freed electron can attain is equal to the difference between the energy of the photons of light and the work function. (Some of the electron's kinetic energy is lost in getting to the surface. In fact, if it loses too much, the electron cannot escape the surface.)

Assume that the work function for a given metal is 3 eV. (One electronvolt equals $1.6 \cdot 10^{-19}$ J, the energy acquired by an electron falling through an electrical potential difference of 1 V. Using this energy unit, $h = 4.14 \cdot 10^{-15}$ eV · s.) A photon of red light (wavelength = 620 nm) has an energy equal to 2 eV. Since all photons of this light have energies of 2 eV, no single photon can provide the 3 eV required to free the electrons from the surface. No matter how intense the red light (no matter how many 2-eV photons are present), the electron will not be freed. If ultraviolet light of wavelength 310 nm impinges on the metal, each photon has a corresponding energy of 4 eV and electrons come flying off the metal. The maximum kinetic energy K these electrons can have is the difference between the two energies:

$$K_{max} = h\nu - \phi.$$

An electron freed by this ultraviolet light may have a kinetic energy as large as 1 eV.

A vending machine provides a useful analogy for what is happening here. Assume we have a vending machine that can't accept multiple coins. You can submit only one coin at a time—a penny, a nickel, a dime, or a quarter. Potato chips cost 10 cents. If you put a penny in the machine, the penny is returned or lost. If you put a nickel in the machine, the nickel is also returned or lost. However, if you put a dime in the machine, a bag of potato chips is released to you. If you put a quarter in the machine, one bag of potato chips is released and up to 15 cents is returned in change. Note that when 25 cents is deposited, *two* bags of potato chips are not released. One coin can yield one bag or none—that's the rule.

The work function of the metal is the price of the potato chips. The energy of the incident photon is the value of the coin dropped into the machine. If an electron is freed, the kinetic energy it attains is represented by the change the machine releases. A low-energy photon of light (a coin less than 10 cents) is unable to free the electron (or a bag of potato chips.) No matter how many nickels you have, you will not be able to free the potato chips. No matter how intense the light (lots and lots of low-energy photons), no electrons will be released. If a high-energy photon is incident on the metal surface, electrons will be emitted. (If a quarter is placed in the machine, a single bag of potato chips will be released. More photons (more quarters), more electrons.

The photon (particle) nature of light satisfactorily explains the experimental results of the photoelectric effect. However, it is not able to explain the wave aspects of light so clearly demonstrated by Young, Maxwell, and others. This leaves us with the question: What is the true nature of light?

More insight into the nature of light was attained in the next decades, as it was shown that light behaves like a particle in an elastic collision between light and an electron, where the momentum of the photon of light is shown to be $h/\lambda = h\nu/c$. Arthur H. Compton earned recognition for these experimental studies in 1922.

Our problems focus on the particle nature of light and the associated energy and momentum of the photons.

A. Monochromatic light sources with a variety of wavelengths were incident on lithium and the maximum kinetic energies of the emitted electrons were recorded by Millikan (of the famous Millikan oil drop experiment, which measured the charge on an electron) as follows:

Wavelength (nm)	Kinetic energy (eV)
433.0	0.55
404.7	0.73
365.0	1.09
312.5	1.67
253.5	2.57

Plot these data and find the numerical values of Planck's constant and the work function for lithium.

B. Show that a free electron cannot completely absorb a photon. (In a metal, the surrounding atoms can participate in the collision, allowing energy and momentum to be simultaneously conserved.)

C. (i) The human eye is so sensitive that it can detect single photons of light. If the pupil of the eye has a diameter of 0.5 cm, at what distance would you place a 50-W light source (of wavelength 500 nm) so that the number of photons reaching the pupil is one per second on average? (ii) At what distance should the light source be placed so that the density of photons is on average 1 photon per cubic centimeter?

Solution

This problem was well received, and excellent responses came from Christopher Rybak (a junior at the Prairie School in Racine, Wisconsin), Jonathan Devor (who represented Israel in the International Physics Olympiad in Australia in 1995), Aaron Manka (a student in Huntsville, Alabama), and Lori Sonderegger and Art Hovey (a student and teacher, respectively, at Amity Regional High School in Woodbridge, Connecticut).

The first part of the problem required readers to interpret data from a photoelectric experiment. The equation relating the kinetic energy of the electron and the frequency of the light is

$$K_{max} = h\nu - \phi.$$

We can substitute $\nu = c/\lambda$, where c is the speed of light in a vacuum and λ is the wavelength. This yields

$$K_{max} = \frac{hc}{\lambda} - \phi.$$

If we plot K_{max} on the y-axis and $1/\lambda$ on the x-axis, the resulting graph should be linear; the slope is hc, and the y-intercept is the negative of the work function ϕ (see the figure below). The slope of this line is $\Delta K_{max}/\Delta(1/\lambda)$, which is equal to 1,242 eV · nm. Therefore,

$$hc = 1{,}242 \text{ eV} \cdot \text{nm},$$

where $c = 3 \cdot 10^8$ m/s $= 3 \cdot 10^{17}$ nm/s and $h = 4.14 \cdot 10^{-15}$ eV · s. The work function for lithium is equal to 2.32 eV.

Part B asked for a proof that a free electron cannot completely absorb a photon. We can assume that absorption can take place, apply conservation of momentum and energy, and then arrive at a paradox.

Momentum conservation requires

$$\frac{h\nu}{c} = mv.$$

Energy conservation requires

$$h\nu = \tfrac{1}{2}mv^2.$$

Dividing these equations requires that the electron's velocity be equal to twice the speed of light—an impossibility.

Part C(i) asked how far away a 50-W bulb would need to be placed so that a human eye, sensitive to single photons, would detect an average of one photon per second.

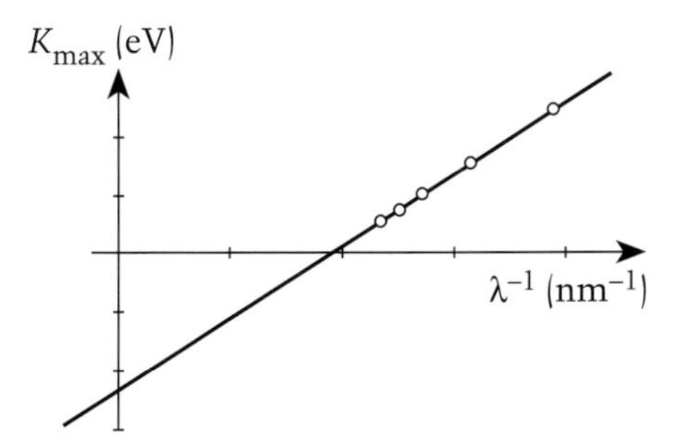

Given a wavelength of 500 nm for the light, the individual photons will have a corresponding energy of $hc/\lambda = 3.97 \cdot 10^{-19}$ J. Therefore, a 50-W light bulb emits on average $1.26 \cdot 10^{20}$ photons/s. These photons spread out isotropically and will intercept the human eye. Since the pupil has a given diameter d of 0.5 cm, its area is $\pi(d/2)^2 = 0.0625\pi$ cm^2.

The ratio of the surface area of the eye to the surface area of the light sphere is equal to the ratio of the number of photons hitting the eye to the total number of photons. The distance at which one photon hits the eye every second is

$$\frac{0.0625\pi \text{ cm}^2}{4\pi R^2} = \frac{1 \text{ photon/s}}{1.26 \cdot 10^{20} \text{ photons/s}}.$$

Thus,

$$R = 14{,}000 \text{ km!}$$

Part C(ii) asked for the distance at which this light source would have to be placed if the density of photons were to be 1 photon/cm^3 on average. If we imagine a spherical shell 1 cm thick, we can assume it is made of 1-cm^3 cubes. The density stated requires one photon to be in the cube at any time. Since the photons travel at $3 \cdot 10^{10}$ cm/s, they will each traverse the cube in $3.33 \cdot 10^{-11}$ s. Therefore, $3 \cdot 10^{10}$ photons will have to travel through the cube in 1 s for the density to be 1 photon/cm^3.

This is identical to having an area density at the shell of $3 \cdot 10^{10}$ photons/cm^2/s. Applying the same procedure as in part C(i), we can solve for the distance R:

$$\frac{1.26 \cdot 10^{20} \text{ photons/s}}{4\pi R^2} = 3 \cdot 10^{10} \text{ photons/cm}^2\text{/s},$$

and

$$R = 183 \text{ m},$$

which is considerably smaller than the previous answer.

Pins and spin

"Nay, sometimes,
Like to a bowl upon a subtle ground,
I have tumbled past the throw . . ."
—The Tragedy of Coriolanus, *Act V, Scene 2*

ALTHOUGH ISAAC NEWTON probably never bowled a perfect 300 game, his physics can be used to analyze the sport of bowling. The collision of the ball with the bowling pins (and the collisions of the pins themselves) must obey the conservation laws. If we ignore the friction of the ball and pins with the floor during the impact, linear momentum in the horizontal directions must be conserved during these collisions. As the pins fly through the air, they conserve linear momentum, angular momentum, and mechanical energy. The path of each center of mass between collisions is the same as those of the projectiles we've all studied in class.

In this problem, let's concentrate on the interaction of the ball with the surface of the lane. Those of us who throw a "curve ball" know that this interaction is very important. Let's imagine that we throw a ball parallel to the fourth board in from the right-hand gutter. This ball would obviously not produce a strike as it will miss the head pin and probably only take out the three pins (6-9-10) on the far right-hand side.

At the time of release, we right-handers can lift our fingers to put spin on the ball so that it curves to the left and hits the "pocket" between the 1 and 3 pins. The curve changes the ball's angle of approach toward the pins and yields a higher percentage of strikes than a "straight ball." This pocket is also more forgiving in that it yields fewer splits.

Analyzing the curve ball involves looking at two rotations: the spin and the rotation down the lane. As we usually do when faced with a problem in physics, let's begin with the simplest case—the straight ball. We assume that the ball of radius r and mass m is thrown horizontally with an initial speed v_0, with no initial rotation, and at a negligible height above the floor. The ball starts with no rotation, loses speed, and picks up

We are in Italy during the Renaissance, the exciting time of the rebirth of classical ideals, of beauty and the highest aspirations for humanity, on top of a roof that consists of three bowling alleys. There are three players and numerous onlookers from different centuries—the curious caveman, Archimedes, Da Vinci, Descartes, all the way to today's postmodernists. The experimenting bowlers are Galileo, forcefully throwing the whole globe at the towers of Pisa; Newton next to him, knocking all the pins with a spinning "curve apple" into space; and an unknown player throwing a huge cylinder into the alley to make sure he takes out all the pins and scores really high. On the street below we see the pins marching back to their jobs, only to be thrashed some more.

—T.B.

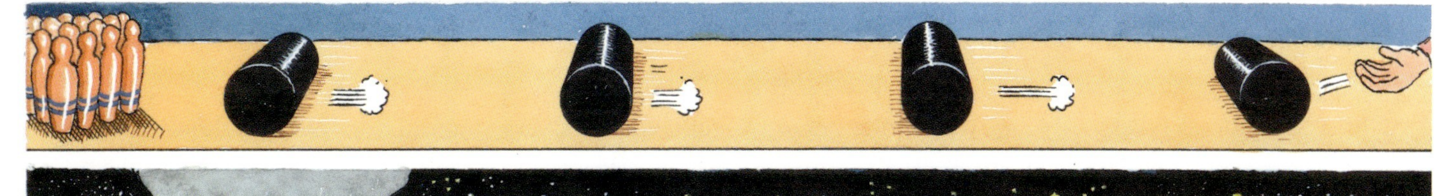

f=ma=
BOWLING
Tom BUNK

rotation due to the friction with the floor, and at some point it rolls without slipping. We are interested in how long this process takes.

Because there is no acceleration in the vertical direction, we know that the force of gravity on the ball is canceled by the normal force of the floor on the ball. Therefore, the net force is due to the frictional force f with the floor, which we assume to have the standard form $f = \mu mg$, where μ is the coefficient of kinetic friction. Therefore, Newton's second law tells us that

$$f = ma = -\mu mg,$$

or the linear acceleration a of the ball is

$$a = -\mu g.$$

The frictional force also exerts a torque $\tau = fr$ on the ball about its center. According to Newton's second law for rotation we have

$$\tau = I\alpha = fr = \mu mgr,$$

where $I = 2/5\ mr^2$ is the moment of inertia of the ball about its center. The resulting angular acceleration is given by

$$a = \frac{5\mu g}{2r}.$$

From the kinematics equations for translational and rotational motion, we have

$$v = v_0 + at = v_0 - \mu gt$$

and

$$\omega = \omega_0 + \alpha t = \frac{5\mu gt}{2r}.$$

Using the condition for rolling without slipping, $v = \omega r$, we can solve for the time when this first occurs:

$$t = \frac{2v_0}{7\mu g}.$$

At the instant the ball starts rolling without slipping, the speed of the ball is $5v_0/7$ and the ball has traveled a distance

$$d = \frac{12v_0^2}{49\mu g}$$

down the alley. Furthermore, we can use the final speed to calculate that the ball loses 2/7 of its initial kinetic energy. However, it is very interesting to note that the loss in kinetic energy is *not* equal to fd. How can you reconcile this?

Let's now return to the curve ball. The ball is initially spinning sideways without any sideways translational motion. This is the inverse problem and leads us to this month's contest problem. Let's use a cylinder instead of a ball to simplify the numbers as we did when this problem was given on the semifinal exam to select the 1995 US Physics Team. Besides, if the cylinder is long enough, bowling would be a lot simpler!

Assume that a uniform cylinder of mass m and radius r has an initial rotational velocity ω_0 about its axis, which is horizontal. Assume further that we drop the spinning cylinder onto the floor from a negligible height.

A. How long is it before the cylinder rolls without slipping?

B. What is the translational speed of the cylinder at this time?

C. How far down the alley does this occur?

D. Use the speed of the cylinder to calculate the fraction of the initial rotational kinetic energy that is lost.

E. Show that this loss of energy can be explained using the work–energy theorem.

We leave it to you to apply the ideas in these two problems to the motion of the curving bowling ball. And we hope that your analysis improves your score!

Solution

Correct solutions were submitted by Noah Bray-Ali from Venice High School in Los Angeles and Canh Nguyen from Little Canada, Michigan, although they had different interpretations of part E.

A. As with the bowling ball, the net force on the cylinder is the usual force of friction $f = \mu mg$. Therefore, Newton's second law tells us that

$$f = ma = \mu mg,$$

and the linear acceleration is

$$a = \mu g.$$

The frictional force also exerts a torque $\tau = rf$ on the cylinder about its center. According to Newton's second law for rotation,

$$\tau = I\alpha = -rf = -r\mu mg,$$

with $I = \frac{1}{2}mr^2$. The resulting angular acceleration is

$$\alpha = \frac{-2\mu g}{r}.$$

Using the kinematic equations for translational and rotational motion, we have

$$v = v_0 + at = \mu gt$$

and

$$\omega = \omega_0 + \alpha t = \omega_0 - \frac{2\mu gt}{r}.$$

Using the condition for rolling without slipping, $v = \omega r$, we solve for the time when this first occurs:

$$\mu gt = r\omega_0 - 2\mu gt,$$

$$t = \frac{r\omega_0}{3\mu g}.$$

B. We can now plug this time into the equation for v,

$$v = \mu gt = \frac{1}{3}r\omega_0,$$

to find the speed when rolling occurs without slipping.

C. We can now obtain the corresponding distance in a variety of ways. We choose

$$d = \frac{1}{2}at^2 = \frac{r^2\omega_0^2}{18\mu g}.$$

D. The initial kinetic energy is only rotational and is given by

$$KE_i = \frac{1}{2}I\omega_0^2 = \frac{1}{4}mr^2\omega_0^2,$$

but the final kinetic energy is a combination of translational and rotational energy:

$$KE_f = \frac{1}{2}mv^2 + \frac{1}{2}I\omega^2,$$

with $v = r\omega$. Therefore,

$$KE_{\text{f}} = \frac{3}{4}mr^2\omega^2 = \frac{1}{12}mr^2\omega_0^2,$$

and the change in kinetic energy is

$$\Delta KE = -\frac{1}{6}mr^2\omega_0^2.$$

This gives a fractional loss of 2/3.

E. The work-energy theorem says that the loss in kinetic energy is equal to the work done by the frictional force. However, in calculating the work, we must use only the distance slipped, not the linear distance traveled. We begin by calculating the total angle through which the cylinder rolls up until the time it quits slipping:

$$\theta = \omega_{\text{ave}}t = \frac{2}{3}\omega_0 t = \frac{2r\omega_0^2}{9\mu g}.$$

If the cylinder had not slipped, this angle would correspond to a linear distance of $r\theta$. The distance slipped is just the difference between this distance and the actual linear distance traveled:

$$d_{\text{slip}} = \left(\frac{2}{9} - \frac{1}{18}\right)\frac{r^2\omega_0^2}{\mu g} = \frac{r^2\omega_0^2}{6\mu g}.$$

The work done by the frictional force is then

$$W = -fd_{\text{slip}} = -\frac{1}{6}mr^2\omega_0^2,$$

which is equal to the change in the kinetic energy calculated above.

Noah Bray-Ali agreed with our interpretation for part E, but decided to calculate the loss in rotational kinetic energy and the gain in translational kinetic energy to get the total change in the kinetic energy. Both Noah and Canh Nguyen showed that the fractional loss in the rotational kinetic energy by itself is 8/9 and that this is due to the rotational work performed by the torque. ◙

Split image

"All these years, your left hand, modest but sinister accompanist, has seen itself in the mirror grown stronger."
—Al Zolynas, "Dream of the Split Man"

AT SPECIAL MOMENTS WE'RE able to bring all of our attention to bear on a single item. We focus our minds, our gaze, and our efforts. Similarly, wouldn't it be interesting to take the light impinging on a surface and focus it to a point? Using the properties of transparent materials and our knowledge of Snell's law and geometry, we can construct an object for this purpose—a lens. A converging lens bends all rays of light parallel to the principal axis (the axis of symmetry of the lens) in such a way that they converge at a single point referred to as the focus (fig. 1).

The lens also takes the light emerging from one point and focuses that light to a point on the other side of the lens. This works whether the light source is on the principal axis (fig. 2) or off axis (fig. 3). This then provides the surprising and technologically vital property of image formation in lenses. All slide projectors, cameras,

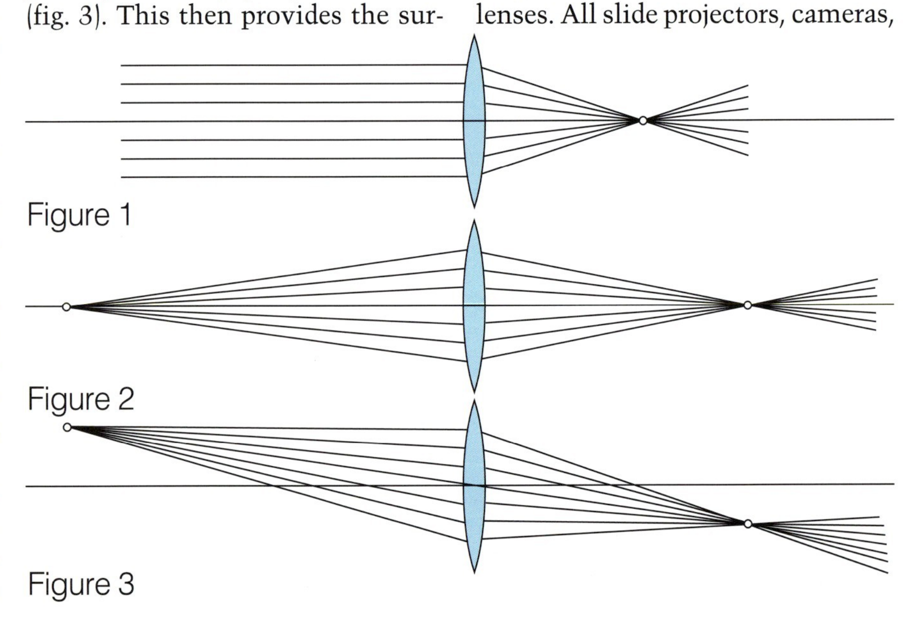
Figure 1

Figure 2

Figure 3

After reading the title of this article, I decided immediately to use it for my illustration. Split is where I was born and where I grew up, a beautiful city on the Mediterranean coast of Croatia. Looking back on my childhood, Split meant always the most perfect place—the temperate climate, the azure blue sea, the town rich in history, founded by the Romans. But in 1995, when I drew this illustration, bloody war was raging, which shattered my image of a perfect paradise. The illustration shows the sunny past, but under the fractured lens we see the sinister side, the destruction and death. Today, ten years later, the fragments of the broken lens have grown together into one piece and the light that passes through is pure and perfect again.

—T.B.

SPLIT IS A BEAUTIFUL OLD TOWN ON THE ADRIATIC COAST IN DALMATIA, WHERE THE ARTIST WAS BORN.

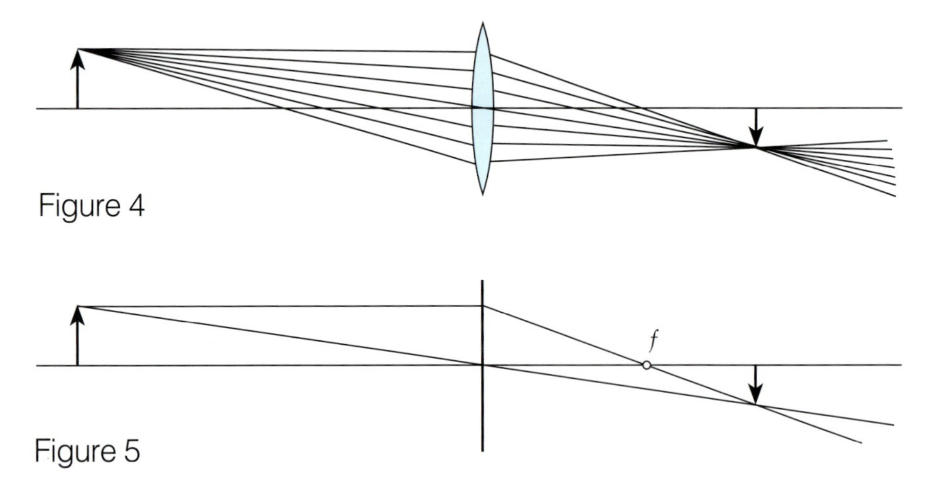

Figure 4

Figure 5

copy machines, microscopes, and binoculars are dependent on a lens being able to produce images.

In forming a real image, all light leaving points of the object and passing through the lens come together on the far side of the lens (fig. 4). Snell's law ($n_1 \sin \theta_1 = n_2 \sin \theta_2$) can be applied to each ray to determine its direction upon leaving the lens. The intersection of any two of the rays determines the location of the image. However, it's much easier to use "special rays" that are easy to draw. In such a ray diagram, a ray of light leaving the "top" of the object parallel to the principal axis must pass through the focal point. A ray through the center of the lens is undeflected. This ray is, in essence, traveling through two parallel sides of a piece of glass. The intersection of these two rays gives the location of the "top" of the image. Ray diagrams are quite an asset in determining the image's location and size (fig. 5).

Beware of the misconceptions that befall those students who begin to believe that this helpful ray diagram describes how the light actually behaves. The ray diagram is a map—it shouldn't be mistaken for the landscape. Those who place all their trust in a ray diagram may begin to believe that if part of the lens is removed, part of the image is also removed. Do our *Quantum* readers understand how the image changes if the top half of the lens is covered?

Those who focus only on the ray diagram may forget that the light diverges after passing through the image location, and may never discover other interesting optical phenomena. The tool of ray diagrams should not limit your understanding nor your curiosity.

Our problem provides you with a broken lens and asks for a description of the resulting light pattern. It was first given at the 1972 International Physics Olympiad in Bucharest, Romania.

A. A lens of focal length f is cut into two parts perpendicular to its plane. The half-lenses are moved apart by a small distance δ (fig. 6—the gap is exaggerated due to typographic considerations). How many interference fringes appear on a screen at a distance L from the lens if a monochromatic light source (wavelength λ) is placed at a distance d ($d > f$) on the other side?

B. If f = 10 cm, d = 20 cm, δ = 0.1 cm, λ = 500 nm, and L = 50 cm, calculate the number of fringes.

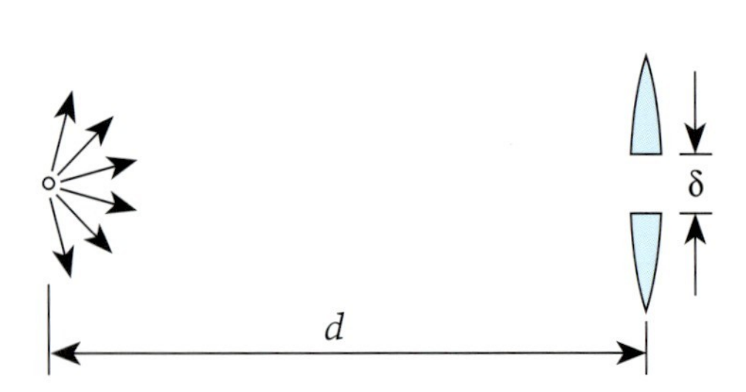

Figure 6

Solution

Because this problem mentions the appearance of interference fringes, the first step in finding the solution is to determine how the multiple sources arise.

The bisection of the lens produces two point images of the point source of light, each displaced from the principal axis, which act as sources to produce the interference pattern. In figure 7, these sources are labeled S_1 and S_2. Our path to a solution is now revealed to us. We can use the lens equation to find the positions of the two new sources. A comparison of similar triangles can provide us with the distance between these sources. This will then allow us to find the spacing between the interference fringes following the analysis of a typical Young's double-slit experiment. A second set of similar triangles will provide us with the size of the overlap region from which we can find the number of fringes. Let's now journey down this solution trail.

We first use the lens equation to find the location of the images:

$$\frac{1}{f} = \frac{1}{d_0} + \frac{1}{d_i},$$

$$d_i = \frac{d_0 f}{d_0 - f}.$$

A comparison of similar triangles yields the distance between S_1 and S_2:

$$\frac{d_s}{\delta} = \frac{d_0 + d_i}{d_0},$$

$$d_s = \delta\left(\frac{d_0 + d_i}{d_0}\right) = \delta\left(\frac{d_0}{d_0 - f}\right).$$

The distance to the screen l can be found in terms of the given dimensions of the problem:

$$l = L - d_i = L - \frac{d_0 f}{d_0 - f}$$

$$= \frac{L(d_0 - f) - d_0 f}{d_0 - f}.$$

The distance x between interference fringes produced by two point sources of wavelength λ separated by a distance d_s on a screen l meters away is

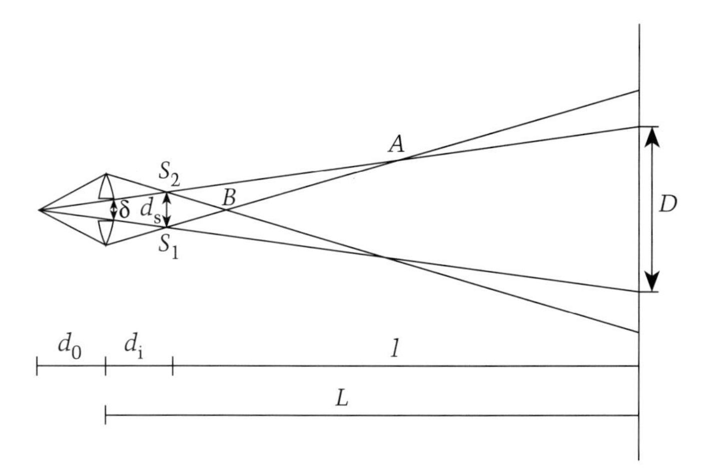

Figure 7

derived in most physics textbooks:

$$x = \frac{\lambda l}{d_s} = \frac{\lambda}{\delta d_0}\left[L(d_0 - f) - d_0 f\right].$$

Interference fringes will appear only in the area where the light beams from S_1 and S_2 overlap. We can determine the size of this region by comparing similar triangles:

$$D = \frac{\delta(L + d_0)}{d_0}.$$

The number of fringes is found simply by dividing the length of the overlap region by the spacing of the fringes:

$$N = \frac{D}{x} = \frac{\delta^2}{\lambda}\frac{(L + d_0)}{L(d_0 - f) - d_0 f}.$$

With the data provided, we can calculate the number of fringes for the specific situation described:

$$f = 10 \text{ cm},\ d_0 = 20 \text{ cm},$$
$$\delta = 0.1 \text{ cm},\ \lambda = 500 \text{ nm},$$
$$\text{and } L = 50 \text{ cm}$$

yield

$$N = 46 \text{ fringes}.$$

A different calculation must be made for D if the screen is nearer than point A in the figure. If the screen is placed nearer than point B, there will be no overlapping region and therefore no interference fringes. ◘

Gravitational redshift

"Impulses from what scarce was matter
Bounced off a shallow platter
Into the realm of number pure . . ."
—Howard Nemerov, "Druidic Rimes"

As we view the state trooper ticketing a speeding motorist, our mind returns to those lectures in physics class where Doppler shifts and radar filled the board. The same type of radar that catches speeding motorists records the speed of pitches in baseball and serves in tennis. At the 1995 US Open Championship, Monica Seles and Steffi Graf both served at speeds in excess of 100 mph, while Pete Sampras recorded serves faster than 120 mph.

The classic example of the Doppler shift is the change in pitch of a train whistle. As the train approaches us, our ears record a higher pitch, which drops as the train passes us. By measuring the change in pitch of the whistle, we can determine the speed of the train.

Doppler shifts occur for all types of wave. Because atoms and ions near the surfaces of stars emit spectral lines characteristic of each element, Doppler shifts are a very important tool in astronomy. For instance, the shift in the frequency of the emission spectrum from a star (or galaxy) tells us how fast the star is approaching or receding from Earth.

Because electromagnetic radiation can propagate through a vacuum, the Doppler shift formula for electromagnetic radiation is simpler than for sound waves. The size of the shift depends only on the relative velocity v of the source and the detector, rather than the velocities of each of these relative to the air. If we use the notation $\beta = v/c$, where c is the speed of light and β is positive when the distance between the source and the detector is increasing, the shifted frequency f is given by

$$\frac{f}{f_0} = \sqrt{\frac{1-\beta}{1+\beta}},$$

where f_0 is the frequency emitted by the source. When $|\beta| \ll 1$, we can use an even simpler expression that can be obtained by approximating

We see old Mr. Doppler standing high up in the universe, on the heads and shoulders of a bunch of the most influential scientists who preceded him and one wondering ape-man on the bottom. Mr. Doppler is listening to cosmic movements, especially to approaching and receding stars and planets, investigating their shifts in pitch or color when passing. Presently a small moonlike planet whooshes by with its potatoheadlike inhabitants, called photons. Some are excitedly bouncing off the surface, some are happily stuck to the planet, and some unfortunate ones get struck by laser beams. A rocket-shaped clown crushes through time and space on his way to nowhere. The watch doesn't have hands to show the time, and time seems to be just another fruit hanging from the tree of knowledge.

—T.B.

SPEED LIMIT 299,000,000 m/SEC
PHOTON
TOM BUNK

$(1 \pm \beta)^n \cong 1 \pm n\beta$. Therefore,

$$\frac{f}{f_0} = (1-\beta)^{1/2}(1+\beta)^{-1/2}$$
$$\cong \left(1-\frac{\beta}{2}\right)\left(1-\frac{\beta}{2}\right)$$
$$\cong 1-\beta,$$

where we've dropped the term in β^2. (The formula used by the police is actually $1 - 2\beta$, because the radar wave is shifted twice. You can think of your car as the first detector and the police car as the second.)

Edwin Hubble showed that most stars (and galaxies) are receding from Earth, which led to the theory that the universe began with a big bang. Because of the expansion, the frequencies of the spectral lines from the stars are shifted to lower values—that is, the light is redshifted. However, this is not the only redshift that occurs. A photon leaving a star is also redshifted as it rises in the gravitational field of the star. The gravitational redshift for our sun is too small to be detected accurately, but the redshifts of photons leaving white dwarfs can be measured and are equivalent to the redshifts corresponding to speeds around 20 km/s.

One of the problems given at the XXVI International Physics Olympiad in Canberra, Australia, in July 1995 combined the effects of these two types of redshift. We use part of that problem here.

Although gravitational redshifts are normally calculated using Einstein's theory of general relativity, we can develop a feeling for the effect by performing a semiclassical calculation. A photon with a frequency f has an effective inertial mass determined by its energy

$$mc^2 = hf.$$

Let's assume that the photon's gravitational mass is the same as its inertial mass and that the photon is emitted from the surface of a star. As the photon travels upward, it loses energy in the form mc^2 as it gains gravitational potential energy.

A. Show that the frequency shift Δf of the photon at an infinite distance from the star is

$$\frac{\Delta f}{f} = -\frac{GM}{Rc^2},$$

where G is the gravitational constant and M and R are the mass and radius of the star.

Let's now imagine launching a probe to a distant star to measure both the mass and radius of the star. Photons emitted from He^+ ions on the surface of the star are monitored through resonant absorption by He^+ ions in the probe. Resonance absorption only occurs if the ions in the probe are given a velocity toward the star that compensates for the gravitational redshift. As the probe approaches the star radially, the velocity v of the He^+ ions relative to the star is measured as a function of the radial distance d from the surface of the star. The experimental data are given in the following table:

$\beta = v/c$ ($\times 10^{-5}$)	d ($\times 10^8$ m)
3.352	38.90
3.279	19.98
3.195	13.32
3.077	8.99
2.955	6.67

B. Utilize these data to determine the mass and radius of the star.

Solution

Essentially correct solutions wre submitted by Christopher Rybak, a senior at the Prairie School in Racine, Wisconsin; Lori Sonderegger, a senior at Amity Regional High School in Woodbridge, Connecticut; and Art Hovey, her AP Physics teacher last year.

A. We use the conservation of energy to find the shift in the frequency of a photon emitted at the surface of the white dwarf:

$$hf - \frac{GMm}{R} = hf' - \frac{GMm}{r},$$

where h is Planck's constant, G is the gravitational constant, M is the mass of the star, m is the effective mass of the photon, and f and f' are the photon's frequencies at radii R (the surface) and $r > R$, respectively. Since $\Delta f \ll f$, the effective mass of the photon does not change appreciably and we have used a single value for m ($m = hf/c^2$). Therefore,

$$h\Delta f = \frac{GMhf}{c^2}\left(\frac{1}{r}-\frac{1}{R}\right),$$

or

$$\frac{\Delta f}{f} = \frac{GM}{c^2}\left(\frac{1}{r}-\frac{1}{R}\right). \quad (1)$$

In the limit $r \gg R$, we get the desired result:

$$\frac{\Delta f}{f} = -\frac{GM}{Rc^2}.$$

The minus sign indicates that the frequency decreases (that is, it is redshifted) as the photon leaves the white dwarf.

B. In the text of the problem, we showed that $f'/f \cong 1 - \beta$, or

$$\frac{\Delta f}{f} \cong -\beta. \quad (2)$$

Equating equations (1) and (2) and setting $r = R + d$, we find that

$$\beta = \frac{GM}{c^2}\left(\frac{1}{R}-\frac{1}{R+d}\right) = \frac{GMd}{c^2R(R+d)}.$$

Inverting both sides of this equation, we obtain

$$\frac{1}{\beta} = \frac{Rc^2}{GM}\left(\frac{R}{d}+1\right).$$

Therefore, if we plot $1/\beta$ versus $1/d$, we should obtain a straight line with a slope equal to $R^2c^2/GM \equiv \alpha R$ and a y-intercept equal to $Rc^2/GM = \alpha$.

Graphing the data gives a slope of $3.2 \cdot 10^{12}$ m and an intercept of $0.29 \cdot 10^5$. These values yield $R = 1.1 \cdot 10^8$ m and $M = 5.1 \cdot 10^{30}$ kg, which are in the right ballpark for a white dwarf.

Focusing fields

"We love to overlook the boundaries which we do not wish to pass."
—Samuel Johnson

HOW UTTERLY DELIGHTFUL is the Lorentz force. When we think of forces, we usually imagine a push or a pull. Such a push or pull is assumed to be in the direction of the line connecting the pusher and the object pushed. Not so with the Lorentz force. A magnetic field acting on a moving charge pulls in a direction perpendicular to the velocity of the particle and perpendicular to the magnetic field. If the particle moves parallel to the field, there is no force. If the particle comes to rest, there is no force. Finally, if the particle were to lose its charge, there would be no force. We can express the Lorentz force mathematically as $\mathbf{F} = q\mathbf{v} \times \mathbf{B}$, where $\mathbf{F}$ is the force, q is the charge, $\mathbf{v}$ is the velocity of the charge, and $\mathbf{B}$ is the strength of the magnetic field. The multiplication sign tells us that this is a vector cross product, which simply means that the force vector is perpendicular to both the velocity vector and the magnetic field vector.

A most important research use of a magnetic field is to control a particle beam. In the cyclotron, the magnetic field causes the charged particle to move in a circle. The magnitude of the Lorentz force acting on the particle is $qvB = mv^2/R$. This can be simplified to an expression that relates the momentum of the particle to the charge, magnetic field, and radius of the particle's path:

$$mv = qBr.$$

This nonrelativistic treatment provides the basic physics. Particles in the cyclotron or a bubble chamber travel at extremely high speeds, and relativistic equations must be used where we replace the momentum mv with the relativistic momentum γmv $\left(\gamma = 1/\sqrt{1-v^2/c^2}\right)$.

Figure 1 (on page 110) is a display of a bubble chamber photograph. An electron is the only charged particle that will make a complete circle in

It is not easy to be a particle pilot for the Lorentz Air Force. After taking off, they have no idea where they are flying to, and when they come near a magnetic field they get pulled and forced to form a straight line leading back and down to Earth, where they crash on the stage floor. This aeronautic show takes place before the backdrop of a Bubble Chamber universe. At the front of the stage we have the master of ceremonies, Pére Ubu, the scandalous scatterbrain with a spiral on his fat belly (modeled by A. Jarry on his high school physics teacher). He is dancing a polka to tunes from an orchestra consisting mostly of distinguished physicists. The theater, a converted hanger, is filled with thousands of spectators watching the cosmic multiple-scattering event unfolding until the last particle has hit the stage floor.

—T.B.

LORENTZ AIR FORCE
P
F
TODAY:
The MULTIPLE SCATTERING EVENT
R
ION
LAF
LAF
CREATION POLKA
EINSTEINWAY
Tom Bunk

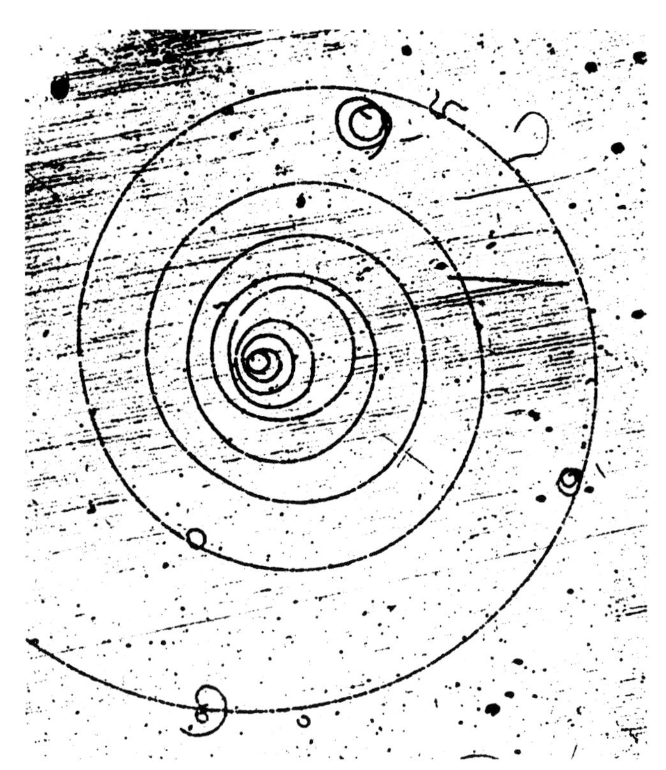

Figure 1

a bubble chamber. Protons, pions, and mesons will trace out circular arcs, but can't trace out the complete circle due to their large momenta and the limited size of the chamber. By investigating a path of an electron, we can find its initial momentum and notice that it loses momentum (spirals in) during its interaction with the hydrogen in the detector. Since the electron is traveling counterclockwise, we can also determine the direction of the external magnetic field.

Similar analyses are able to determine the momenta of other particles and to discover new particles as they interact with known particles. An example of a multiple scattering event that can be analyzed is shown in figure 2.

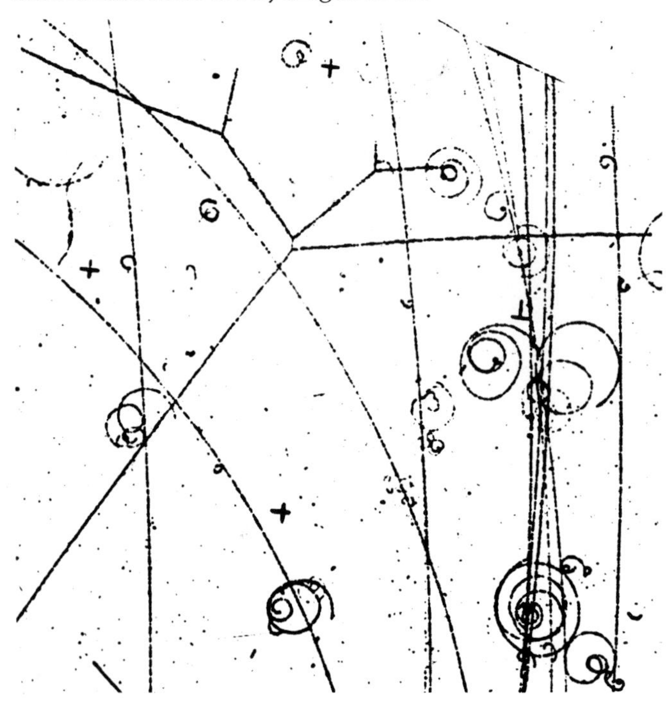

Figure 2

Our problem is from the VIII International Physics Olympiad that was hosted by East Germany in 1975. It requires us to find a magnetic field that can focus charged particles.

Ions of identical mass m, charge q, and speed v diverge from a point P (fig. 3). A uniform magnetic field B perpendicular to the plane of the page focuses them to a point R located at a distance $PR = 2a$ away from P. Their trajectories have to be symmetrical to the axis that is the perpendicular bisector of PR. Determine the boundaries of the magnetic field.

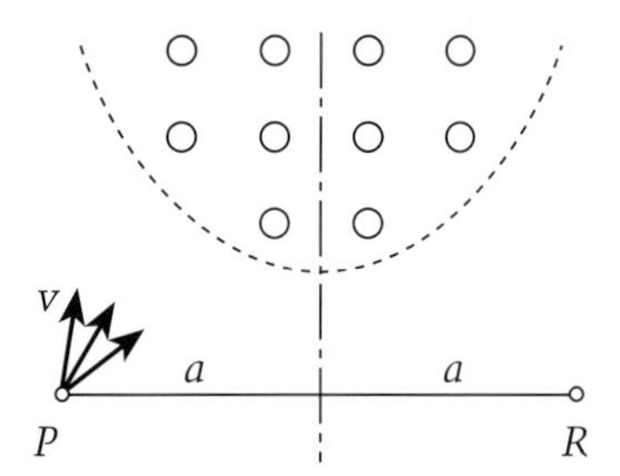

Figure 3

Solution

The best solutions came from the wide range of readers that *Quantum* reached. The best solution by a high school student came from Christopher Rybak, a senior at The Prairie School in Racine, Wisconsin. The best solution by a college student was written by Joseph Hermann, an undergraduate at the University of Missouri. Arthur Hovey, a physics teacher at Amity Regional High School in Woodbridge, Connecticut, and Daniel Dempsey of Canisius College in Buffalo, New York, also contributed excellent solutions. Prof. Dempsey sent along an exciting research paper he had published in *The Review of Scientific Instruments* (Vol. 26, No. 12, 1141–45, Dec. 1955). In this paper, Prof. Dempsey provides solutions to the contest problem where the magnetic field boundaries are limited to straight lines and circles.

We know that the particles will travel in straight lines when they are not in the magnetic field and will travel in circular paths when they are in the magnetic field. These circular paths are defined by the relation

$$qvB = \frac{mv^2}{R}.$$

Since all particles in this problem have the same velocity, mass, and charge, they will all travel along circular paths of a specific radius:

$$r = \frac{mv}{qB}.$$

Upon leaving the magnetic field, the particles will travel along the tangent at the point where they leave the magnetic field. The solution to the problem consists of finding a boundary line where all circles of radius r have tangent lines that meet at point R. The centers of the circles of radius r lie on the y-axis.

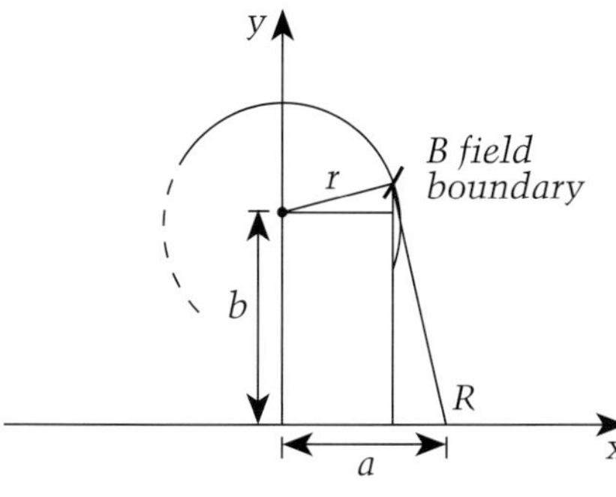

Figure 4

Figure 4 shows the geometry of the solution. The equation of the circle is

$$x^2 + (y - b)^2 = r^2.$$

The similarity of the triangles yields

$$\frac{y-b}{x} = \frac{a-x}{y}.$$

Combining these equations by eliminating $(y - b)$, we find that the locus of points is defined by the relation

$$y = \frac{x(a-x)}{\sqrt{r^2 - x^2}}.$$

This function can most easily be viewed by using a graphing calculator or a spreadsheet and graphics program. It is symmetrical with respect to the y-axis.

The solution depends on the relationship between the radius of the circle and a. If $r < a$, the boundaries of the field extend to infinity, and all ions entering the field can be focused. See figure 5 for the case $a = {}^5\!/_3 r$. If $r = a$, the boundaries have finite positions, and once again all ions entering the field can be focused (fig. 6). If $r > a$, the boundaries begin to "flatten" and therefore restrict the angles at which ions can leave point P and arrive at R. See figure 7 for the case $a = {}^5\!/_8 r$.

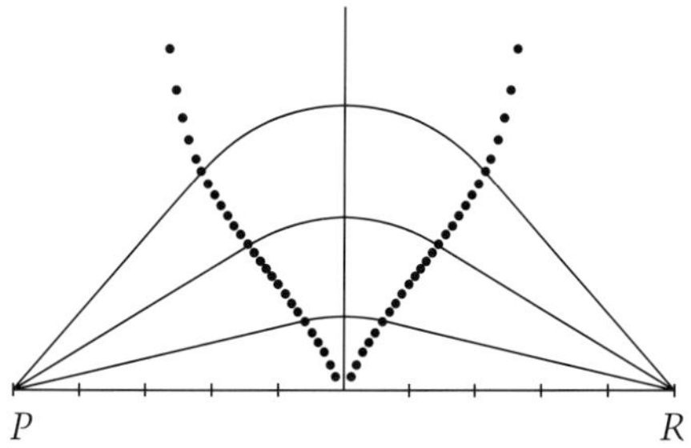

Figure 5

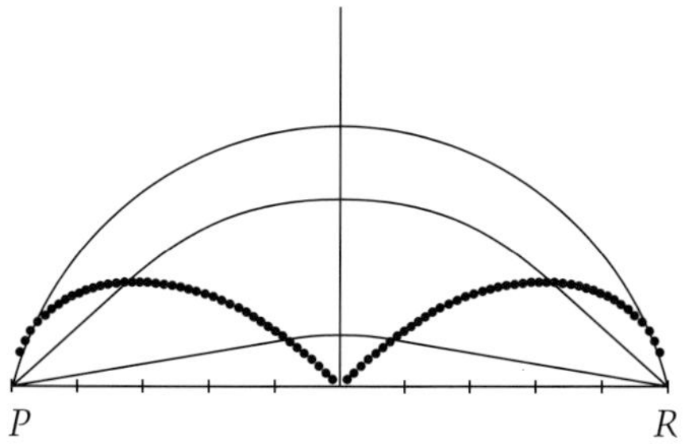

Figure 6

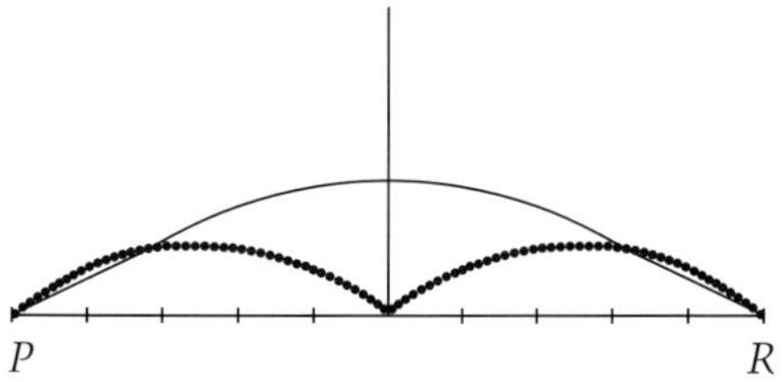

Figure 7

Sea sounds

"Come here Uncle John's band by the riverside,
Got some things to talk about, here beside the rising tide."
—Garcia–Hunter

REFRACTION IS RESPONSIBLE for distorting and correcting our vision. The shimmering of light as it passes over a hot barbecue requires a knowledge of refraction to be understood. Refraction of light at the surfaces of lenses benefits those of us who wear glasses or contact lenses.

Refraction was known as far back as the time of Ptolemy in the 2nd century A.D. It wasn't until 1621, however, that the correct relationship between the incident and refracted angles was discovered. Snell determined that

$$n_1 \sin \theta_1 = n_2 \sin \theta_2,$$

where θ_1 and θ_2 are the angles of the light measured relative to the normal to the surface in materials with indices of refraction n_1 and n_2, respectively. When Foucault measured the speed of light in water and other transparent materials in the 1850s, it was shown that the index of refraction is inversely proportional to the speed of light in the material—that is, $n \propto 1/v$.

Modern manufacturing techniques have allowed the development of a new type of lens that can be made out of a flat piece of glass. The graded index lens has an index of refraction that varies from the center to the edge. This variation causes the light to change direction as it would in a regular lens.

This technique also finds applications in fiber optics. The graded index fiber has a core with an index of refraction that decreases with distance from the center. This helps keep the rays traveling along the axis of the fiber.

Of course, Nature has exhibited the effects of a variable index of refraction for a very long time. The index of refraction of air varies with its density. Therefore, the index of refraction of the Earth's atmosphere decreases with altitude, and light rays bend toward the vertical as they approach the ground. As a con-

Above sea level are three men in a tub, playing a dashing tango. Einstein, Newton, and Ptolemy have tied themselves to the mast like Odysseus so as not to succumb to the irresistible and seductive voices of the three sirens on the tiny island nearby. Meanwhile a party is in progress deep down on the bottom of the sea, the flamboyant Mr. Shark gracefully leading the charming Miss Octopus around the tiny dance floor while the other guests enjoy themselves eating fish and drinking water. While the adults are having a good time, the little fish have to attend school, learning the hard facts of refracted sound rays. A simple light show in the darkness of the deep sea is provided by Dr. Nemo's submarine, while a technician at the bottom-right tries to get the sound system to work in accordance with the laws of physics.

—T.B.

FISH SCHOOL PS1
REFRACTED LIGHT RAYS
TODAY: FISH & CHIPS
TUNA
TONIGHT: the GRAND FISH BALL
REFRACTED SOUND RAY SPECIAL
TOMAS BUNK

sequence, the Moon appears about one diameter higher than it actually is when it is near the horizon. Temperature also affects the density of air, and this dependence creates the mirages we see on road surfaces as we drive down highways on hot days.

Identical effects occur with other types of wave—for instance, sound waves. This formed the basis for a theoretical problem given at the XXVI International Physics Olympiad held in Canberra, Australia, in July 1995.

The speed of propagation of sound in the ocean varies with depth, temperature, and salinity. Let's assume that the speed has a minimum v_0 midway between the surface and the ocean floor. For convenience, we choose the origin $z = 0$ at the level of the minimum, $z = z_s$ at the surface, and $z = -z_f$ at the floor. Let's also assume that the speed of sound increases linearly above and below $z = 0$ according to

$$v = v_0 + b|z|,$$

where b is a positive constant.

Let x be a horizontal direction, and let's place a source of sound S at the position $x = 0$, $z = 0$. A ray of sound is emitted from S at an angle $\theta_0 < \pi/2$ measured relative to the positive z-axis—that is, vertically upward (fig. 1).

A. Show that the trajectory of the ray (constrained to the xz-plane) is a circle with radius

$$R = \frac{v_0}{b \sin\theta_0}.$$

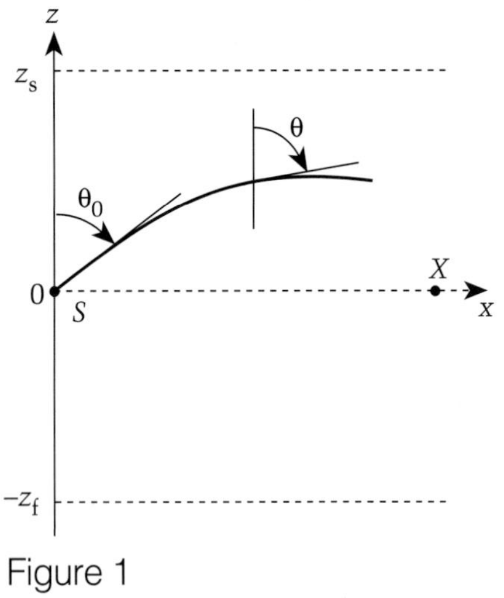

Figure 1

This can be derived by using calculus or demonstrated by means of a spreadsheet. In using the spreadsheet, assume that the ocean is divided up into a large number of horizontal sheets, each with a speed of sound equal to that in the mid-depth of the sheet. You can then apply Snell's law at the interfaces between sheets to obtain the trajectory of the ray. (It's sufficient to show that the path follows a circular arc at the beginning of the trajectory. The spreadsheet will have difficulties when the ray is horizontal.

B. Derive an expression for the smallest angle of $\theta_0(z_s, b, v_0)$ that can be transmitted without the sound ray hitting the surface.

C. Assume that you have a microphone at a position $x = X$, $z = 0$. Find the series of values for $\theta_0(X, b, v_0)$ required for the sound ray emerging from S to reach the microphone. Assume that z_s and z_f are sufficiently large to remove the possibility of reflection from the ocean surface or floor.

D. Calculate the smallest four values of θ_0 for these rays given that $X = 10{,}000$ m, $v_0 = 1{,}500$ m/s, and $b = 0.02000$ s^{-1}.

E. Let's now compare the times required for sound to travel along two different paths. The first path is the direct horizontal, or axial, path. The second path is the one corresponding to the smallest angle for θ_0. The time required for the second path can be obtained by integrating ds/v along the path—that is, by adding up the times that it takes to move small distances along the path. (The integral $\int dx/\sin x = \ln\tan(x/2)$ should prove useful.) This result can also be obtained numerically with a spreadsheet or a computer program. Which path takes the shorter time?

Solution

A very good solution was jointly submitted by André Cury Maiali and Gualter José Biscuola, who are physics teachers in Jundiaí, São Paulo, Brazil. We will follow their reasoning in our solution.

A. We can show that the sound rays follow circular paths by utilizing a spreadsheet or resorting to calculus. Knowing that the general formula for a circle in the xz-plane is given by

$$(x - x_c)^2 + (z - z_c)^2 = R^2, \qquad (1)$$

where the center of the circle is located at (x_c, z_c) and R is the radius of the circle, and recognizing that Snell's law is given in terms of the angle θ, we decide to find the coordinates x and z as functions of θ.

Let's restrict ourselves to an upward-pointing ray in the region above the minimum sound speed (that is, $z > 0$ and $\theta < \pi/2$). From the symmetry of the problem, we will obtain the same results for $z < 0$. We substitute the relationship for the variation of the sound speed into Snell's law—

$$\frac{\sin\theta_0}{v_0} = \frac{\sin\theta}{v} = \frac{\sin\theta}{v_0 + bz} \qquad (2)$$

—and solve for z to obtain

$$z = \frac{v_0}{b}\left(\frac{\sin\theta}{\sin\theta_0} - 1\right).$$

Anticipating that we will eventually need the derivative, we have

$$\frac{dz}{d\theta} = \frac{v_0}{b}\left(\frac{\cos\theta}{\sin\theta_0}\right). \qquad (3)$$

Let's now look at the slope of the curve:

$$\frac{dz}{dx} = \tan\left(\frac{\pi}{2} - \theta\right) = \cot\theta = \frac{\cos\theta}{\sin\theta}. \qquad (4)$$

We can also use the chain rule to write

$$\frac{dz}{dx} = \frac{dz}{d\theta}\frac{d\theta}{dx}. \qquad (5)$$

Substituting equations (3) and (4) into equation (5), we obtain

$$\frac{\cos\theta}{\sin\theta} = \frac{v_0}{b}\frac{\cos\theta}{\sin\theta_0}\frac{d\theta}{dx}.$$

We now solve for dx—

$$dx = \frac{v_0}{b\sin\theta_0}\sin\theta d\theta$$

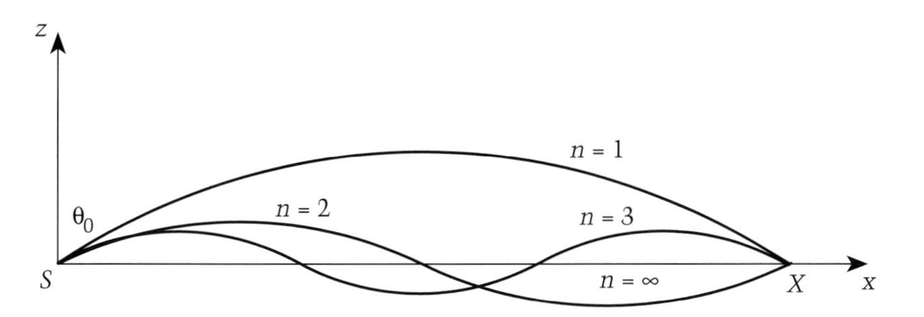

Figure 2

—and integrate:

$$x = \frac{-v_0 \cos\theta}{b \sin\theta_0} + C.$$

For $x = 0$, $\theta = \theta_0$. Therefore, $C = v_0/b \tan\theta_0$, and finally

$$x = \frac{v_0}{b \tan\theta_0}\left(1 - \frac{\cos\theta}{\cos\theta_0}\right). \quad (6)$$

In a similar fashion we obtain

$$z = \frac{v_0}{b}\left(\frac{\sin\theta}{\sin\theta_0} - 1\right). \quad (7)$$

Moving the constant terms to the left-hand sides of equations (6) and (7), squaring them, and adding them together, we find that

$$\left(x - \frac{v_0}{b \tan\theta_0}\right)^2 + \left(z + \frac{v_0}{b}\right)^2 = \left(\frac{v_0}{b \sin\theta_0}\right)^2,$$

which has the form of equation (1). Therefore, the path is that of a circle of radius

$$R = \frac{v_0}{b \sin\theta_0}$$

centered at

$$\left(\frac{v_0}{b \tan\theta_0}, \frac{-v_0}{b}\right).$$

B. The smallest value of θ_0 that can occur without the sound ray hitting the surface is obtained when the circular path is tangent to the surface of the sea. This requires that

$$R = z_s + |z_c|,$$

or

$$\sin\theta = \frac{v_0}{b z_s + v_0}.$$

C. In figure 2 we have drawn four possible paths (including the direct path) the sound rays could take from S to X. For each case the length of the chord of the circle must be X/n. Therefore,

$$2R \cos\theta_0 = \frac{X}{n} = 2\frac{v_0 \cos\theta_0}{b \sin\theta_0},$$

yielding a series of values for θ_0:

$$\tan\theta_0 = n\frac{2v_0}{bX}, \quad n = 1, 2, 3, \ldots .$$

D. For the given data we have

$\theta_0 = 86.19°$ for $n = 1$,
$\theta_0 = 88.09°$ for $n = 2$,
$\theta_0 = 88.73°$ for $n = 3$,
$\theta_0 = 89.05°$ for $n = 4$.

Note that the limiting value is $\theta_0 = 90°$, as expected.

E. For the axial path, we simply divide the distance by the speed to obtain the time taken:

$$\Delta t_\infty = \frac{X}{v_0} = 6.667 \text{ s}.$$

For the circular arc we need to add up the times for a large number of small pieces of the path. This can be done using a spreadsheet or by integrating. Let's do the latter:

$$dt = \frac{ds}{v} = \frac{R d\theta}{v}.$$

Rather than simply plugging in the expressions for R and v, we note that

$$R = \frac{v_0}{b \sin\theta_0} = \frac{v}{b \sin\theta},$$

according to Snell's law (equation (2)). Therefore,

$$dt = \frac{d\theta}{b \sin\theta}.$$

To simplify the integration we take advantage of the symmetry and only integrate to the top of the path and then multiply by 2:

$$\Delta t_1 = 2\int_{\theta_0}^{\pi/2} \frac{d\theta}{b \sin\theta} = \frac{2}{b}\left[\ln \tan\frac{\theta}{2}\right]_{\theta_0}^{\pi/2}$$
$$= -\frac{2}{b} \ln \tan\frac{\theta_0}{2}.$$

This gives a time $\Delta t_1 = 6.6546$ s, which is a *shorter* time than for the direct ray. ◘

Moving matter

"God . . . created matter with motion and rest in its parts, and . . . now conserves in the universe, by His ordinary operations, as much of motion and of rest as He originally created."
—René Descartes, Principia Philosophiae *(1644)*

HOW FAST CAN YOU THROW a baseball? How fast is a speeding bullet? Restricted to simple tools in the laboratory, both measurements can be completed with a clever approach and some elementary physics.

Although the baseball's speed would be difficult to measure directly, you can throw it into a box and measure the movement of the box across the table. Such an arrangement requires a box that will slow the ball down appreciably. The ball is thrown into the box. The box slides across the table and comes to a stop. Measuring the distance the box travels allows us to find the work done by friction. This then allows us to find the combined velocity of the box and ball when they began sliding. Since momentum is conserved in all collisions, we can back up one more step and determine the initial speed of the ball before it got embedded in the box.

Let's assume that a 0.5-kg ball is hurled into a 10-kg box that is resting on the table. If the box slides 2 m before coming to rest, we know that the work W done on the box is Fd, where F is the force of friction and d is the distance the box travels along the table. We can determine F independently by pulling the box at a constant speed along the table with a spring scale or by first determining the coefficient of friction μ and the normal force F_n. Let's assume that the force of friction is 20 N. The momentum of the ball before the collision with the box is equal to the combined momentum of the ball and box after the impact:

$$mv_0 = (m + M)V,$$

where m is the mass of the ball, M is the mass of the box, v_0 is the velocity of the ball, and V is the velocity of the ball and box. The kinetic energy of the ball and box is lost due to the work done by the

To measure the speed of a ball is a difficult task. On the top floor of the cosmic dollhouse we see a huge apparatus that can throw a ball like a bullet at a pendulum, which hits a spring and causes a bell to ring. Somehow the man in the top hat can then figure out the speed of the ball. On the middle floor we have similar arrangements to measure the speed of a bullet by using a suspended pendulum or an ordinary box sliding on the floor. The audience admires the poetry in motion, while God plays marbles by elegantly moving his divine finger. On the ground floor we see a dramatic rescue mission where Tarzan uses the momentum of the swinging pendulum to save Jane from the wild beasts. Does he have enough speed to safely reach the balcony, full of monkeys? Meanwhile, the somewhat tired Einstein rests on a sofa, sipping coffee and sawing away at some inspirational tunes.

—T.B.

POETRY IN MOTION

TOM BUNK

frictional force:

$$\frac{1}{2}(m+M)V^2 = f \cdot d.$$

In our selected example, the ball's velocity is approximately 60 m/s, or 120 mph. (Does this agree with your calculated value?)

For bullets, it's preferable to suspend the box as a pendulum. This method was first used by Benjamin Robins in 1742. Professor A. P. French (MIT) shared with us a description by Robins in the *Philosophical Transactions of the Royal Society* for 1742–1743 (Vol. 42, pp. 437–56). In this journal, Robins described how he deliberately designed the pendulum to provide the first reliable way of measuring the speed of bullets.

With the bullet embedded in the pendulum, the pendulum rises and the initial kinetic energy of the pendulum becomes gravitational potential energy. It is then only necessary to measure the change in the vertical height of the pendulum to determine the speed of the bullet.

In both the baseball and bullet approaches, a common error is to assume that the initial kinetic energy of the object is equal to the kinetic energy after the collision. Since the object and the target stick together, we should recognize that this is an inelastic collision, in which kinetic energy is not conserved.

In some classroom demonstrations, a bullet is shot into the pendulum. In alternative demonstrations, an arrow is shot into the pendulum. The speed of the arrow in this latter experiment can be determined by a second, unrelated approach. The kinetic energy of the arrow is equal to the work performed by the bowstring, which can be determined by measuring the average force for every centimeter of pull. When two unrelated approaches provide an equivalent measure for the speed of the arrow, our confidence in the value and utility of the technique rises.

We'll look at a fun Tarzan-and-Jane problem that we used on the first screening test for the US Physics Olympiad Team. Then we'll challenge you to find the speed of a bullet using a ballistic pendulum. Finally, we'll ask you to design an apparatus to be used in school physics laboratories.

A. Tarzan (mass = 80 kg) is standing on a small hill (height = 10 m) when he spots Jane (mass = 40 kg) in danger in the valley below. He grabs a vine and swings toward Jane. Grabbing her at the bottom of the arc, he hopes to make it to the 5-meter-high hill on the other side. Is Tarzan successful?

B. A bullet of mass 5 g hits a target of mass 5 kg that is free to swing as a pendulum. The cord holding the target is 5 m long. The target captures the bullet in a very short time and moves 30 cm horizontally. Calculate the velocity of the bullet.

C. You wish to design a laboratory apparatus that propels a ball by compressing a spring. The ball then travels a short distance into a target that is free to swing as a pendulum. What are the relative masses of the ball and target that produce the maximum movement of the target for a given initial compression of the spring?

D. A second variation of your laboratory apparatus has the ball stick (with a Velcro™-type fastener) to the bottom of a uniform rod. How is the final angle dependent on the mass of the ball and mass of the rod?

E. After using your propulsion device for the ballistic pendulum, you decide to have some real fun by inventing a target game. The propulsion device is mounted on a table and propels a marble horizontally in order to hit a target on the floor below. In your first attempt you compress the spring 1.0 cm and the marble falls 30 cm short of the target, which is 3.0 m horizontally from the edge of the table. How much should you compress the spring in your second attempt to get a perfect hit?

Solution

Two *Quantum* readers sent in excellent solutions to this problem: Noah Bray-Ali, a junior at Venice High School in Los Angeles, California, and Wolfgang Wais of Tübingen, Germany. We will follow along with Noah as he describes his approach.

A. There are three parts to this problem: the jump, the collision, and the upswing. Conservation of energy during Tarzan's (mass M) swing from height h_0 gives his speed v_0 at the bottom of the valley:

$$Mgh_0 = \frac{1}{2}Mv_0^2, \qquad v_0^2 = 2gh_0. \tag{1}$$

During the ensuing perfectly inelastic collision with Jane (mass m), momentum is conserved, and we can solve for the couple's final velocity v':

$$Mv_0 = (M+m)v',$$

$$v' = \frac{Mv_0}{M+m},$$

or, with equation (1),

$$v'^2 = \frac{M^2}{(M+m)^2}v_0^2 = 2gh_0\frac{M^2}{(M+m)^2}. \tag{2}$$

We can now find the height h to which they rise using conservation of energy during their upswing:

$$\frac{1}{2}(M+m)v'^2 = (M+m)gh'. \tag{3}$$

Solving for h' using equations (2) and (3) yields

$$h' = \frac{M^2}{(M+m)^2}h_0 = 4.44 \text{ m}.$$

Alas, Tarzan and Jane end up more than half a meter short of the crest of the hill they had hoped to reach.

B. When the bullet (mass m) and pendulum (mass M) collide, momentum is conserved:

$$mv_0 = (m + M)v',$$

where v_0 is the original velocity of the bullet and v' is the velocity of the system after the collision. Thus

$$v'^2 = \frac{m^2}{(m+M)^2} v_0^2. \qquad (4)$$

After the collision, the bullet–pendulum system rises to a height h, which we can find from the given parameters: the length L of the cord and the horizontal distance s traveled by the system. If we denote by θ the angle the cord makes with the vertical, we have

$$\theta = \sin^{-1} \frac{s}{L} = 3.44°,$$

$$h = L - L\cos\theta = L(1-\cos\theta) = 0.009 \text{ m},$$

From conservation of energy we can solve for the bullet's speed:

$$\frac{1}{2}(M+m)v'^2 = (M+m)gh.$$

Replacing v'^2 using equation (4), we get

$$v_0^2 = 2gh\frac{(M+m)^2}{m^2}, \qquad (5)$$

or

$$v_0 = 420 \text{ m/s}.$$

C. If the spring of force constant k is initially compressed a distance d, from conservation of energy during compression we can solve for the speed v_0 at which the ball (mass m) is ejected:

$$\frac{1}{2}kd^2 = \frac{1}{2}mv_0^2,$$

$$v_0^2 = \frac{kd^2}{m}. \qquad (6)$$

All of the work in part B is still true, so we can combine equations (6) and (5) and solve for h, where M is the mass of the pendulum:

$$h = \frac{kd^2}{2g}\frac{m}{(M+m)^3}.$$

To solve this for the relative masses that optimize h, we take the derivative of h with respect to m, set it equal to zero, and solve for m in terms of M. Noting that the term $kd^2/2g$ is constant and can be factored out, this reduces to

$$0 = \frac{dh}{dm} = \frac{1}{(M+m)^2} - \frac{2m}{(M+m)^2},$$

and finally

$$M = m.$$

D. This question can be broken into two parts: the collision and the upswing. From the conservation of angular momentum about the pivot point of the rod, we can calculate the final angular speed ω of the system after the collision in terms of the ball's speed v_0 after ejection, which we know from part C:

$$mv_0L = I'\omega,$$

where m is the mass of the ball, L is the length of the rod, and I' is the moment of inertia of the system after the collision. We can calculate I by the superposition principle, knowing that I for the rod is $^1/_3ML^2$:

$$I' = I + mL^2 = \frac{1}{3}ML^2 + mL^2.$$

Replacing and solving for θ gives

$$\omega = \frac{mv_0}{\left(\frac{1}{3}M + m\right)L}.$$

During the upswing, the rotational kinetic energy from this collision is converted into gravitational potential energy. The ball rises a height $L - L\cos\theta$, and the center of mass of the rod rises $L/2 - L/2\cos\theta$, where θ is the final angle we're trying to find. So conservation of energy gives

$$\frac{1}{2}I'\omega^2 = \left(m + \frac{M}{2}\right)gL(1-\cos\theta).$$

Replacing I' and ω and solving gives

$$1 - \cos\theta = \frac{3m^2v_0^2}{(M+2m)(M+3m)gL}.$$

Equation (6) is still valid, so we can replace and solve:

$$\theta = \cos^{-1}\left(1 - \frac{3kd^2m}{gL(M+2m)(M+3m)}\right),$$

which expresses the final angle in terms of the masses of rod and ball.

E. From equation (6) we know the velocity v_0 of the marble is proportional to the compression d of the spring. This horizontal velocity is constant during the flight, so the final distance traveled in the x-direction will simply depend on the time of flight, which is the same for both trials: the time it takes to fall from the table to the floor. The distance traveled in the x-direction without air resistance is thus directly proportional to v_x and, therefore, to the compression distance. If we denote the values of the first attempt by d_0 and x_0 and those of the second attempt by d and x, this proportionality gives

$$d = d_0\frac{x}{x_0} = 1.11 \text{ cm}.$$

Boing, boing, boing . . .

"Nature is an endless combination and repetition of a very few laws. She hums the old well-known air through innumerable variations."
—Ralph Waldo Emerson

WHILE WATCHING THE Olympic Games every four years, you may be reminded once again of the versatility of physics. The equations for projectile motion can be used to analyze many different track and field events at the Olympic Games—shotput, discus, hammer throw, javelin, high jump, long jump, triple jump, and pole vault. Now, the athletes are not required to understand all of this physics, but we know that the information is important. Coaches study the physics principles behind the events, and sports physicists analyze the events to improve the athlete's performance.

When we begin our study of projectile motion, we simplify the mathematics by assuming that there is no air resistance, which is definitely not true for the javelin and the discus. Under this assumption, we are fortunate that the motions in the vertical and horizontal directions can be analyzed separately. The horizontal motion is one with a constant velocity because there is no horizontal force acting—that is,

$$x = x_0 + v_{0x}t.$$

The vertical motion is one with a constant acceleration due to the force of gravity:

$$y = y_0 + v_{0y}t - \tfrac{1}{2}gt^2.$$

We then proceed to analyze motion on a flat plane, where we are given a launch speed v_0 and a launch angle θ. If we choose the origin of our coordinate system at the launch site so that $x_0 = y_0 = 0$, our equations reduce to

$$x = v_0 \cos\theta\, t,$$
$$0 = v_0 \sin\theta - \tfrac{1}{2}gt.$$

We can then solve these simultaneous equations for the range x and the time of flight t:

$$t = \frac{2v_0 \sin\theta}{g},$$

$$x = \frac{2v_0^2 \cos\theta \sin\theta}{g}.$$

An enormous moon rises behind classical Greek scenery, which is very busy this time of year. It's the time of the Olympic games, when people from all over the world gather peacefully to throw things around. Demonstrating the disciplines of projectile motion, Diogenes plays marbles in his barrel, Pythagoras throws numbers around, and Plato hits a cricket ball, while on top of a pillar Aristotle drops a rather heavy ball down a ramp. Meanwhile the Moon is just as busy, where people compete by tossing, hurling, and pitching all kinds of objects and even themselves. The dervish whirls a globe, Newton throws an apple, Einstein his bow, and Galileo a javelin, while an unrefined Viking, not knowing anything about the Olympic's noble spirit, warms up his hot dog over the Olympic flame.

—T.B.

E=mc²
NO BALL PLAYING
TOM FUNK

These equations can be used as a first analysis of the long jump. Because the maximum range occurs for $\theta = 45°$, the jumper should leave the ground at 45° (provided v_0 is the same independent of angle).

The analysis of the shotput is more difficult because the shot is launched at a different height than it lands. Our second equation is then quadratic. We often analyze the simpler case in which the projectile is launched horizontally.

One of the problems on the preliminary exam used to select the 1996 US Physics Team was an interesting example of this last type of projectile motion problem. A ball is dropped vertically, falls a distance h, and strikes a ramp inclined at 45° to the horizontal. The ball undergoes a perfectly elastic collision. This means that the velocity component parallel to the surface of the ramp remains the same, while the perpendicular component reverses directions. How far down the ramp does the ball land after the first bounce?

For the free-fall portion of the motion, our second equation becomes

$$-y = -\tfrac{1}{2}gt^2,$$

and therefore the time t_0 to reach the ramp is

$$t_0 = \sqrt{\frac{2h}{g}}.$$

We can use the conservation of energy or the kinematic equation $v = -gt$ to obtain the speed v_0 at impact:

$$v_0 = \sqrt{2gh}.$$

The choice of 45° for the ramp angle makes the problem easy, because the ball leaves the first bounce in the horizontal direction. This choice (and setting the new origin at the location of the first bounce) also leads to simple coordinates for the location of the second bounce, $(L_1/\sqrt{2}, -L_1/\sqrt{2})$, where L_1 is the distance measured down the ramp. Therefore, our equations for the projectile motion become

$$\frac{L_1}{\sqrt{2}} = v_0 t_1,$$
$$\frac{-L_1}{\sqrt{2}} = -gt_1^2.$$

Solving for L_1, we obtain our answer:

$$L_1 = 4\sqrt{2}h.$$

In trying to find a suitably challenging problem for our readers, we played around with changing the angle of the ramp, but things get messy and nothing very interesting emerged. However, when we looked at the second bounce, something very interesting emerged that prompted us to look at the third bounce, and the fourth . . .

A. Complete the analysis of the first bounce by showing that the time t_1 in the air equals $2t_0$, the speed of impact is $\sqrt{5}v_0$, and the angle θ_1 that the ball makes with the vertical is given by $\tan\theta = 1/2$.

B. For the second and third bounces, find the distance the ball travels down the ramp (in terms of L_1), the time in the air (in terms of t_1), the speed at impact (in terms of v_0), and the tangent of the angle with the vertical at impact.

C. Generalize your answers to the nth bounce and give physical reasons for the existence of the observed patterns.

Solution

This problem produced a very good response from *Quantum* readers. Excellent solutions were submitted by Noah Bray-Ali from Venice High School in Los Angeles, California; Richard Burstein from the Commonwealth School in Boston; André Cury Maiali and Gualter José Biscuola, physics teachers in São Paulo, Brazil; Mary Mogge, a professor at Cal Poly Pomona; and Charles Thiel, a student at Montana State University–Bozeman.

Dr. Mogge points out that the problem is easiest to solve if we use the standard coordinate system for solving inclined plane problems, the x-axis pointing down the plane and the y-axis normal to the plane. She then writes that we should think of the problem as the projectile motion we all know and love, with the extra twist of an x-acceleration. We can then solve the problem in general and do not need to consider each bounce separately.

The motion in the y-direction is that of a ball bouncing with totally elastic collisions with the "floor" and a constant acceleration equal to the component of g perpendicular to the plane, $g/\sqrt{2}$. Therefore, the time for each bounce is always the same. Because of the symmetry of the motion in the y-direction, the time for each bounce is twice the time it took the ball to hit the plane the first time—that is, $t_1 = 2t_0$, where $t_0 = \sqrt{2g/h}$ as calculated in the article. Likewise, the magnitude of the y-component of velocity just before each bounce is always the same—it is the component of v_0 perpendicular to the plane, $v_y = v_0/\sqrt{2}$, where $v_0 = \sqrt{2gh}$ from conservation of energy. If we use n to denote the bounce number, counting the initial bounce as number zero, we can write this as

$$v_{yn} = \frac{v_0}{\sqrt{2}}.$$

The motion parallel to the plane has a constant acceleration equal to the component of g along the plane, $g/\sqrt{2}$. Therefore, the x-component of the velocity is given by

$$v_x = \frac{gt}{\sqrt{2}},$$

where t is measured from the time the ball is dropped. The zeroth bounce occurs at $t = t_0$, the first bounce at $t = 3t_0$, and the second bounce at $t = 5t_0$. Therefore, the x-component of the velocity just before (and after) each bounce is

$$v_{xn} = \frac{(2n+1)gt_0}{\sqrt{2}} = (2n+1)\sqrt{gh}$$
$$= \frac{(2n+1)v_0}{\sqrt{2}}.$$

This immediately tells us the speed of the ball at each bounce:

$$v_n^2 = \frac{\left[(2n+1)^2 + 1\right]v_0^2}{2}.$$

It's easiest in this coordinate system to calculate the tangent of the angle with respect to the plane. Just before each bounce, we have

$$\tan\phi_n = \frac{v_{yn}}{v_{xn}} = \frac{1}{2n+1}.$$

Note that both the tangent and the angle approach zero. As the x-component of the velocity increases, the ball hits closer and closer to parallel with the plane. You can obtain the tangent of the angle relative to the true vertical by rotating the coordinate system by 45° using the trigonometric identity

$$\tan(x-y) = \frac{\tan x - \tan y}{1 + \tan x \tan y},$$

with $x = \phi$ and $y = 45°$, yielding

$$\tan\theta_n = \frac{n}{n+1}.$$

The coordinate of the ball along the plane is given by

$$x = \frac{gt^2}{2\sqrt{2}}$$

and

$$x_n = \frac{(2n+1)^2 h}{\sqrt{2}}.$$

Therefore, the distance traveled down the plane during each bounce is

$$L_n = x_n - x_{n-1} = \frac{8nh}{\sqrt{2}} = nL_1.$$

Noah points out a nifty demonstration as an extension of this problem. Release a frictionless puck at $x = 0$ at time $t = 0$. Every time the ball hits the plane, the puck will be directly under the ball. If you cover the ball with ink, there will be no marks on the plane! Noah goes on to point out that the physics behind this demonstration is the same as that of the "monkey-shoot" described in most textbooks.

The bombs bursting in air

"Education is the art of making man ethical."
—Georg Hegel

IS EDUCATION INTENDED TO expand our horizons or to make us conform? Can the knowledge we acquire be a tool to maintain the status quo and instill specific values in us?

When we learn arithmetic, we assume that the information is value free. What hidden message could be sheltered in the equation 3 + 4 = 7? In some children's books, this problem is illustrated with three apples and four apples. We can imagine another primer illustrating the problem with three machine guns and four machine guns. Does the choice of illustration promote values?

There is such a history in physics texts, as well. As we peruse the introductory physics texts on our shelves, we find the following trajectory problems:

A rescue plane is flying at a constant elevation of 1200 m with a speed of 430 km/h toward a point directly over a person struggling in the water. At what angle of sight ϕ should the pilot release a rescue capsule if it is to strike (very close to) the person in the water? (Halliday, Resnick, and Walker, *Fundamentals of Physics*, Wiley, 1993)

In the 1968 Olympics in Mexico City, Bob Beamon shattered the record for the long jump with a jump of 8.90 m. Assume that the speed on takeoff was 9.5 m/s. How close did this world-class athlete come to the maximum possible range in the absence of air resistance? The value of g in Mexico City is 9.78 m/s^2. (Halliday, Resnick, and Walker, *Fundamentals of Physics*, Wiley, 1993)

A golf ball hit with a 7-iron soars into the air at 40.0 degrees with a speed of 54.86 m/s. Overlooking the effect of the atmosphere on the ball, determine the range and where it will strike the ground. (Hecht, *Physics*, Brooks/Cole, 1994)

In contrast, the predominant problems in the older texts appear to

The stage suggests a place between Pisa and Stockholm, and the science is shown as something between good and evil. It can create amusement parks and it can create bombs to annihilate entire cities. A curious mind can drop apples from the Leaning Tower of Pisa, or a life-saving boat for a drowning person, or a bomb. The answer should be easy, yet mankind seems to be more fascinated with destruction. The picture shows Da Vinci, one of the most brilliant minds and artists of the Renaissance, using his genius to construct sophisticated deadly weapons. And Alfred Nobel, whose name we associate with the world-renowned peace prize, actually made his fortune by inventing and patenting dynamite—not a very peaceful creation. The Thinker at the front of the stage is sitting on dynamite, thinking—not thinking too long before starting to act, we hope. The audience can't wait to see the fireworks.

—T.B.

BOOM!
VILLA NOBEL
R.I.P.
TNT
NOBEL
TOM BUNK

be illustrated by these examples:

A bomber is flying at a constant horizontal velocity of 820 miles/hr at an elevation of 52,000 feet toward a point directly above its target. At what angle of sight ϕ should a bomb be released to strike the target? (Halliday and Resnick, *Physics*, Wiley, 1966)

The projectile of a trench mortar has a muzzle velocity of 300 ft/s. Find the two angles of elevation to hit a target at the same level as the mortar and 300 yd distant. (Sears and Zemansky, *College Physics*, Addison-Wesley Press, 1948)

The angle of elevation of an anti-aircraft gun is 70° and the muzzle velocity is 2700 ft/s. For what time after firing should the fuse be set, if the shell is to explode at an altitude of 5000 ft? Neglect air resistance. (Sears and Zemansky, *College Physics*, Addison-Wesley Press, 1948)

All of these problems have similar solutions. We can analyze the trajectory of any object (without air resistance) by recognizing that the horizontal and vertical motions are independent of one another. The horizontal motion has a constant velocity and the vertical motion has a constant acceleration. The standard kinematic equations for motion in one dimension—

$$s = v_{av}t,$$
$$s = \tfrac{1}{2}at^2 + v_i t,$$
$$v_f^2 = 2as + v_i^2,$$
$$v_f = at + v_i$$

—permit us to find the time in the air, the range of the projectile, or whatever else is required in the problem.

As an example, let's solve the rescue plane problem given above. The initial velocity of the capsule is the same as that of the plane. That is, the initial velocity v_0 is horizontal and has a magnitude of 430 km/h. We can find the time of flight of the capsule:

$$y - y_0 = v_{0y}t - \frac{1}{2}gt^2,$$

$$t = \sqrt{2\frac{(y_0 - y)}{g}} = \sqrt{2\frac{(1200\text{ m})}{9.8\text{ m/s}^2}} = 15.6\text{ s}.$$

The horizontal distance covered by the capsule in that time is

$$\begin{aligned} x - x_0 &= v_{0x}t \\ &= (430\text{ km/h})(15.6\text{ s})(1\text{ h}/3600\text{ s}) \\ &= 1{,}860\text{ m}. \end{aligned}$$

The angle of sight can be calculated by comparing the horizontal and vertical displacements:

$$\phi = \arctan\frac{x}{h} = 57.2°.$$

The pedagogical/social question is more difficult than the physics: is the selection of examples and problems of concern? Does it make a difference if we learn to solve projectile problems using sports and rescue planes or mortar shells and bombs? What is your opinion?

As the second part of our problem, we ask you to solve a difficult projectile problem. A fireworks aerial display is shot into the air and explodes isotropically (uniformly in all directions) into a very large number of fragments. At some time t_1, the first fragment(s) will begin to hit the ground. At some later time t_2, the final fragment(s) will hit the ground. Neglecting the effects of air resistance, at what time will the frequency of fragments hitting the ground be the greatest?

We pose the problem with a fireworks display. One can easily imagine how the same problem could have military applications in terms of bomb fragments or interference with ground radar. There is an answer to the second part of our problem. As students and instructors of physics, we should reflect and discuss the first part as well.

Solution

We follow the solution given a few years ago by Tainan Wang, an undergraduate student at SUNY Stony Brook and former member of the Chinese Olympiad Team.

Assuming that all the fragments have the same speed relative to the center of mass, then the fragments will form a sphere with the center of mass as its center. The radius of this sphere will increase in proportion to the time, as the entire sphere descends toward the ground with acceleration g.

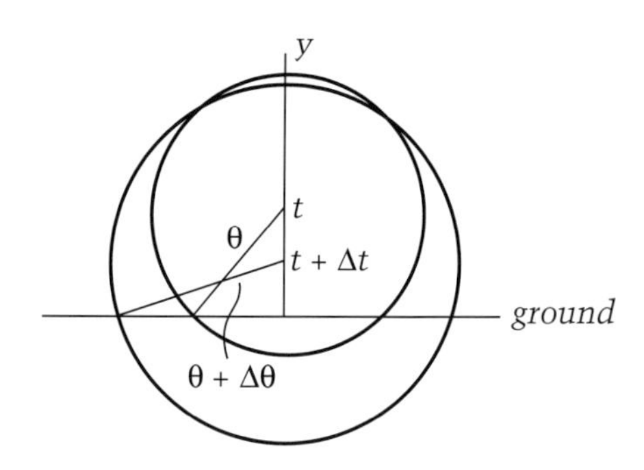

Figure 1

Figure 1 shows the expanding sphere at times t and $t + \Delta t$. During the time interval Δt, all fragments within $\Delta\theta$ have hit the ground.

The motion of the fragments is described by the following simultaneous equations:

$$y(t) = h - \frac{1}{2}gt^2 + v_0 t,$$
$$r(t) = v_1 t,$$

where v_0 is the initial velocity of the center of mass and v_1 is the speed of the expanding sphere.

Since

$$r\cos\theta = y,$$

$$\cos\theta = \frac{h + v_0 t - \frac{1}{2}gt^2}{v_1 t}.$$

the fragments within $\Delta\theta$ can be found by comparing the area of this surface fragment with the total surface area of the sphere. The area of the surface fragment is the product of the circumference of the surface

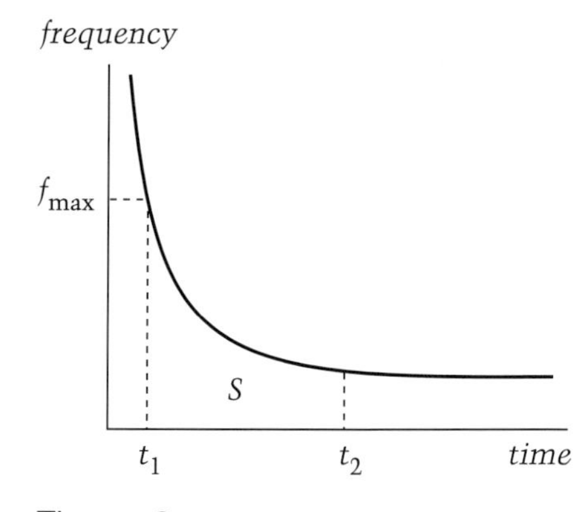

Figure 2

and the arc length of that fragment $(r\Delta\theta)$, assuming that Δt is very small.

The number of fragments is therefore

$$\Delta n = \frac{(2\pi r \sin\theta)(r\Delta\theta)}{4\pi r^2} N$$
$$= \frac{1}{2} N \sin\theta \Delta\theta,$$

where N is the total number of fragments.

The frequency can now be found:

$$f = \frac{dn}{dt} = \lim_{\Delta t \to 0} \frac{\Delta n}{\Delta t}$$
$$= \lim_{\Delta t \to 0} \frac{N}{2} \sin\theta \frac{\Delta\theta}{\Delta t}$$
$$= \frac{1}{2} N \sin\theta \frac{d\theta}{dt}$$
$$= -\frac{1}{2} N \frac{d\cos\theta}{dt}$$
$$= \frac{1}{2} N \left(\frac{g}{2v_1} + \frac{h}{v_1} \frac{1}{t^2} \right).$$

The graph of this equation is shown in figure 2. In the figure, t_1 is the time at which the first fragment hits the ground and is the time where the frequency of particles hitting the ground is the greatest.

As a check on our work, the area S beneath the curve must equal the total number of particles N:

$$\int_{t_1}^{t_2} f\, dt = \int dn = \int_0^{\pi} \frac{N}{2} \sin\theta d\theta = N.$$

Does this make physical sense? Imagine a limiting case where the velocity of the fragments is very, very large. We can see that the particles that were shot downward will have a small difference in their vertical velocities and will all hit the ground at almost the same time. The particles that were shot straight up will also have a small difference in vertical velocities. This small difference will lead to a large lag time due to the long time in flight.

We also asked the more provocative question about whether it makes a difference if we learn to solve projectile problems using sports and rescue planes or morter shells and bombs. We now wonder why most of our readers did not bother to express their opinion on this pedagogical/social question.

The nature of light

"The invisible influences of gravitation and electromagnetic fields remain magic; describable, but nevertheless implacable, nonhuman, alien, magic."
—*B. K. Ridley*

LIGHT PLAYS SUCH A CRUCIAL role in our lives that it's very hard to imagine a universe without light. Almost all of the information that we receive from outside our solar system comes to us in the form of light. Observations of the heavenly bodies and attempts to find regularities in their motions led to tremendous advancements in science and the modern-day scientific method. Studies of light and color revolutionized painting and the fine arts. The invention of the electric light allowed us to work and study at night. More recently the invention of lasers has had profound effects on our abilities to understand the world around us and to make great advances in technology used in such areas as surgery, cutting, welding, surveying, communication, the arts, advertising, and manufacturing.

But what is light? How do we describe its behavior? We have two basic models that we can use to describe light—particle behavior and light behavior. The debate over the best way to describe light has been waged for centuries. Newton felt that light was composed of tiny particles that traveled very fast. He used this idea to predict that light would travel faster in transparent materials like water and glass than in air. By the time the speed of light in water was measured by Jean Foucault in 1862 as slower than in air, the particle model for light was already out of favor. By 1801 Thomas Young had demonstrated the interference of light, a uniquely wavelike effect, and the wave model for light dominated for the next century.

As we discussed in The First Photon (p. 92), Albert Einstein reintroduced the particle aspect of light in 1905 to explain the observations of the photoelectric effect. Einstein said the light can behave like a particle (known as a photon) that has an energy $E = hf$, where $h = 6.63 \cdot 10^{-34}$ J-s is Planck's constant and f is the frequency of the light.

On a dark night in the northern countryside, a jolly photon is given the boot at a bar next to the lighthouse. Somewhat woozy, the particle drives off down the lane, zigzagging in short wavelengths in accordance with the wave model of light, when he suddenly collides with an electron sleeping happily in his bed on four wheels. While the befuddled photon makes a left turn, increasing his wavelength, the electron wakes up, gets energized from the collision, and starts speeding down the curvy road, where canyons fall off steeply into the bottomless universe. Einstein, who likes to travel with light, tries to hitch a ride with the speeding electron. Meanwhile, on top of the wall we can join the X-ray party with some luminaries from all walks of life. The light show demonstrates that inside we all look alike.

—T.B.

BAR
RADIOLOGY
DR. ROENTGEN
PRIVATE
photon
photon
ELECTRON
ELECTRON
TOMAS
BUNK

Additional verification of these revolutionary ideas was provided by Arthur Holly Compton, the son of a Presbyterian minister. Compton was "turned on" to the study of X rays by his older brother and friend Karl. It was known that X rays are another form of electromagnetic radiation similar to visible light and radio, infrared, and ultraviolet waves. Therefore, the puzzle of light was the puzzle of X rays.

Compton's investigations began with the study of the angular distribution of X rays from crystals, for which he was awarded a Ph.D. from Princeton University in 1916. In his research Compton learned about Bragg scattering and was able to measure the wavelengths of X rays rather accurately. He discovered that the wavelengths of some of the X rays scattered by matter were lengthened.

After rejecting classical explanations for these observations, Compton combined Einstein's ideas about photons and relativity and emerged with a simple explanation for his observations. He assumed that X rays consisted of photons with energy $E = hf$ and momentum $p = E/c$. When a photon undergoes a collision with an electron, some of the energy and momentum of the photon is transferred to the electron, reducing the energy of the photon and consequently increasing its wavelength. Using the relationship $\lambda f = c$ for electromagnetic waves, where λ is the wavelength and c is the speed of light, Compton was able to calculate that

$$\lambda' - \lambda = \frac{h}{mc}(1 - \cos\theta), \qquad (1)$$

where λ' and θ are the wavelength and scattering angle of the scattered photon and m is the mass of the electron. Compton verified this effect experimentally and it is now known as the Compton effect. It is interesting that Compton was the one who suggested the name "photon" for light when it acts like a particle. Compton shared the 1927 Nobel prize in physics with Charles Wilson, the inventor of the cloud chamber.

Later Compton studied cosmic rays and helped to establish that cosmic rays are charged particles rather than high-energy electromagnetic waves. After working on the Manhattan Project during World War II, Compton became the chancellor of Washington University in St. Louis. His brothers became presidents of the Massachusetts Institute of Technology and Washington State University, both alma maters of one of your authors (LDK).

By the way, how does the debate about the nature of light come out? We now believe that light exhibits both particle and wave aspects depending on the types of measurement that we make. This is known as wave–particle duality and is a property of all particles at the subatomic level, including electrons and protons.

One of the problems on the semifinal exam used to select the 1996 US Physics Team was a one-dimensional, nonrelativistic derivation of the Compton effect and is the basis for the problem we offer below.

A. Consider the one-dimensional collision of a photon with a free electron initially at rest. Assume that the energy of the photon is much less than the rest energy mc^2 of the electron and the photon recoils in the backward direction with frequency f'. Write expressions for the conservation of energy and linear momentum.

B. Neglecting additive terms of order v^2/c^2, show that

$$h^2 ff' = \left(\frac{1}{2}mv^2\right)\left(\frac{1}{2}mc^2\right), \qquad (2)$$

where v is the electron's speed after the collision.

C. Show that equation (2) can be written in the form

$$\lambda' - \lambda = \frac{2h}{mc},$$

in agreement with the formula for the Compton effect. Note that the change in wavelength does not depend on the original wavelength.

The quantity h/mc is known as the Compton wavelength and has the value $2.43 \cdot 10^{-12}$ m = 2.43 pm—a very small change. This change is difficult to measure unless λ is also small. Compton used X rays with a wavelength of 71.1 pm.

D. What is the energy of these X rays? Can the electrons in matter be treated as if they were free? Does the recoil energy of the electron satisfy the condition for a nonrelativistic treatment?

E. If you would like to work more with this effect, try obtaining the two-dimensional result given in equation (1). Let ϕ be the angle of the electron leaving the collision. Write down the equations for the conservation of energy and the two components of momentum. Use these three equations to eliminate ϕ and v. You will need to use the fact that the collision is nonrelativistic to neglect small additive terms. (Most textbooks on modern physics give the relativisitic derivation.)

Solution

Correct solutions were submitted by Robert Marasco (a junior at North Penn High School in Lansdale, Pennsylvania), Timothy Spegar (a graduate student in mechanical engineering at Penn State University), and jointly by André Cury Maiali and Gualater José Biscuola (engineers and physics teachers in Jundiaí, São Paulo, Brazil). For the most part, we will follow Maiali and Biscuola's solution.

A. Knowing that the energy and momentum of a photon are given by $E = hf$ and $p = hf/c$, respectfully, we can write down the equations for conservation of energy and momentum in one dimension:

$$hf = hf' + \frac{1}{2}mv^2, \qquad (3)$$

$$\frac{hf}{c} = -\frac{hf'}{c} + mv, \qquad (4)$$

where f and f' are the initial and final frequencies of the photons, m is the mass of the electron, and v is the speed of the electron after the collision.

B. We now collect the frequency terms in equations (3) and (4) and square the equations to obtain

$$f^2 - 2ff' + f'^2 = \frac{m^2v^4}{4h^2}, \quad (5)$$

$$f^2 + 2ff' + f'^2 = \frac{m^2v^2c^2}{h^2}. \quad (6)$$

Subtracting equation (5) from equation (6) yields

$$4ff' = \frac{m^2v^2c^2}{h^2}\left(1 - \frac{v^2}{4c^2}\right).$$

Neglecting the last term in the parentheses, we have

$$h^2ff' = \left(\frac{1}{2}mv^2\right)\left(\frac{1}{2}mc^2\right). \quad (7)$$

C. Replace the term in the first parentheses in equation (7) with its equivalent from equation (3) and use the wave relationship $c = \lambda f$ to obtain

$$\frac{h^2c^2}{\lambda\lambda'} = h\left(\frac{c}{\lambda} - \frac{c}{\lambda'}\right)\left(\frac{1}{2}mc^2\right),$$

or

$$\lambda' - \lambda = \frac{2h}{mc}. \quad (8)$$

D. We can get the energies of the X rays from

$$E = hf = \frac{hc}{\lambda} = 2.80 \cdot 10^{-15}\ \text{J}.$$

Because this energy (17.5 keV) is much, much larger than the binding energies of the electrons to their atoms (~10 eV), the electrons can be treated as being free.

The kinetic energy K of the recoil electron can be found from

$$K = h(f - f') = hc\left(\frac{1}{\lambda} - \frac{1}{\lambda'}\right).$$

Using equation (8) and defining $\lambda_C = h/mc$, we obtain

$$K = hf\left(\frac{2\lambda_C}{\lambda + 2\lambda_C}\right) = 1.12\ \text{keV}.$$

Because this is very small compared to the rest energy of the electron (0.511 MeV), the electron can be treated nonrelativistically.

E. We obtain the two-dimensional equation by starting with the two components of the equation for the conservation of momentum.

$$p = p'\cos\theta + p_e\sin\phi, \quad (9)$$

$$0 = p'\sin\theta - p_e\cos\phi, \quad (10)$$

where $p_e = mv$ is the electron's momentum and $p = hf/c = h/\lambda$ and $p' = h/\lambda'$ are the momenta of the photons. In equations (9) and (10), put the electron terms on the left-hand side and the photon terms on the right-hand side, square both equations, and add them together to obtain

$$p_e^2 = p^2 - 2pp'\cos\theta + p'^2. \quad (11)$$

The equation for the conservation of kinetic energy is

$$K = pc - p'c. \quad (12)$$

Divide equation (12) by c, square it, and subtract it from equation (11):

$$p_e^2 - \frac{K^2}{c^2} = 2pp'(1 - \cos\theta). \quad (13)$$

Now let's look at the left-hand side of equation (13) in more conventional terms:

$$m^2v^2 - \frac{m^2v^4}{4c^2} = m^2v^2\left(1 - \frac{v^2}{4c^2}\right)$$

$$\cong m^2v^2 = 2mK = 2mc(p - p'),$$

where we've neglected the term in v^2/c^2. Therefore,

$$mc(p - p') = pp'(1 - \cos\theta).$$

Dividing through by pp' and expressing the momentum in terms of the wavelengths, we get the desired result:

$$\lambda' - \lambda = \frac{h}{mc}(1 - \cos\theta).$$

Do you promise not to tell?

"Everything that happens is the message: you read an event and be one and wait, like breasting a wave, all the while knowing by living, though not knowing how to live."
—William Stafford

CAN YOU KEEP A SECRET? On the television show Seinfeld, George refused to tell his fiancé his code number to withdraw money from his bank with his ATM (automated teller machine) card. "They told me not to tell anyone." It's only when a person's life is in danger that George finally blurts out his secret code—Bosco. During World War II, the public was constantly reminded that "loose lips sink ships." It was through the work of Alan Turing (of computer fame) and others that the British were able to break the code of the Axis powers and win the war for the Allies. In the Pacific, the Japanese met defeat because they thought their code could not possibly be broken.

As we have more conversations electronically by e-mail or electronic bank transactions through the Internet, we assume somebody has successfully dealt with the question of security and secrecy. When the security is breached, we read about it in the newspapers and wonder how safe the whole system is. New interest has surfaced in the areas of random numbers and quantum properties of matter in hopes that more secrecy can be maintained.

How would you go about sending a message so that only one recipient can understand it? Although code books and number sequences have been somewhat successful in the past, we can look at the properties of waves for our secret transmissions. Is it possible to send a signal out so that one person will receive the signal but another will not? Let's rephrase the question as a physicist might: can we emit a wave that will be localized?

All waves emanating from a point source diverge from that source. A light bulb can be seen from all directions. We know that it is possible to send out a beam of light that is directional because we have used flashlights or observed the beam from a lighthouse. A collimated beam is produced by the careful placement of a lens or a mirror. If all the light emanates from the focal point of a lens, the transmitted beam will be parallel to the principal axis. Similarly, if the light is emitted from the focal point of a concave mirror, all of the light

On top of the image the Second World War is raging, while further below it has become the Cold War, with spies lurking everywhere, especially around the Berlin Wall. Coded messages are sent and received, spies spying on spies who are being spied on by other spies, all trying to decode the secret messages. Here we have also a case where a singing lady spy sends out two signals in order to produce for the enemy a destructive interference and for the ally a constructive one. Unfortunately the latter falls in love with her seductive voice and totally forgets to decode the message. On the snowy road Einstein himself is dreaming how to decipher some of God's mysteries, knowing that the answer entails another question, which in turn begs to be answered, and so on.

—T.B.

LOOSE LIPS SINK SHIPS
CODE
Sound of Silence
TOM BUNK
TOP SECRET

reflected from the mirror will emerge parallel to the principal axis. In both of these cases, the light that moves away from the lens or mirror without reflection will diverge, and spies could pick up our weaker signal from other positions.

What happens if we send out two signals? Two signals will produce places of constructive and destructive interference. If we know the location of our ally and the location of our enemy, we can, in theory, set up our pair of emitters such that our ally gets constructive interference of the two signals and our enemy gets destructive interference of the two signals.

As an example, we can look at a simple version of Young's double slit experiment. If the light emerges from two sources, we can calculate the positions of the constructive interference (antinodes) and the destructive interference (nodes). The conditions for the antinodes are that the distances from the two sources must differ by an integral number of wavelengths. The nodes must have distances that differ by an odd integral number of half-wavelengths ($^1/_2\lambda$, $^3/_2\lambda$, and so on).

Many people have experienced the destructive interference of radio waves when they are stopped at a red light near a large building and notice that the car radio now has lots of static. The direct signal from the transmitter and the reflected signal from the building produce the unpleasant static. Moving a short distance forward can bring your car to an antinode and away from the node so that you can hear your tunes.

In both of these examples, the two sources are in phase. When we send our secret messages, they needn't be. This gives us more flexibility in the placement of our transmitters.

Our problem is adapted from one used at the XVI International Physics Olympiad held in Portoroz, Yugoslavia, in 1985.

A young radio amateur maintains a link with two friends living in two towns. The two antennas are positioned such that when one friend, living in town A, receives a maximum signal, the other friend, living in town B, receives no signal, and vice versa. The two antennas transmit with equal intensities uniformly in all directions in the horizontal plane.

A. Find the distance between the antennas and the orientation of the antennas such that the electrical signals provide a maximum signal for one friend and no signal for the other. Assume that the two antennas transmit the signals in phase.

B. Find the parameters of the array (that is, the distance between the antennas, their orientation, and the phase shift between the signals supplied to the antennas) such that the distance between the antennas is a minimum.

C. Find the numerical solution if the radio station broadcasts at 27 MHz and the angles between north and the directions to town A and town B are 72° and 157°, respectively.

Solution

A wonderful solution was submitted by our colleagues André Cury Maiali and Gualter José Biscuola (jointly) and Flavis Pakianathan.

In part A of the problem the two radio sources were in phase. To solve this problem, we must realize that the path difference between the antennas S_1 and S_2 and town A must be equal to an integral number of wavelengths. The geometry, as shown in the figure, leads to the familiar equation

$$n_1\lambda = d \sin \theta_1.$$

The path difference between the antennas S_1 and S_2 and town B must be equal to an odd integral number of half-wavelengths. This leads to a similar equation

$$\left(n_2 + \frac{1}{2}\right)\lambda = d \sin\theta_2.$$

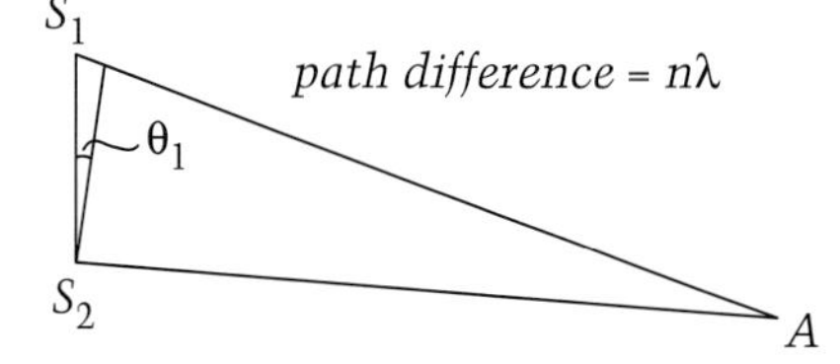

Subtracting these equations and solving for the distance d between the sources, we get

$$d = \frac{\left[n_1 - (n_2 + 1/2)\right]\lambda}{\sin\theta_1 - \sin\theta_2}.$$

We can see that there are an infinite number of selections for the distance between the antennas and their orientation that allows for an antinode at A and a node at B. If we take the simplest situation where town A lies on the perpendicular bisector of the line connecting the antennas, the distance d is defined by the orientation of the angle θ_2:

$$d = \frac{(n_2 + 1/2)\lambda}{\sin\theta_2}.$$

Part B of the problem asked for the parameters of the array (including the phase shift) such that the distance between the antennas is a minimum.

The total path difference when there is a phase delay δ between the sources is

$$d \sin\theta + \frac{\delta\lambda}{2\pi}.$$

The constructive and destructive interference equations thus become

$$n_1\lambda = d\sin\theta_1 + \frac{\delta\lambda}{2\pi},$$

$$\left(n_2 + \frac{1}{2}\right)\lambda = d\sin\theta_2 + \frac{\delta\lambda}{2\pi}.$$

Subtracting these equations and solving for the distance between the sources d, we get

$$d = \frac{\left[n_1 - (n_2 + 1/2)\right]\lambda}{\sin\theta_1 - \sin\theta_2}.$$

We can minimize d by minimizing the numerator and maximizing the denominator. The numerator would be a minimum if

$$n_1 = n_2 + \frac{1}{2}.$$

But since n_1 and n_2 are integers, this is impossible. The smallest value of the numerator occurs when $n_1 = n_2$.

We can find the maximum value of the denominator by taking the derivative of the denominator and setting it equal to zero:

$$f = \sin \theta_1 - \sin \theta_2.$$

Since θ_1 and θ_2 are both dependent on the orientation of the antennas, let's replace θ_2 with $\theta_1 - \phi$, where ϕ is the angle between the two locations:

$$f' = \cos \theta_1 - \cos (\theta_1 - \phi),$$
$$\cos \theta_1 = \cos (\theta_1 - \phi).$$

For the cosine of the two angles to be equal and $\phi \neq 0, 2\pi, 4\pi, \ldots$, we must have

$$\theta_1 = -(\theta_1 - \phi),$$
$$\theta_1 = \frac{\phi}{2}.$$

This makes sense, because in setting up the array we found from our analysis of the numerator that n_1 and n_2 are equal. Therefore, we expect that the perpendicular bisector of the array should pass between the two towns.

The equation for the minimum distance between the sources then becomes

$$d = \frac{\lambda}{4 \sin \frac{\phi}{2}}.$$

The corresponding phase shift between the antennas can now be found:

$$n_1 \lambda = d \sin \theta_1 + \frac{\delta \lambda}{2\pi},$$
$$\delta = \frac{-\pi}{2} - 2\pi n_1 = \frac{3\pi}{2} - 2\pi n_1.$$

Part C of the problem asked for a numerical solution for a broadcast frequency of 27 MHz and angles between north and the directions to the towns of 72° and 157°, respectively:

$$\lambda = \frac{c}{\nu} = 11.1 \text{ m},$$
$$d = \frac{11.1 \text{ m}}{4 \sin 42.5°} = 4.1 \text{ m}.$$

The orientation of the antenna is such that the perpendicular bisector of the line connecting the antennas makes an angle of 72° + 42.5° with the north. ◘

Mars or bust!

"I've always wanted to see a Martian," said Michael. "Where are they, Dad? You promised."

"There they are," said Dad, and he shifted Michael on his shoulder and pointed straight down.

—*Ray Bradbury,* The Martian Chronicles

HAVE YOU EVER WANTED TO go to Mars? Mars is the next frontier. Those of you who are too young to have watched the Herculean efforts to send the first humans to the Moon may be able to participate in the next big space exploration. You may be an astronaut, an engineer, or a computer analyst helping with the mission. Thousands of people will be required. In 1996 the public's interest in Mars was heightened by NASA's announcement that scientists may have found evidence for the existence of primitive life on ancient Mars.

Sending humans to Mars will require a lot of preparation. The work has already begun. Besides the crucial work on studying how humans live in space for long periods of time, two recent launches have sent satellites to our nearest planetary neighbor.

On November 7, 1996, NASA launched the Mars Global Surveyor (MGS), which reached Mars on September, 12, 1997, to begin a two-year survey of the atmosphere and surface of Mars from orbit. The MGS required 309 days to make its journey. (You can learn more about the Mars Global Surveyor mission on the web at *marsprogram.jpl.nasa.gov/mgs/.*)

On December 4, 1996, NASA launched Mars Pathfinder on a trajectory designed to have it land on the surface of Mars on Independence Day 1997, a trip of 212 days. Mars Pathfinder contained a microrover that was used to deploy scientific instruments and explore the terrain around the landing site. (You can learn more about Mars Pathfinder at *marsprogram.jpl.nasa.gov/MPF/.*)

It's interesting that the satellite that was launched last arrived at Mars first. We can get some understanding of this by looking at a simplified orbital problem.

Let's assume that we have a satellite in a circular orbit around the Sun with a radius equal to the average radius of the Earth's orbit. Let's fire rockets in the forward direction

Leaving the overcrowded blue planet, two scouts and one scientist arrive on Mars aboard Pathfinder I. Immediately behind them follows an endless procession of wagons filled with settlers and interplanetary explorers. On the ground a Martian family watches the invasion, wondering what kind of primitive life is coming to their shore. Are the humans coming in peace, or will they decimate the natives and destroy the planet bit by bit? The curtains on each side of the picture suggest that we are in a theater, and indeed there is a balcony with an audience, mainly astronomers like Ptolemy, Copernicus, Galileo, Newton, Hubble, the curious caveman, all watching the new frontier musical unfolding, while the great stage designer is creating new planets like bubbles for everybody to enjoy.

—T.B.

VIKING
MARS 1
PATHFINDER 1
USA
CCCP
MARS
1
TOM
BUNK

tangent to the orbit. If we increase the speed of the satellite by the correct amount, the satellite will be placed into an elliptical orbit that has its greatest distance from the Sun equal to the average radius of Mars's orbit. If Earth and Mars are in the proper relative positions, this would allow the satellite to orbit or land on Mars. In our calculations, we neglect the gravitational effects of Mars and the Earth and consider only the Sun's gravity.

We can find the required speed of the satellite using conservation of energy

$$\frac{1}{2}mv_E^2 - \frac{Gmm_S}{r_E} = \frac{1}{2}mv_M^2 - \frac{Gmm_S}{r_M}$$

and conservation of angular momentum

$$mv_E r_E = mv_M r_M,$$

where m and m_S are the masses of the satellite and the Sun, respectively; r_E and r_M are the orbital radii of Earth and Mars, respectively; and v_E and v_M are the orbital speeds of Earth and Mars. Note that these two velocities occur at the ends of the ellipse—that is, when $r = r_E$ and $r = r_M$, respectively, and that the velocities are perpendicular to the radii. Solving these two equations for v_E, we obtain

$$v_E = v_0\sqrt{\frac{2r_M}{r_E + r_M}},$$

where

$$v_0 = \sqrt{\frac{Gm_S}{r_E}}$$

is the orbital speed of the Earth. Using $r_M = 1.53\ r_E$, we find that $v_E = 1.10\ v_0$. Knowing that $G = 6.67 \cdot 10^{-11}\ \mathrm{N \cdot m^2/kg^2}$, $m_S = 2.0 \cdot 10^{30}$ kg, and $r_E = 1.5 \cdot 10^{11}$ m, we obtain numerical values of $v_0 = 29.7$ km/s and $v_E = 32.7$ km/s. Therefore, we must increase the satellite's forward speed by 3.0 km/s.

We can do a similar calculation at the other end of the ellipse to find out how much we need to accelerate the satellite to match the orbital speed of Mars. Conservation of angular momentum tells us that $v_M = 21.4$ km/s and that Mars's orbital speed is 24.1 km/s. Therefore, the satellite needs to speed up by 2.7 km/s.

Because Kepler's laws are applicable for any object orbiting the Sun, we can use Kepler's third law to find out how long it takes the satellite to reach Mars. Let's compare the circular orbit of the Earth to this transfer ellipse connecting Earth and Mars with a major axis equal to $r_E + r_M$:

$$\left(\frac{T_t}{T_E}\right)^2 = \left(\frac{r_E + r_M}{2r_E}\right)^3.$$

Therefore, $T_t = 1.42\ T_E = 1.42$ years. Because the satellite only executes one half of the elliptical orbit, the time is 0.71 years = 260 days. Longer and shorter periods can be obtained by using different ways of leaving Earth orbit and entering Mars orbit. For a better calculation, we also need to take into account the gravitational fields of Mars and Earth.

Our problem is based on a problem that appeared on the second exam used to select the members of the 1995 US Physics Team, which won four gold medals and one silver medal at the International Physics Olympiad in Australia.

Let's assume that the Mars Global Surveyor is in a circular orbit about Mars at the designed height of 367 km above the surface. We also assume that we can neglect the effects of Mars's atmosphere and that Mars has a radius R = 3,400 km and a surface gravity $g = 3.72\ \mathrm{m/s^2}$.

A. Find the speed of the satellite in its circular orbit about Mars in terms of the values given above.

Although the Global Surveyor is not designed for this purpose, let's assume that we want to send the satellite down to the Martian surface. The satellite could reach the surface by firing its rocket engines for a short period of time. We will consider two special cases.

B. In the first method, the retrorockets are fired at point X tangent to the orbit to slow the satellite. The circular path becomes an elliptical path that brings the satellite to a landing strip on the Martian surface at point A on the side opposite to point X, as shown in figure 1.

(i) Determine the speed of the satellite immediately after the retrorockets have been fired.

(ii) Determine the speed of the satellite as it reaches Mars's surface at point A.

C. In the second method the rockets are fired at point X perpendicular to the orbit, giving the satellite a momentum directed toward Mars. The circular path becomes an elliptical path that brings the satellite to a landing strip on the lunar surface at point B one quarter of the way around Mars, as shown in figure 2.

(i) Determine the speed of the satellite as it reaches Mars's surface at point B.

(ii) Determine the velocity of the satellite immediately after the rockets have been fired.

D. How do the magnitudes of the changes in velocity at point X compare for the two methods?

E. How do the speeds of the satellite at Mars's surface compare for the two methods?

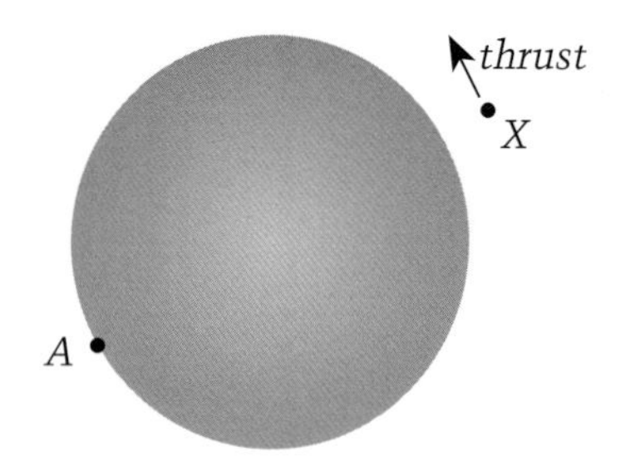

Figure 1

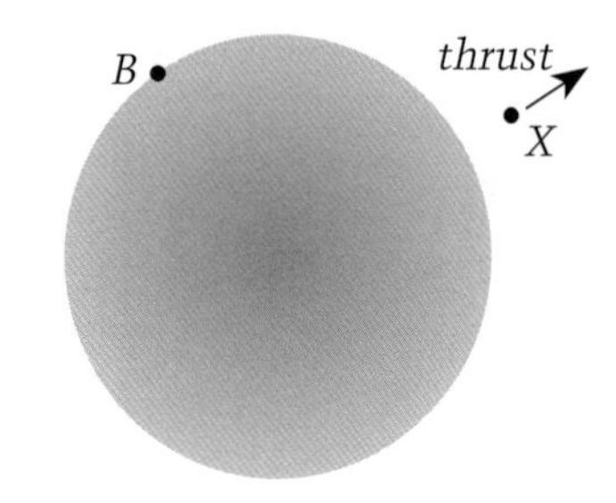

Figure 2

Solution

Excellent solutions to this problem were submitted by André Cury Maiali and Gualter José Biscuola from Brazil.

A. The gravitational force acting on a satellite of mass m orbiting Mars is given by

$$F = \frac{GMm}{(R+h)^2},$$

where G is the gravitational constant, M and R are the mass and radius of Mars, and h is the altitude of the orbit above the surface. Because the surface gravity on Mars is given by

$$g = \frac{GM}{R^2},$$

this force can also be written as

$$F = \frac{mgR^2}{(R+h)^2}.$$

Equating either of these expressions for the gravitational force to the centripetal force $mv_0^2/(R+h)$ allows us to solve for the orbital velocity v_0:

$$v_0 = R\sqrt{\frac{g}{R+h}} = \sqrt{\frac{GM}{R+h}}.$$

Using the numerical values given in the problem, $v_0 = 3.38$ km/s.

B. According to the statement describing the first method of landing, the path is an ellipse that is tangent to the orbit and to the Martian surface at the two ends of the ellipse. Let the speed of the satellite be v_X after the retrorockets have fired and v_A at the surface and write down the expressions for the conservation of angular momentum

$$mv_AR = mv_X(R+h)$$

and the conservation of mechanical energy

$$\frac{1}{2}mv_A^2 - \frac{GMm}{R} = \frac{1}{2}mv_X^2 - \frac{GMm}{R+h}.$$

Solving for v_X we obtain

$$v_X = v_0\sqrt{\frac{2R}{2R+h}} = 3.29 \text{ km/s}.$$

Conservation of angular momentum now tells us that

$$v_A = v_X\frac{R+h}{R} = 3.65 \text{ km/s}.$$

This answer makes sense, because the speed must increase as the satellite descends to the surface.

C. This time the rockets fire in the radial direction. Therefore, the rockets do not change the angular momentum of the satellite. Conservation of angular momentum tells us that

$$mv_BR = mv_0(R+h).$$

Thus

$$v_B = v_0\frac{R+h}{R} = 3.74 \text{ km/s}.$$

We can now use conservation of mechanical energy to find the speed at point X. Let's call this speed v_Y to distinguish it from the speed we obtained for the first method. Then

$$\frac{1}{2}mv_Y^2 - \frac{GMm}{R+h} = \frac{1}{2}mv_B^2 - \frac{GMm}{R}.$$

Solving for v_Y and using our expression for v_B, we obtain

$$v_Y = v_0\sqrt{1+\frac{h^2}{R^2}} = 3.40 \text{ km/s}.$$

D. Our Brazilian readers point out that we can get much better comparisons by calculating the changes in the velocities at point X algebraically rather than using our numerical results. In the first case, v_0 and v_X lie along the same direction and we can simply subtract them to obtain

$$\Delta v_A = v_0 - v_X = v_0\left(1-\sqrt{\frac{2R}{2R+h}}\right) = 87.7 \text{ m/s}.$$

In the second case, the two velocity vectors $\mathbf{v}_0$ and $\mathbf{v}_Y$ are not parallel, but we know that $\mathbf{v}_Y = \mathbf{v}_0 + \mathbf{v}_r$, where $\mathbf{v}_r$ is the radial component of velocity imparted by the rocket engines. Therefore, we can use the Pythagorean theorem to find

$$\Delta v_B = v_r = v_0\frac{h}{R} = 365 \text{ m/s}.$$

Then $\Delta v_A/\Delta v_B = 0.240$, or about one-fourth as much.

E. We now compare the landing speeds:

$$v_A = v_X\frac{R+h}{R} = v_0\left(\frac{R+h}{R}\right)\sqrt{\frac{2R}{2R+h}},$$

$$v_B = v_0\left(\frac{R+h}{R}\right).$$

Thus

$$\frac{v_A}{v_B} = \sqrt{\frac{2R}{2R+h}} = 0.974.$$

Color creation

*"She comes in colors everywhere,
she combs her hair,
she's like a rainbow."*
—Mick Jagger/Keith Richards

WE LOVE COLORS—THE colors of spring and summer, the colors of butterfly wings and rainbows, the colors of soap bubbles, and the colors from a CD. How are these the same? How are they different? Should we look to the same cause for what appears to be the same effect?

Isaac Newton, in his study of colors, devised some wonderful investigations. Described in his book *Opticks*, published in 1704, he details a series of experiments. In one, he shined light from his window through a prism and observed the colors of the spectrum that had tantalized so many others before him. He then brought these colors back together again with a second prism and saw, as no one before him had, that the white light returned. Newton then surmised that white light is a combination of all the colors of the spectrum. In school we learn the name of Mr. Roy G. Biv to help us remember the order of these colors (**r**ed, **o**range, **y**ellow, **g**reen, **b**lue, **i**ndigo, **v**iolet). These spectral colors are observed through diffraction gratings, through prisms, and in rainbows.

The rainbow is arguably Nature's most beautiful optical display. After a rainfall, the bow of colors can extend from horizon to horizon. The creation of the rainbow involves the physics of refraction and reflection and a geometry first explained by Descartes. The light rays from the sun refract as they enter the raindrops (fig. 1). This refraction causes the different colors of the white light to bend by different amounts, producing a spectrum. Upon hitting the back side of the water droplets, the light is partially reflected back toward the general direction of the sun. These light

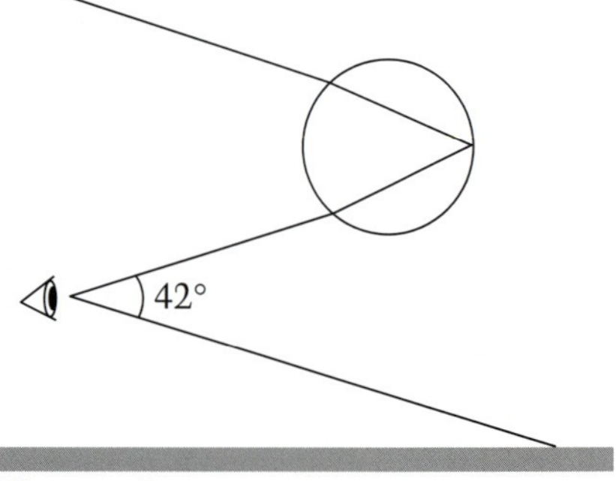

Figure 1

On a dull and dreary afternoon somewhere in England, a speeding oil truck loses control on a slippery road and hits a tree. The oil spills all over the gray landscape and displays wonderful spectral colors. Newton, looking out the window, gets inspired to add some more colors to the bleak scenery and creates kaleidoscopic bubbles, sending them up into the gloomy sky where a young girl, dressed in psychedelic colors, dances gracefully over a rainbow, chasing the pretty bubbles. Meanwhile, Einstein drifts down to Earth suspended by a multicolored parachute, fiddling a colorful tune, and in his attic an artist dabs vivid colors on his canvas to brighten up this dull and dreary afternoon. The diagram at the very bottom shows that all these beautiful colors are in reality nothing but refractions and reflections of plain white light.

—T.B.

ROYAL
SOCIETY
OIL
TOMAS
BUNK

rays then refract again upon leaving the water droplet. The light rays emerge at many angles depending on where the light enters that rain drop. But each color tends to be concentrated at a special angle. With your back to the sun, the special angle between the shadow of your head and the red light is 42°. If you look up at 42°, you see red light. This also occurs at 42° to the left or right. In fact, it occurs at 42° in every direction. The locus of points that is 42° from the shadow of your head is a cone. This explains why you see a bow of red light across the sky—a bow at 42°. The special angle for orange light is a little less than 42°. The arc of orange light is, therefore, seen just below the red light. Similarly, the arcs for the other spectral colors are unique and together they form the rainbow in Roy G. Biv order from outside to inside.

The colors from soap bubbles and oil slicks are not the Roy G. Biv colors of the rainbow. If you have occasion to look at soap bubbles or to notice the colors in oil puddles after a rain, you will recognize the colors as muted reds and blues and not the pure vibrancy of the rainbow colors. The way in which these colors are produced is quite different. We refer to the creation of these colors as due to thin film interference. In thin film interference, light reflected from the top and bottom surfaces of the film interfere, enhancing some colors and diminishing other colors.

Let's take a closer look at the creation of colors from thin film interference (fig. 2). Imagine the thin layer of oil that rests on a puddle of water. If only red light was shining on the thin layer of oil, some of the light would reflect off the surface and some of the light would refract into the oil layer. Some of the refracted light would then hit the inner surface and some would reflect back into the oil film. Most of this light would then exit the film (undergoing another refraction) and combine with the originally reflected ray of light. If the two light beams had traveled equal distances, they would constructively interfere with one another, and lots of red light would be reflected. But the light ray that traveled into the film and back again traversed an additional distance. If the total distance were equal to one wavelength of red light, then the two rays of red light would, once again, constructively interfere. If the total distance were equal to one-half wavelength of light or $1\frac{1}{2}$ wavelengths of light, the rays would destructively interfere. Destructive interference essentially means that there is no red light reflected.

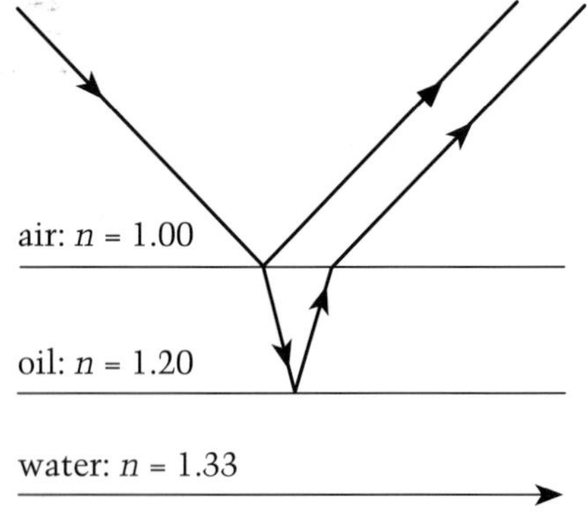

Figure 2

Let's imagine the more complex situation where white light is shining on the thin film of oil. If the thickness of the film is such that the red light undergoes destructive interference, you would see the complete spectrum minus red light—Oy G. Biv. This light is a muted purple, which we do see in oil slicks and in soap bubbles. If the thickness of the film is such that the violet light undergoes destructive interference, you would see Roy G. Bi, which looks like a muted red.

The actual situation has one more complication that must be taken into account. As many of us have seen, a wave reflecting off a boundary can undergo a phase shift. A crest of a wave can hit a boundary and reflect as a trough. This occurs when the wave is reflected from a stiffer medium or one with a higher index of refraction. A phase shift can be mathematically expressed as a shift of one-half wavelength. To determine the correct thickness of the film for destructive interference, one must calculate the total path difference due to the thickness of the oil and any phase shifts that may occur at the boundaries. Since the index of refraction of oil is 1.20 and the index of refraction of water is 1.33, there is a phase shift of the ray reflected off the oil *and* a phase shift reflected off the water surface.

The phenomenon of thin film interference has industrial applications. One of the problems in building sophisticated lens systems is that the internal reflections can cause stray light in the photograph. By coating the lens with a thin film of magnesium fluoride, the unwanted reflection can be eliminated. Let's assume that the coating of magnesium fluoride has an index of refraction of 1.36 and a 100-nanometer layer is evaporated onto a lens of index of refraction 1.60. Which wavelength of light will not be reflected from the surface? We will assume that the light is incident perpendicular to the surface.

The light must travel through the thin film and back again, a total distance of 200 nm. Since the film has a higher index of refraction than the air (1.36 > 1.00), there is a phase shift of one-half wavelength upon reflection. The light which reflects from the film-glass interface also undergoes a phase shift of one-half wavelength, since once again the light is reflecting from a material with a higher index of refraction (1.60 > 1.36). The light traveling along path 1 reflected from the first surface and underwent a phase shift of $\frac{1}{2}$ wavelength. The light traveling along path 2 entered the film, traveled 100 nm, reflected off the second surface and underwent a phase shift of $\frac{1}{2}$ wavelength, and then traveled an additional 100 nm to the first surface. But the 100 nm in the film is not like 100 nm in the air. The wavelength of light is shorter in the film by a factor equal to the index of refraction. The light from these two paths will destructively interfere if the total path difference is a multiple of $\frac{1}{2}$ wavelength in the film. For the thinnest coating of magnesium fluoride, the total path length in the film must be equal to $\frac{1}{2}\lambda$ in the film:

$$\frac{1}{2}\lambda_{film} = 200 \text{ nm},$$

or

$$\lambda_{film} = 400 \text{ nm}.$$

Therefore, the wavelength in air is

$$\lambda = n\lambda_{film} = 544 \text{ nm}.$$

And now we offer two problems. One is adapted from a problem first given at the International Physics Olympiad (IPhO) in Czechoslovakia in 1977, and the other is a problem from *Fundamentals of Physics* by Halliday, Resnick, and Walker.

A. White light falls on a soap film at an angle of 30° with the normal. The reflected light displays a predominantly bright green color of wavelength 500 nm. The index of refraction of the liquid is 1.33. (i) What is the minimum thickness of the film? (ii) What color would be seen if the light source fell on the same soap film from the vertical direction.

B. A thin film of acetone (n = 1.25) coats a thick glass plate (n = 1.5). White light is incident normal to the film. In the reflection, fully destructive interference occurs at 600 nm and fully constructive interference at 700 nm. Calculate the minimum thickness of the acetone film.

Solution

Readers sending in solutions chose to make some simplifying assumptions to ease the derivation. Specifically, they assumed that the total distance through the soap film was equal to twice its thickness, ignoring the angle at which the light traveled. They also assumed that the ray that reflects from the top surface and the ray that enters the film and reflects from the bottom surface interfere with one another even though they are displaced from one another. We will do a more thorough analysis of this standard problem.

As shown in figure 3, the beam of light incident at point A refracts into the film, reflects at point B, and then reemerges at point C. A second ray partially reflects at C, and these two rays interfere. The path difference between these two rays determines whether there is constructive or destructive interference. The beam of light arrives along the line AD in phase. We see that the refracted and reflected rays travel additional distances $AB + BC$ and DC, respectively. From the geometry of the problem, we see that

$$AB + BC = \frac{2d}{\cos\beta},$$

where β is the incident angle of the ray at the lower surface.

The wavelength in the soap film is smaller than the wavelength in air by a factor of n, the index of refraction. The optical path length is therefore

$$\frac{2d}{\cos\beta} \div \frac{\lambda_0}{n} = \frac{2dn}{\lambda_0 \cos\beta}.$$

The path length DC is given by

$$DC = AC \sin\alpha = 2d \tan\beta \sin\alpha,$$

where α is the angle of incidence at the top surface. The number of wavelengths in this distance is

$$\frac{2d \tan\beta \sin\alpha}{\lambda_0}.$$

Because this wave reflects from a medium with a higher index of refraction, it undergoes a phase shift of 180°. This makes the total number of wavelengths equal to

$$\frac{2d \sin\beta \sin\alpha}{\lambda_0 \cos\beta} + \frac{1}{2}.$$

Constructive interference will occur when the path difference is equal to an integral number of wavelengths k:

$$k = \frac{2dn}{\lambda_0 \cos\beta} - \frac{2d \sin\beta \sin\alpha}{\lambda_0 \cos\beta} - \frac{1}{2}.$$

Because Snell's law applies to these rays of light, we know that $\sin\alpha = n \sin\beta$. Therefore,

$$\begin{aligned} k &= \frac{2dn}{\lambda_0 \cos\beta}\left(1 - \sin^2\beta\right) - \frac{1}{2} \\ &= \frac{2dn}{\lambda_0 \cos\beta}\left(\cos^2\beta\right) - \frac{1}{2} \\ &= \frac{2dn\cos\beta}{\lambda_0} - \frac{1}{2}. \end{aligned}$$

Using Snell's law and the relationship $\sin^2\alpha + \cos^2\alpha = 1$, we can write the equation in terms of α:

$$2k + 1 = \frac{4d}{\lambda_0}\sqrt{n^2 - \sin^2\alpha}.$$

If k = 0,

$$d = \frac{\lambda}{4}\frac{1}{\sqrt{n^2 - \sin^2\alpha}} = 0.1\ \mu m.$$

For a light ray incident on the film along the vertical direction,

$$\lambda_0 = 4d\sqrt{n^2 - \sin^2 0} = 4dn.$$

Because d = 0.1 μm, λ_0 = 0.53 μm. This is a greenish-yellow light.

Part B of the problem asked for the minimum thickness of a film of acetone (n = 1.25) placed on glass (n = 1.50) such that light coming from the vertical would have destructive interference at 600 nm and constructive interference at 700 nm.

The path difference for the ray reflected off the top surface of the acetone and the ray reflected off the bottom surface of the acetone is simply $2dn$. Because both reflections occur for materials with higher indices of refraction, a phase shift occurs at both surfaces and therefore the 1/2 does not appear in our equation.

For constructive interference, the optical path difference $2dn/\lambda_{0c}$ should equal an integral number of wavelengths:

$$k = \frac{2dn}{\lambda_{0c}}.$$

For destructive interference, the optical path difference $2dn/\lambda_{0d}$ should equal a half integral number of wavelengths:

$$k + \frac{1}{2} = \frac{2dn}{\lambda_{0d}}.$$

Eliminating k and solving for the thicknesses, we find

$$d = \frac{1/2}{2n\left(\frac{1}{\lambda_{0d}} - \frac{1}{\lambda_{0c}}\right)} = 840\ \text{nm}.$$

This is the minimum thickness. There are other possible thicknesses. What are some of them?

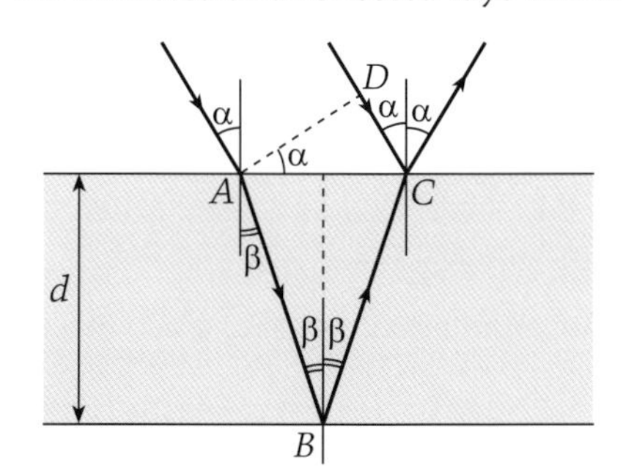

Figure 3

A physics soufflé

"Enough! or Too Much."
—*William Blake,* The Marriage of Heaven and Hell

WHAT DISTINGUISHES THE world's great chefs from the millions of adequate cooks is an understanding of the concepts of cooking. We strive for a similar appreciation of physics concepts in our students. Most of the time the problems in physics textbooks do not require much understanding to obtain the answer in the back of the book. If the problem gives the mass m and acceleration a of an object and asks for the value of the net force F acting on the object, it is not too difficult to find a formula containing m, a, and F and plug in the numbers. To enhance such a cookbook problem, we may provide superfluous information like the velocity v of the object or its color λ. Students who simply look for an expression containing m, a, v, F, and λ will not succeed with this approach. Students must understand the concepts well enough to understand that the velocity and color are not needed. Only the capable chef can ensure that the soufflé will rise.

As an example of giving extra information, consider the following problem: "How much work does the gravitational force perform on a satellite in a circular orbit around Earth? The mass of the satellite is 4,500 kg, the color of the satellite is blue ($\lambda = 450$ nm), the circumference of its orbit is 42,000 km, the Earth's radius is 3,760 km, and the gravitational field at the Earth's surface is 9.8 N/kg." All of these numbers are extra information for students who realize that the gravitational force on the satellite is perpendicular to the satellite's displacement and, therefore, the work is zero!

Is this fair or is it a "trick" question? Recognizing what information is useful and what is extraneous is important in life and in physics problems. We have had students complain about problems in which additional, superfluous information was provided. In an attempt to gen-

A divine soufflé of cosmic proportions floats like a star in the vast universe, waiting to be consumed by hungry extraterrestrial gourmands. While some chefs are fluffing up their soufflés with atmospheric pressure using conventional recipes, Einstein, the world's most inspired chef, is mixing up secret ingredients to revolutionize the commonly held conceptions of previous centuries. The lower image displays a huge kitchen area and control center where the cooks manage shuttle take-offs and prepare soufflés at the same time. It takes a real master cook to understand this concept, and that's why the soufflés sometimes blast off and rise into space and shuttles descend, crashing through the restaurant windows. On the right side an experiment is taking place to show the incredible power of the vacuum, for no apparent reason.

—T.B.

РАСЧЕТНОЕ ВРЕМЯ СТАРТА 20
СПУТНИКА МАРС1 - ОБОС-1
ГОТОВНОСТЬ : 30 МИН
OUR
ESE
SALT
ITES
S FLOUR
4 EGG YOCKS
1 CUP MILK
1 TSP BUTTER
PRESSURE VS TIME DATA
TOMAS BUNK

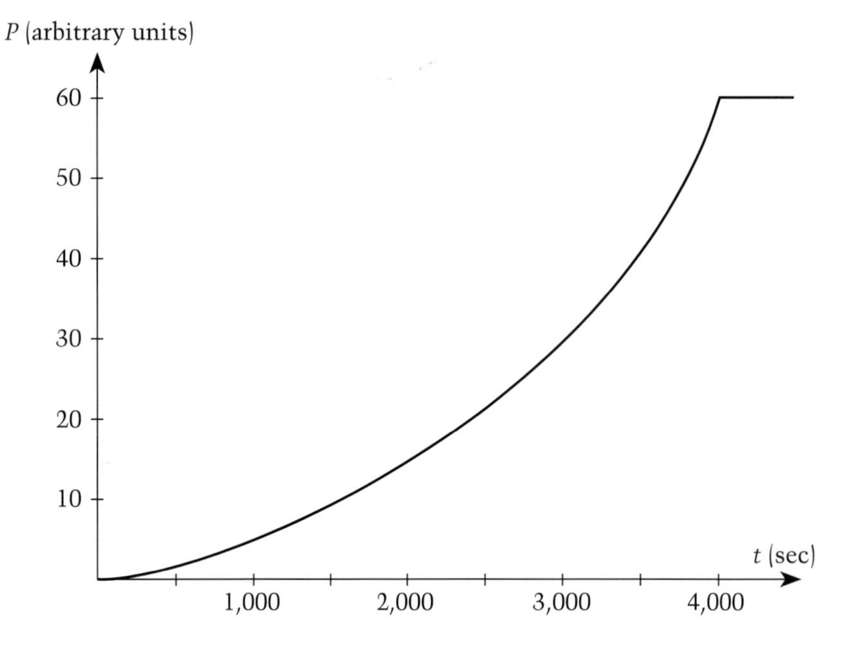

erate discussion, one of us (LDK) gave the students a problem and deliberately omitted one vital piece of information. He then asked, "What do you need to be able to solve this problem?" The problem was not well received, but asking such problems should be commonplace. In the real world, physicists and engineers often have to figure out what measurements are required to obtain the data needed to solve problems.

Sometimes, the physicist works through the theoretical aspects of a problem only to discover that a number is not needed, or that a calibration is not needed. This brings us to the problem at hand. It is based on one of the problems on the second examination used to select members of the US Physics Team that competed in the International Physics Olympiad hosted by Canada in Sudbury, Ontario, during July 1997. The problem originally appeared in *Kvant* (*Quantum*'s sister publication) many years ago.

Upon entering the atmosphere of a planet, a probe descended straight down to the surface. Along the way it recorded the atmospheric pressure as a function of time as shown in the figure. Unfortunately the calibration of the pressure gauge has been lost and the units on the pressure axis are not known. Your mission, should you choose to accept it, is to compensate for this lack of calibration.

The atmosphere is mostly carbon dioxide with a molecular mass of 44 g/mol and can be treated locally as an ideal gas. The surface temperature T_s is 400 K, the gravitational field g at the surface is 9.9 N/kg, and the radius R of the planet is 5,000 km.

A. Apply Newton's second law to a small slab of the atmosphere of vertical thickness Δy to show that the change in pressure ΔP between the top and bottom of the slab is given by

$$\Delta P = \rho g \Delta y,$$

where ρ is the atmosphere's density and g is the local gravitational field.

B. Using this formula for the change in pressure with altitude and the graph, estimate the probe's speed v_0 just before it strikes the surface. Why is the calibration data not needed?

C. Under the simplifying assumption that the probe's speed is constant during its travel through the lower atmosphere, estimate the temperature of the atmosphere at a height of 15 km above the surface.

D. Estimate the uncertainty in your determination of this temperature. How confident are you that your value for the temperature is meaningful?

Solution

We begin by examining a small rectangular slice of the atmosphere with a horizontal cross-sectional area A and thickness Δy. The force due to the pressure differences on the slice in the vertical direction must support the weight of the air within the slice. Therefore,

$$P_l A - P_u A = \rho A \Delta y g,$$

where P_u and P_l are the pressures at the upper and lower surfaces, ρ is the mean density of the air, and g is the value of the local gravitational field. Thus,

$$\Delta P = \rho g \Delta y.$$

If we divide both sides by a small increment of time Δt, we can find how the pressure changes with time:

$$\frac{\Delta P}{\Delta t} = \rho g \frac{\Delta y}{\Delta t} = \rho g v,$$

where v is the speed of the probe.

Because we do not know the density of the gas, we use the ideal gas law

$$PV = nRT$$

to find that

$$\rho = \frac{nM}{V} = \frac{PM}{RT},$$

where n is the number of moles of gas, M is the molecular weight, R is the gas constant, and T is the absolute temperature of the gas. This leads us to

$$v = \frac{\Delta P}{\Delta t} \cdot \frac{RT}{PMg}.$$

We use this relationship to find the speed of the probe just before it hits the surface. We find the surface value of $\Delta P/\Delta t$ by calculating the slope of the pressure-time curve just before it flattens out. We estimate a value of 0.060 ± 0.006 units/s. Using values of T_s = 400 K, P_s = 60 units from the graph, $M = 44 \cdot 10^{-3}$ kg/mol, g_s = 9.9 N/kg and R = 8.3 J/mol·K, we obtain $v_s = 7.6 \pm 0.8$ m/s, where the uncertainty in the speed is due to our uncertainty in the slope. We note that the unit of pressure, and therefore the calibration, does not matter because the units in the ratio $\Delta P/P$ cancel. Therefore, the pressure calibration is not needed for this measurement.

At a constant speed it takes the probe

$$t = \frac{h}{v_s} = 2000 \pm 200 \text{ s}$$

to fall from a height h = 15 km. At t = 2000 s, P = 15.0 units, g = 9.8 N/kg at altitude, and $\Delta P/\Delta t$ = 0.012 units/s, we get

$$T = \frac{PMgv_s}{R\frac{\Delta P}{\Delta t}} = 490 \text{ K}.$$

Using the other times and the corresponding pressures and slopes, we estimate that there is an uncertainty of 90 K or more in the temperature. Therefore, we are not able to determine the temperature very accurately.

A significant factor in the analysis is the assumption that the probe has a constant speed for the entire 15 km. If the probe had reached terminal speed before it reached an altitude of 15 km, the probe should slow down due to the increased air resistance as the atmospheric density increases. If the probe takes 20% longer to complete the descent from 15 km, the calculated temperature decreases to about 430 K. If the probe had not reached terminal speed at an altitude of 15 km, the descent time could be quite a bit shorter, resulting in a much higher temperature. We conclude that we really do not know the temperature very well and that if the mission is to be repeated, the mission scientists should be capable of measuring the probe's speed profile.

Finally, we note that we obtained a higher temperature at altitude than at the surface. This differs from Earth's atmosphere, where the temperature at an altitude of 15 km is approximately –50°C, very much lower than typical surface temperatures of 20°C. ◘

Cool vibrations

"The world is never quiet, even its silence eternally resounds with the same notes, in vibrations which escape our ears."
—Albert Camus

A LOW RUMBLE THROUGH the Earth convulses a highway like a fish gasping for air. A child in a distant playground gracefully moves his body, propelling the swing to new heights. An operatic singer shatters a crystal glass with the precision of her voice.

The Earth, the child, and the soprano play with oscillations. Not content with the simple vibrations of sound, or a mass on a string, or a screen door swinging to and fro, this active cast of characters forces the systems and produces fascinating results. To understand what is happening, we will review the simplest vibrating system before exploring the more complex activities of our players.

A mass hangs from a massless spring. In its stable position, the force of gravity on the mass must be equal and opposite to the force of the spring. This defines the equilibrium position of the system x_0:

$$mg - kx_0 = ma = 0,$$

$$x_0 = \frac{mg}{k}.$$

If the system is pulled below its equilibrium position, there is a net force pulling the mass upward. We can define the stretch of the spring x as the distance beyond the equilibrium position:

$$-kx = ma = m\frac{d^2x}{dt^2}.$$

We can "guess" at the solution to this differential equation:

$$x = A\cos(\omega t + \phi).$$

Taking derivatives, we find

$$v = \frac{dx}{dt} = -A\omega\sin(\omega t + \phi),$$

$$a = \frac{d^2x}{dt^2} = -A\omega^2\cos(\omega t + \phi).$$

To find out if we have guessed successfully, we substitute these solu-

Within the frame of this broken mirror we find ourselves in the Germany of the late 1930s. A voluminous singer of Wagnerian proportions is blasting her high-pitched voice into her surroundings, creating vibrations that cause crystal goblets to shatter. In this case the image hints at the so-called *Kristallnacht*, when Jews were beaten and killed by organized thugs all over Germany. Shortly after this incident, Hitler's *Autobahn* was convulsed by marching boots rumbling through the earth leading straight into the hell of World War II, the destruction of cities and millions of lives. Some, like Einstein, managed to escape in time to reach the boat to safety. The innocent little girl is swinging happily to new heights, creating good vibrations, but they could not compete with the poisonous ones that ruled the times.

—T.B.

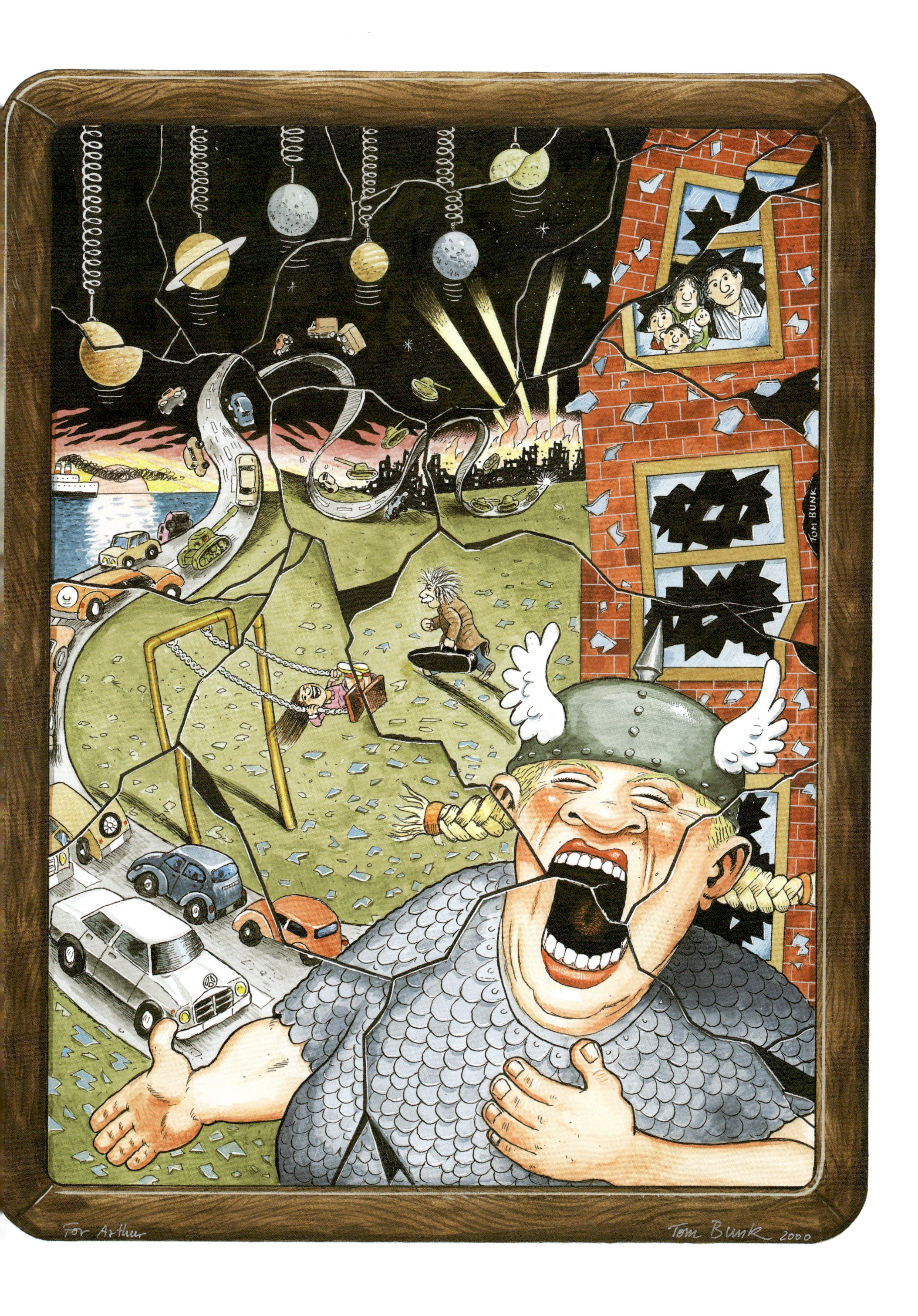
TOM BUNK
For Arthur
Tom Bunk 2000

tions for x and d^2x/dt^2 into our original equation $-kx = md^2x/dt^2$ and find that this is indeed a solution when

$$\omega = \sqrt{\frac{k}{m}},$$

where $\omega/2\pi$ is equal to the frequency of vibration ν of the oscillator.

The situation gets more complicated when it is made more realistic. All oscillating systems have a retarding force. Often this retarding or damping force is proportional to the velocity of the mass. The amplitude of this oscillation will decrease and decrease until the mass eventually comes to rest. But does the frequency change under this force? The equation of motion is now

$$-kx - b\frac{dx}{dt} = m\frac{d^2x}{dt^2}.$$

A solution to this equation if b (the coefficient of the damping force) is small is

$$x = Ae^{-bt/2m}\cos(\omega' t + \phi),$$

$$\omega' = 2\pi\nu' = \sqrt{\frac{k}{m} - \left(\frac{b}{2m}\right)^2}.$$

A plot of this equation shows an oscillation of constant frequency, which damps out exponentially. Exploring this solution will be part of the problem offered below.

A more interesting motion occurs when a varying external force also drives the oscillating mass with a damping force. This external force can be the positioning of a child's body to make the swing reach new heights or the soprano's voice driving the molecules in the glass crystal. In this case, the equation of motion is

$$-kx - b\frac{dx}{dt} + F_m\cos(\omega'' t) = m\frac{d^2t}{dt^2}.$$

The solution of this equation is

$$x = \frac{F_m}{G}\sin(\omega'' t - \phi),$$

where

$$G = \sqrt{m^2\left(\omega''^2 - \omega^2\right)^2 + b^2\omega''^2}$$

and

$$\phi = \cos^{-1}\frac{b\omega''}{G}.$$

We can see that the frequency of oscillation is now that of the external driving frequency and not the natural frequency. If the driving frequency is equal to the natural frequency and the damping force is zero ($b = 0$), we see that G becomes zero and the displacement x will get infinitely large. This is called resonance. Of course, there is never a situation where the damping force is exactly zero, and we find that the resonance does produce very large displacements, although not infinite. This is the explanation for the collapsing highway, the crystal being broken, and the child being able to swing to such new heights.

Our problem oscillates from some simple problems to some graphical and mathematical analysis.

A. A mass hangs from a massless spring and oscillates with a frequency of 1 Hz. If the spring is cut in half, what is the new oscillation frequency?

B. A mass m hangs from 3 massless springs as shown in figure 1. The springs have spring constants k_1, k_2, and k_3. When m is displaced from its equilibrium position, what is the period of oscillation?

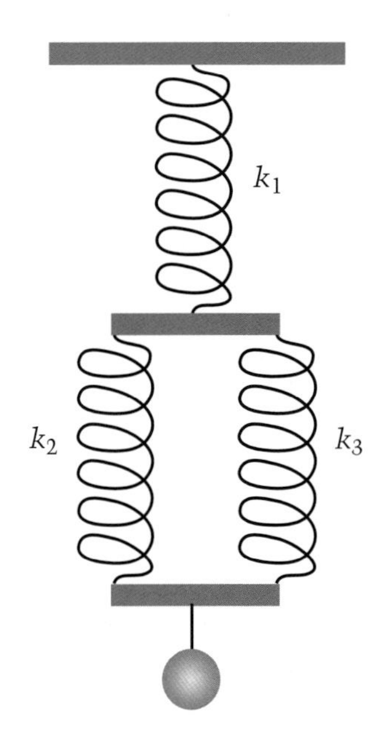

Figure 1

C. (1) Sketch the solution for the damped oscillator. (2) The mean lifetime is defined as the time it takes for the oscillator's amplitude to reach $1/e$ of its initial value. Derive an equation for the mean lifetime. (3) A hanging block with a mass of 2 kg is attached to a massless spring with spring constant equal to 10 N/m. The mass is displaced from its equilibrium position by 12 cm. If the damping force has a b value of 0.18 kg/s, find the number of oscillations made by the block during the time interval in which the amplitude falls to 1/4 of its original value. (4) Derive an expression for the velocity of the mass at any given time.

D. A forced oscillator can have substantially different effects on a mass. (1) Show graphically how the amplitude depends on the ratio of the driving frequency ω'' and the natural frequency ω for the following values of the damping coefficient b: $b = 0$, $m\omega/4$, $m\omega/2$, $m\omega$, and $2m\omega$. (2) Derive an expression for the velocity of the mass at any given time.

Solution

Most readers found the problem of two springs in parallel to be straightforward. Two identical springs in parallel will each have to support half the weight of the suspended mass. Each spring will stretch half as much in order to apply half the force. The equivalent spring constant of the pair of two identical springs would therefore be $2k$ since the pair stretches only half as much:

$$F = kx = 2k\frac{x}{2}.$$

A general rule for springs in parallel is

$$k' = k_1 + k_2.$$

In contrast, springs in series must each support the entire weight of the suspended mass. The total stretch of the two identical springs would be twice the stretch of one spring alone.

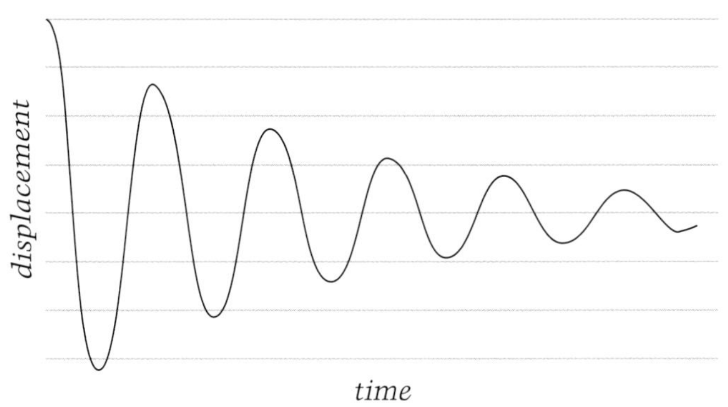

Figure 2

The combined spring constant is $k/2$:

$$F = kx = \frac{k}{2}2x.$$

A general rule for springs in series is

$$\frac{1}{k'} = \frac{1}{k_1} + \frac{1}{k_2}.$$

In part A of the problem, a spring is cut in half. With only half the spring, the new spring constant must be $2k$. Two identical half springs of $2k$ and $2k$ in series would have the combined (original) spring constant of k. With k being doubled as a result of cutting the spring in half, the frequency is now

$$\nu = \frac{1}{2\pi\sqrt{\frac{2k}{m}}} = \nu_0\sqrt{2},$$

where ν_0 is the original frequency.

In part B, springs k_2 and k_3 are in parallel. The combined spring constant of k_2 and k_3 is $k_2 + k_3$, following our first rule. Adding k_1 in series makes the equivalent resistance of the combination

$$\frac{1}{k} = \frac{1}{k_1} + \frac{1}{k_2 + k_3},$$

or

$$k = \frac{k_1(k_2 + k_3)}{k_1 + k_2 + k_3}.$$

The corresponding period T is

$$T = 2\pi\sqrt{\frac{m}{k}} = 2\pi\sqrt{\frac{m(k_1 + k_2 + k_3)}{k_1(k_2 + k_3)}}.$$

C. Using a spreadsheet and corresponding graphing program, we can generate a graph of the solution for the damped oscillator (fig. 2). The mean lifetime can be found by substituting A/e for the displacement in the equation

$$\frac{1}{e}A = Ae^{-\frac{bt}{2m}},$$

$$\ln\frac{1}{e} = -\frac{bt}{2m},$$

$$t = \frac{2m}{b}.$$

Finding the number of oscillations requires us to first calculate the frequency and the elapsed time when the maximum displacement is $A/4$:

$$\frac{1}{4}A = Ae^{-\frac{bt}{2m}},$$

$$t = (\ln 4)\left(\frac{2m}{b}\right) = 30.8 \text{ s}.$$

The ω, using the values given, is 2.236 rad/s, which corresponds to a frequency of 0.356 Hz. Then

(0.356 oscillations/s)(30.8 s) = 11 oscillations.

The corresponding velocity can be found by differentiating the displacement equation:

$$v = \frac{dx}{dt} = Ae^{-\frac{bt}{2m}}\left[-\sin(\omega' t + \phi)\right]\omega' + x\left(-\frac{b}{2m}\right),$$

$$v = -Ae^{-\frac{bt}{2m}}\left[\omega'\sin(\omega' t + \phi) + \left(\frac{b}{2m}\right)\cos(\omega' t + \phi)\right].$$

D. Using a spreadsheet and corresponding graphing program, we can generate a graph of the solution for the forced oscillator for different values of the damping coefficient b (fig. 3). One notices that if $b = 0$ and the driving frequency is equal to the natural frequency, we have a resonance effect gone wild, and the amplitude grows without bound. As the damping coefficient b gets larger, we notice that the resonance effects seem to diminish in size.

The corresponding velocity can be found by differentiating the displacement equation:

$$v = \frac{dx}{dt} = \frac{F_m}{G}\omega''\cos(\omega'' t - \phi).$$

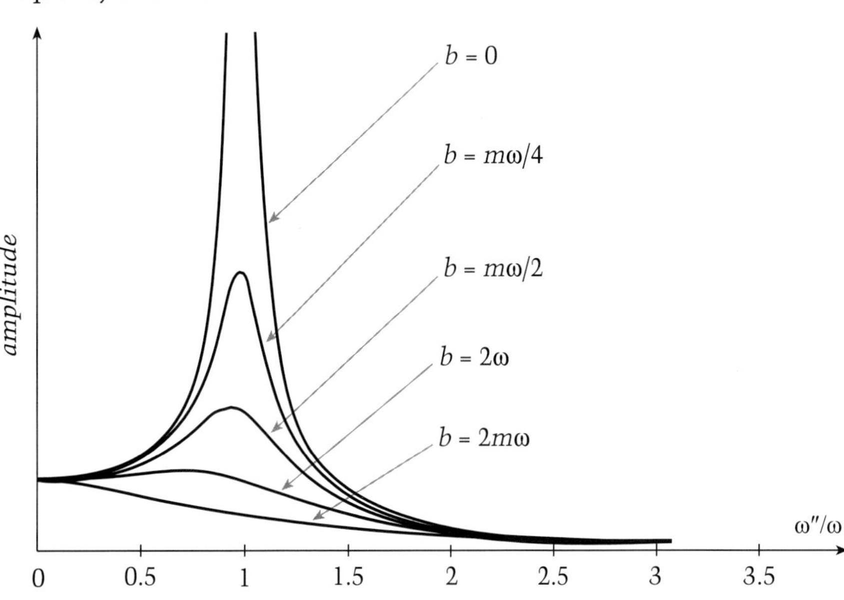

Figure 3

Elephant ears

"Sir Isaac Newton was very much smaller than a hippopotamus, but we do not on that account value him less."
—Bertrand Russell (1872–1970)

WHY DO ELEPHANTS HAVE such big ears? And why do they have such thick legs? In other words, why do elephants have different shapes than horses? These questions and more can be answered using the laws of scaling that we learn in physics.

Elephant bones are made from the same basic material as human bones. Therefore, the bones must be thicker to support the extra mass of the elephant. But how much thicker? Let's compare an elephant to a horse. A typical horse has a mass around 600 kg and a typical elephant has a mass around 4200 kg, or some 7 times larger. Because all mammals have a density near that of water, the elephant must have 7 times the volume of the horse. If we assume that the two have the same shape (they both have four legs!), the linear dimensions of the elephant must be $\sqrt[3]{7} = 1.9$ times the corresponding dimensions of the horse.

Each elephant leg must support 7 times as much weight as a horse leg. Because the compression strength of a beam depends on its cross-sectional area, an elephant leg bone must have 7 times the cross-sectional area of a horse leg bone. In other words, the elephant leg bone must have 2.6 times the diameter of a horse leg bone. Notice that the elephant and the horse cannot have the same shape; the legs must be proportionately larger than the other dimensions. The comparison would be even more dramatic if we compared the elephant to a mouse!

This explains why elephants have such thick legs, but what about the ears? Let's assume for the moment that an elephant eats 7 times as much as a horse because it has 7 times as much mass. As this food is used by the body, it generates heat. Therefore, an elephant must dissipate 7 times as much heat as a horse. We know that the thermal loss is proportional to the difference in the temperature across the skin and to the area of the skin. The surface area of any solid depends on the square of its linear dimensions, so the elephant only has $1.9^2 = 3.6$ times the surface area. This means that the

Tonight at the circus an elephant is being measured to find out if it makes sense to be so big or to be different from a horse or piglet, for example. A handful of specially appointed clowns are investigating with scientific unseriousness his ears, legs, memory, blood pressure, weight, height, inside mechanism that makes him tick, and the important capacity of his trunk to hold water. The results will be collected, may be analyzed, but will definitely and thoroughly be ignored. Even Einstein and Galileo stepped down from the heights of their scientific status to join the fun of circus life, performing daring stunts on the tightrope while a mouse tries to get hold of a horse on the flying trapeze. Not an easy task compared to the little piglet dancing on top of the world.

—T.B.

CIRCUS
SCIENTIFICUS
(T=287K)=
TOM BUNK

elephant must have a much higher body temperature or some other way of getting rid of the thermal energy. This is one of the roles of the big ears. They increase the surface area and, by moving in the air, keep the air temperature near the skin from climbing very much. The elephant also eats less per unit mass than a horse.

It is interesting to note that although elephants communicate by ultrasound, it is not necessary for them to have big ears for this purpose.

Not all scaling deals with lengths. We can use any factor as a scaling parameter. For instance, the Bohr radius for the hydrogen atom is given by

$$a_0 = \frac{\hbar^2}{mke^2} = 0.0529 \text{ nm},$$

where $\hbar$ is Planck's constant divided by 2π, m is the mass of the electron, k is Coulomb's constant, and e is the electronic charge. What would be the new radius if the electron were replaced by a muon with a mass 207 times as large? (We assume that the mass of the proton is large compared to the mass of the muon.) We do not need to solve for the new radius from scratch; all we need to know is that the radius scales inversely with mass. Therefore, the radius of the muonic hydrogen atom is

$$a_\mu = a_0\left(\frac{m_e}{m_\mu}\right) = \frac{a_0}{207} = 0.256 \text{ pm}.$$

This was one of a series of five problems on scaling that made up one of the three theory questions at the International Physics Olympiad held in Sudbury, Canada, in July 1997. The theory problems were developed under the direction of Chris Waltham, who is a faculty member at the University of British Columbia. Three of the other scaling problems make up the problem we offer below.

A. The mean temperature on the Earth is $T = 287$ K. What would the new mean temperature T' be if the mean distance between the Earth and the Sun were reduced by 1%?

B. On a given day, the air is dry and has a density $\rho = 1.2500$ kg/m^3. The next day the humidity has increased and the air contains 2% water vapor by mass. The pressure and temperature are the same as the day before. What is the new air density ρ'? Assume ideal gas behavior. The mean molecular weight of dry air is 28.8 g/mol, and the molecular weight of water is 18 g/mol.

C. A type of helicopter can hover if the mechanical power output of its engine is P. If another helicopter is produced that is an exact half-scale replica (in all linear dimensions) of the first, what mechanical power P' is required for it to hover?

Solution

A correct solution to the first question was submitted by Tal Carmon from Technion, Israel.

A. The first question asked what would happen to the mean temperature T of Earth if the mean distance R between Earth and the Sun decreased by 1%. To do this we match the input radiation to the output radiation because Earth is in thermal equilibrium. If the power output of the Sun is P, the radiation reaching Earth per unit area is $P/4\pi R^2$. If we denote Earth's radius by R_E and its reflectance by r, the input power P_{in} to Earth is

$$P_{in} = (1-r)\frac{P}{4\pi R^2}\pi R_E^2.$$

Stefan's Law gives the output power

$$P_{out} = 4\pi R_E^2 \varepsilon\sigma T^4,$$

where ε is the Earth's emissivity and σ is Stefan's constant. Although the emissivity is a function of temperature, the change in temperature is expected to be small, and we can neglect this dependence. Therefore,

$$T \propto \sqrt{\frac{1}{R}},$$

and a reduction of 1% in R gives a 0.5% rise in T. For a mean temperature of 287 K, we get a rise of 1.4 K.

B. The second question asked about the change in the density of dry air with an increase in humidity when the temperature and pressure remain the same. Let's use the subscripts d and m for dry and moist air, respectively. Then the number of molecules N_d in the dry air is

$$N_d \propto \frac{M_d}{28.8},$$

where M_d is the mass of dry air in a unit volume and the mean molecular mass of dry air is 28.8 g/mol. For moist air, we must account for the proportions of dry air and water vapor. For 2% humidity, we have

$$N_m \propto 0.02\frac{M_m}{18} + 0.98\frac{M_m}{28.8},$$

where the mean molecular mass of water is 18 g/mol.

We know that identical volumes of ideal gases with the same temperature and pressure have the same number of molecules. Therefore,

$$M_d = 1.012 M_m.$$

Because the densities of equal volumes are proportional to the respective masses,

$$\frac{\rho_m}{\rho_d} = \frac{M_m}{M_d} = 0.988,$$

and using $\rho_d = 1.25$ kg/m^3, we get our answer:

$$\rho_m = 1.235 \text{ kg/m}^3.$$

C. The last question asked how the power required for a helicopter to hover depends on the size of the helicopter. The mechanical power P of the helicopter is equal to the thrust T times the downward velocity component v of the air below the blades. The thrust is given by the change in momentum of the air per unit time

$$T = v\frac{dm}{dt},$$

with

$$\frac{dm}{dt} = \rho A v,$$

where ρ is the density of the air and A is the cross-sectional area covered by the blades. Thus,

$$T = \rho A v^2.$$

When the helicopter is hovering, the thrust must be equal to the helicopter's weight. Therefore,

$$v^2 = \frac{T}{\rho A} = \frac{W}{\rho A}.$$

If the size of the helicopter is characterized by a linear dimension L, then $W \propto L^3$, $A \propto L^2$, and $v \propto L^{0.5}$. Thus,

$$P = Tv = Wv \propto L^{3.5}.$$

For a half-scale helicopter, the required power is $0.5^{3.5}P = 0.0884P$.

Local fields forever

"What doth gravity out of his bed at midnight?"
—Shakespeare, Henry IV

AN ASTRONAUT'S DAY MAY be more intense than yours, but the physics is basically the same. You get in an elevator and feel a bit heavier as the elevator accelerates upward. The astronaut's shuttle accelerates upward and the astronaut feels three times as heavy. You jump off a step and experience weightlessness for less than a second. The astronaut lives on the space shuttle or *Mir* and experiences weightlessness for days or months.

How do we comprehend these sensations of feeling heavier or lighter? Can we distinguish between accelerations and gravitational fields? The elevator problem is a classic in elementary physics: If an elevator accelerates a 60 kg student upward at 3 m/s^2, what is the student's weight? If you are standing on a bathroom scale in a stationary elevator (as only a dedicated, socially secure physics student would), the scale supports you by applying a force equal in magnitude to Earth's gravitational force. The force exerted by the scale is what we call your "weight." When the elevator begins to ascend, the scale pushes on you with an additional force—enough to accelerate you:

$$\begin{aligned}\sum F &= ma,\\ F_s - F_g &= ma,\\ F_s &= ma + F_g\\ &= m(a + ag)\\ &= (60\ \text{kg})(3\ \text{m/s}^2 + 9.8\ \text{m/s}^2)\\ &= 768\ \text{N},\end{aligned}$$

where F_s is the force of the scale and F_g is the gravitational force.

The scale reads 768 N, or 180 N more than your normal weight. But it's not just the scale's reading. You really do feel heavier while the elevator accelerates upward. You experience the same sensations that you would have on a planet with a stronger gravitational pull.

If we move to two dimensions, we have a more interesting effect. As your car accelerates, you are pressed against the seat. It is almost as if another planet appeared behind your car. You now experience the gravity of Earth and the apparent gravity connected to the acceleration. If the car accelerates at 9.8 m/s^2, you will experience the vector sum of the effects of the gravity and the acceleration. Ig-

We are at another theater tonight, this time the famous Globe. The ghost of Hamlet's father emerges at midnight, scaring the wits out of Shakespeare himself, who tumbles out of bed and, thanks to gravity's pull, lands plop in Yorick's grave, where Newton wonders about the apple that fell on his wig. Around the bed other players from Shakespeare's plays give in to gravity, dropping down to the floor. While people in elevators going up feel heavy and pushed down, the ones going down feel light and elevated. The red car is accelerating, pushing the passengers to the back, creating additional gravitational pressure and causing a policeman to automatically follow the speeding car. The audience is thrilled and excited about the notable gravity of the drama unfolding.

—T.B.

THE
GLOBE
Elevator
Elevator
TOMAS
BUNK

noring the prior knowledge that cars are horizontal, the two accelerations are indistinguishable.

An interesting experiment to perform involves a helium balloon tied to the seat in your car. As the car accelerates forward, the helium balloon will lean forward. There are two distinct ways of explaining why. The first involves the inertia of the air. An acceleration forward compresses the air in the rear of the car. The increased pressure in the rear of the car forces the balloon forward to the area of lower pressure. Alternatively, we can imagine the acceleration of the car as being equivalent to a gravitational field pointing backward. We can find the effective gravity by finding the vector sum of the two gravitational forces. The helium balloon will point opposite this vector sum in the same way that it points opposite the Earth's downward pull. We can call the vector sum of the gravitational fields the "local field."

Both approaches, pressure differences and local fields, can be used to explain the motion of the helium balloon. What's so nice about the field interpretation is that the angle of the balloon can provide an instantaneous calculation of the car's acceleration. If the angle is 45°, the car must be accelerating at 9.8 m/s^2.

We live on an accelerating planet. Earth is rotating on its axis, and all objects on Earth must have a centripetal force pulling them toward the axis of rotation. Objects at the equator require a large centripetal force in comparison to objects in the United States. Objects at the North Pole require no centripetal force at all. What force is responsible for this centripetal force? It must be gravity since gravity is the only force present. If a component of gravity is required for the centripetal acceleration, then the remaining gravitational force must be considered the object's weight. Assuming a spherical Earth, the weight of an object at the equator (what a bathroom scale would read) would be less than the weight of that same object at the North Pole.

The bathroom scale in the elevator had a different reading because of its acceleration. We can use this weight as a measure of the local field. Similarly, we can consider the local field of Earth at each latitude. The local field is equal to the vector difference of that gravitational field and the centripetal acceleration. These local field effects are quite real. The astronauts sense them, we sense them, and Earth senses them. The local field defines the direction an object falls and the perpendicular to the surfaces of liquids. Over time, the local field has actually changed the shape of Earth!

Our problem begins by asking you to find some local fields on an idealized spherical Earth.

A. Calculate the local field at the equator, the North Pole, and at 40° latitude.

B. Determine the angular deviation of the local field at 40° latitude from the radial line toward the center of Earth.

C. The local field at the equator is along a radial line; the local field at the North Pole is also along a radial line. For all other latitudes, the local field deviates from the radial direction. For which latitude is the deviation of the local field from the radial line the greatest? Calculate this deviation.

Solution

A. At the North Pole, the local field is due only to the gravitational force, since the angular velocity there is zero:

$$\sum F = \frac{Gm_1m_2}{R^2} = mg_N,$$

$$g_N = \frac{GM_E}{R_E^2} = 9.804 \text{ m/s}^2,$$

where

$$G = 6.6726 \cdot 10^{-11} \text{ N} \cdot \text{kg}^2/\text{m}^2,$$
$$M_E = 5.977 \cdot 10^{24} \text{ kg},$$
$$R_E = 6.378 \cdot 10^6 \text{ m}.$$

At the Equator, part of the gravitational force is needed to provide the centripetal force on the rotating Earth. The local field is reduced by this amount:

$$\sum F = \frac{GM_E m}{R_E^2} - m\omega^2 R_E = mg_E,$$

$$g_E = \frac{GM_E}{R_E^2} - \omega^2 R_E$$
$$= (9.804 - 0.034) \text{ m/s}^2$$
$$= 9.770 \text{ m/s}^2.$$

At 40° latitude, a component of the gravitational force is needed to provide the centripetal force on the rotating Earth. The local field is once again reduced by this amount. In this case, we resolve the gravitational force into components parallel and perpendicular to Earth's axis of rotation and reduce the perpendicular component by the centripetal force.

In the perpendicular direction,

$$\sum F_\perp = \frac{GM_E m}{R_E^2}\cos\theta - m\omega^2 R_E$$
$$= mg_\perp,$$

where $R_E \cos\theta$ is the radius of the circle that objects at 40° latitude rotate. Therefore,

$$g_\perp = \left(\frac{GM_E}{R_E^2} - \omega^2 R_E\right)\cos\theta$$
$$= 7.484 \text{ m/s}^2.$$

In the parallel direction,

$$\sum F_\parallel = mg_\parallel,$$

$$g_\parallel = \frac{GM_E}{R_E^2}\sin\theta = 6.302 \text{ m/s}^2.$$

The vector sum of the two components is therefore

$$g_{40} = \sqrt{7.484^2 + 6.302^2} \text{ m/s}^2$$
$$= 9.784 \text{ m/s}^2,$$

$$\theta = \tan^{-1}\frac{6.302}{7.484} = 40.10°.$$

B. The angular deviation between the local field at 40° latitude and the radial line toward the center of Earth is 0.10°.

C. The local field is along the radial line at the Equator (both the gravitational force and centripetal force are along the same line) and along the radial line at the North Pole (no centripetal force). There must be a latitude for which the deviation is greatest. Finding this latitude and its corresponding deviation requires us to use the equations derived in part A:

$$g_{\perp} = \left(\frac{GM_E}{R_E^2} - \omega^2 R_E \right) \cos\theta,$$

$$g_{\parallel} = \frac{GM_E}{R_E^2} \sin\theta,$$

$$\theta' = \tan^{-1} \frac{\frac{GM_E}{R_E^2} \sin\theta}{\left(\frac{GM_E}{R_E^2} - \omega^2 R_E \right) \cos\theta}.$$

To find where the deviation of this angle from θ is a maximum, we can plot the equation $(\theta' - \theta)$ versus θ to find the maximum. Alternatively, we can find the maximum on a spreadsheet or take the derivative and set it equal to zero:

$$\theta' - \theta = \tan^{-1} \frac{\frac{GM_E}{R_E^2} \sin\theta}{\left(\frac{GM_E}{R_E^2} - \omega^2 R_E \right) \cos\theta} - \theta$$
$$= \tan^{-1} K \tan\theta - \theta,$$

where

$$K = \frac{\frac{GM_E}{R_E^2}}{\left(\frac{GM_E}{R_E^2} - \omega^2 R_E \right)},$$

$$\frac{d(\theta' - \theta)}{d\theta} = \frac{K \sec^2\theta}{1 + K^2 \tan^2\theta} - 1 = 0,$$

$$\sin\theta = \sqrt{\frac{1}{K+1}}.$$

On the rotating Earth, K is very close to 1 (that is, 9.804/9.770 = 1.0035), and the maximum deviation occurs at an angle slightly less than 45°. On objects where the rotational speed is much greater, we find that the maximum deviation occurs at even smaller latitudes. ◘

Around and around she goes

"Revolutions are celebrated when they are no longer dangerous."
—Pierre Boulez (b. 1925)

MERRY GO-ROUNDS ARE the most egalitarian ride in that everybody can have a good time. As our personal thrill tolerances increase, we can try such rides as the Loop-the-Loop, the Tilt-A-Whirl, and the Rotor, where the floor is pulled out from under us when the cylinder is spinning fast enough to "pin" us to the wall. The most deceptive ride is the Tea Cups at Disneyland. The Tea Cups subjects us to three simultaneous circular motions. It looks tame enough, but the ride can be a quite dizzying experience depending on how your friends spin the cups and platform.

Circular motions can also be entertaining in the physics class, though we wouldn't suggest there will ever be two-hour lines of people waiting to learn about centripetal forces and angular momentum.

Let's consider some classical examples of circular motion before we move on to some interesting variations. We choose problems that are most easily solved using the conservation laws of linear momentum, angular momentum, and kinetic energy.

Consider a circular disk of radius R_1, mass M_1, and moment of inertia I_1 rotating at a constant angular speed w1 about its axis of symmetry. Its angular momentum $L_1 = I_1\omega_1$, its rotational kinetic energy $KE_1 = (^1/_2)I_1\omega_1^2$, and if it's a uniform disk, its momentum of inertia $I_1 = (^1/_2)M_1R_1^2$, all about the center of mass of the disk.

Now let's assume that we drop a similar disk so that it lands exactly on top of the first disk, face-to-face, as shown in figure 1 on page 162. Then conservation of angular momentum tells us that

$$L_f = L_1 + L_2 = I_1\omega_1 + I_2\omega_2$$
$$= (I_1 + I_2)\omega_f,$$

where the subscript f refers to the final conditions. As a simple example, let $\omega_2 = 0$ and $I_1 = I_2$. Then $\omega = {}^1/_2\omega_1$, as we might expect.

Revolutions are not everyone's cup of tea. They are violent, heads sometimes roll, new social structures replace old ones only to become old themselves. Here on the not-so-merry-go-round we see a couple of revolutions enacted by the underclasses: the Lutheran peasant revolution in Germany, the American and French revolutions, the October revolution, the Chinese, Cuban, and Iranian revolutions. Sitting in the cups are the upper classes—kings, the aristocracy, the wealthy bourgeoisie. In some cups we find unorganized revolutionaries like A. Jarry, Don Quixote, Pére Ubu, and a pirate; in other cups, scientific revolutionaries like Einstein, Galileo, Newton, and the Grim Reaper, who of course was the busiest player in all the revolutions. The whole social-historical circus is put in circular motion by a divine cosmic engine, controlled by an intelligent computer being (one hopes, at least).

—T.B.

TOM BUNK
LEFT - RIGHT
ON
OFF
UP DOWN

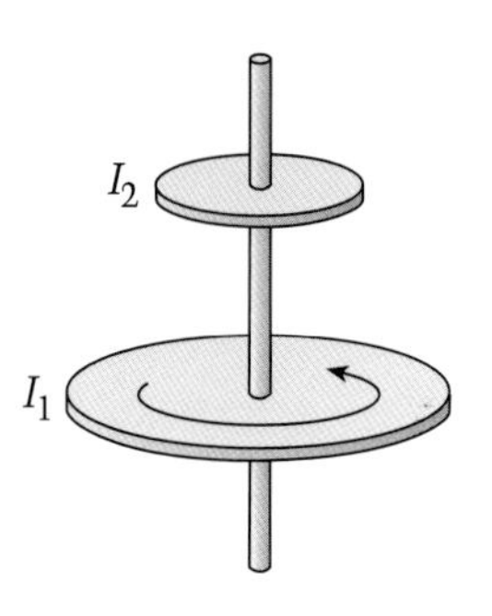

Figure 1

Although angular momentum is conserved, kinetic energy is not conserved. Let's remain with the case $\omega_2 = 0$. Then

$$KE_i = \frac{1}{2} I_1 \omega_1^2$$

and

$$KE_f = \frac{1}{2}(I_1 + I_2)\omega_f^2 = \frac{1}{2}\left(\frac{I_1^2}{I_1 + I_2}\right)\omega_1^2$$
$$= KE_i\left(\frac{I_1}{I_1 + I_2}\right),$$

where the subscript i refers to the initial conditions. Note that KE_f is always less than KE_i and that the collision is inelastic. If $I_1 = I_2$, half of the original kinetic energy is lost.

As a second example, let's consider a small ball with mass m and speed v colliding with the rim of the circular disk shown in figure 2. We assume that the disk is initially not rotating, the disk is free to rotate about a fixed axis through its center, and that the ball sticks to the rim. With what angular speed does the wheel rotate?

We use conservation of angular momentum. You might think that there is no initial angular momentum, but even a ball moving in a straight line has angular momentum about all points not on the line of its motion:

$$L_i = mvR_1.$$

We also know that the final angular momentum is

$$L_f = I_f \omega_f,$$

where I_f is the combined moment of inertia of the disk and the ball. If we assume a uniform disk, we have

$$I_f = I_1 + mR_1^2 = \left(\frac{1}{2}M_1 + m\right)R_1^2.$$

Because $I_f > I_1$,

$$\omega_f < \frac{v}{R_1},$$

and the ball slows down. For the case of equal masses,

$$\omega_f = \frac{2v}{3R_1}.$$

Convince yourself that one-third of the original kinetic energy of the ball is lost in this collision.

The first problem below is based on part of a problem on the preliminary exam that was given nationwide in January 1998 to select the US Physics Team that competed in Iceland the next summer.

A. A disk of radius R spins with angular speed ω_0 about its axis, which is held vertically in frictionless bearings. The disk's moment of inertia about the spin axis is I_0. At a certain instant, a small chip of mass m breaks off the rim of the disk and flies away moving tangent to the disk as shown in figure 3. What is the angular speed of the disk after the chip breaks off?

B. A ball of mass m and speed v strikes the end of a thin rod of mass M and length a as shown in figure 4. Assume that the rod lies on a frictionless table and that the ball stops after the collision. What are the velocity of the center of mass of the rod and its rotational speed about its center of mass? For what range of mass m is such a collision possible? (The moment of inertia for the thin rod is $Ma^2/3$ about one end and $Ma^2/12$ about its center of mass.)

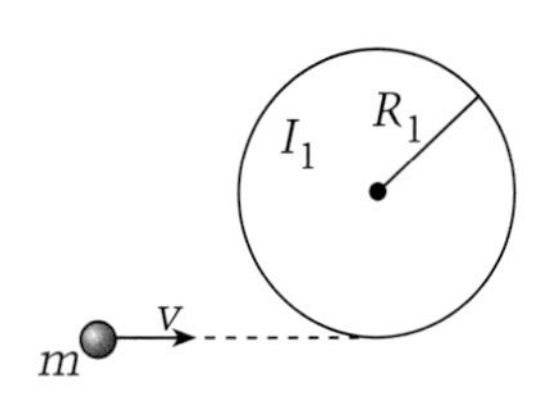

Figure 2

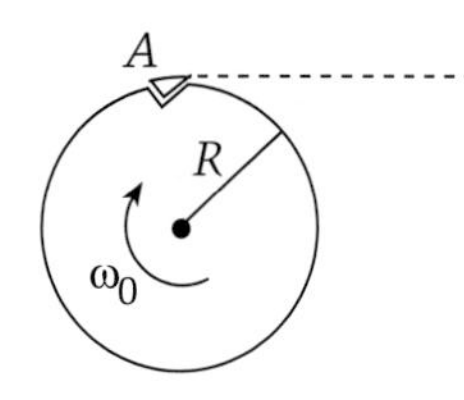

Figure 3

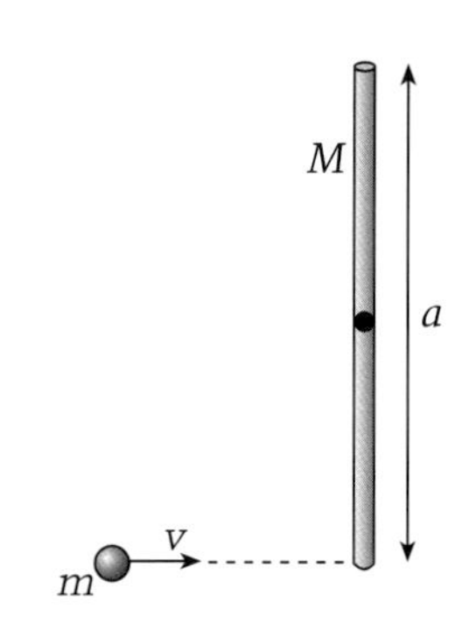

Figure 4

Solution

Three readers sent in different approaches to the solution of the first part of this problem. Art Hovey, a teacher at Amity Regional High School in Woodbridge, Connecticut, stated that when the chip breaks off the rim of a rotating disk, it exerts no impulse on the disk, so the angular momentum (and, thus, the angular speed) of the disk does not change. David Heller, his student, noted that the angular momentum of the particle is initially $mR^2\omega_0$ and finally mvR. But $v = \omega_0 R$, and the angular momentum of the particle doesn't change. By the conservation of angular momentum, the angular momentum of the disk doesn't change. Rob Morasco from Hatfield, Pennsylvania, used a more mathematical approach. Conservation of angular momentum requires

$$I_0\omega_0 = (I_0 - mR^2)\omega_f + mR^2\omega_0.$$

Rearranging terms,

$$(I_0 - mR^2)\omega_0 = (I_0 - mR^2)\omega_f.$$

Therefore, $\omega_f = \omega_0$.

The problem in which the ball of mass m and speed v hits a stick of mass M and length a proved to be a bit more difficult. As with all collision problems, we must conserve momentum. Conservation of linear momentum gives us

$$mv = MV,$$

where V is the speed of the stick's center of mass and the final speed of the ball is zero. Therefore,

$$V = \frac{m}{M}v.$$

Because the ball strikes the stick near one end and perpendicular to the stick, the angular momentum of the ball about the center of mass of the stick is

$$L_i = mv\frac{a}{2}.$$

The angular momentum of the stick about its center of mass is

$$L_f = I\omega = \frac{1}{12}Ma^2\omega,$$

where ω is the stick's angular speed about its center of mass.

Conservation of angular momentum then yields

$$\omega = \frac{6mv}{Ma}.$$

Art Hovey was able to find the condition on the masses for such a collision to be possible. We know that the kinetic energy after the collision cannot exceed the kinetic energy available before the collision. Therefore,

$$\frac{1}{2}mv^2 \geq \frac{1}{2}MV^2 + \frac{1}{2}I\omega^2$$
$$= \frac{1}{2}M\left(\frac{m}{M}v\right)^2 + \frac{1}{2}\left(\frac{1}{12}Ma^2\right)\left(\frac{6mv}{Ma}\right)^2$$
$$= \frac{1}{2}mv^2\left(\frac{4m}{M}\right),$$

and

$$M \geq 4m.$$

Depth of knowledge

"Adaptability is not imitation. It means power of resistance and assimilation."
—Mahatma Gandhi

IT IS A COMMON SIGHT DURing the summer months to see a car crowned with camping gear, canoes, or beach chairs as the family heads off on vacation. Such a sight causes automobile engineers responsible for aerodynamics to convulse.

Millions of dollars of research and experimentation have been invested in sleek car designs to minimize the adverse effects of air resistance. And here is the American household negating all efforts to maximize gas mileage with a bramble bush of outdoor gear strapped to the roof.

A home study of air resistance requires only a stopwatch and some coffee filters. A single coffee filter is dropped from a selected height, and its descent time is recorded. The experiment is repeated with two nested filters, three nested filters, and so on. In this way, we can find the relationship between mass and descent time. Since the filters fall with constant velocity, having reached a terminal velocity quite quickly, this is a measure of the effect of mass on terminal velocity as well. (Proving that terminal velocity is reached quickly is an important digression.)

A second experiment can be conducted to determine the effect of surface area on terminal velocity. Two coffee filters can be taped side by side, and their descent time can be compared with the time for two nested filters. This is followed by three taped filters and three nested filters. Continuing in this manner, the desired relationship is derived.

Theoretically, we can look at the effect of air resistance on an air cart moving horizontally with a piece of cardboard providing the resistive force. The cart would continue to move forward at constant velocity with out this cardboard obstruction. For slow speeds the force of air resistance on the cardboard is proportional to the velocity of the cart:

People can't wait to leave our stressful civilization and go someplace in nature to relax. To get there they not only create unnerving traffic jams, they drive aerodynamically designed vehicles to minimize air resistance and save gas, and at the same time they counteract this same gas-saving scheme by piling up tons of vacation gear on their roofs. Gandhi, on the other hand, is not only using free and clean energy like wind and gravity, he is already enjoying his vacation. Right behind him is the fiddling Einstein enjoying himself by letting gravity do all the work. The idyllic beach scene on the right is the artist's childhood backyard, and all he needs to get back to this paradise is to use some ink and paper.

—T.B.

TEXACO
MOBIL
GAS
SKIPPY
TOM BUNK

$$\sum F = ma,$$
$$-kv = ma,$$
$$-kv = m\frac{dv}{dt},$$
$$\frac{dv}{v} = -\frac{k}{m}dt,$$
$$\ln\frac{v}{v_0} = -\frac{k}{m}t,$$
$$v = v_0 e^{-\frac{k}{m}t}.$$

Therefore, the velocity decreases exponentially from its initial value. The proportionality constant k determines the rate of velocity decrease. It seems from the equation that the cart will never reach zero velocity in a finite time. Of course, when t is large enough, we can have a velocity that is effectively zero.

This air resistance costs us money. If there were no air resistance, the only force slowing our car on the highway would be the friction between the tires and the road. This small force would hardly retard our motion. We would glide along the ribbon of road at 60 miles per hour with little need for additional fuel.

Most of the fuel our cars consume is to counteract the air resistance. Open the car window as the car cruises at a fast clip and feel the wind on your hand. Your car is heading into quite a storm! Recognizing that the air resistance is proportional to the velocity of the car (or the square of the velocity of the car), it is easy to see that a small decrease in the speed of the car reduces the required fuel. Lower speed limits not only save lives but also save fuel.

An automobile design that lowers the air resistance so that cars could get an additional 1 mile per gallon saves an extraordinary sum of money. Let's try a quick "Fermi" calculation. There are approximately 60 million passenger cars in the United States (one for every four people). If each car travels approximately 20,000 miles per year and gets 20 miles per gallon of gas, then the fuel consumption is 1000 gallons per car or 60 billion gallons of gas. At $1 per gallon, this is 60 billion dollars. If these cars could get 21 miles per gallon, we could save 5% of the total, or 3 billion dollars per year. How valuable is an automobile design that can save this much money every year?

Air resistance also affects sports. A baseball hit at 50 m/s (110 mph) at an angle of 45° would be able to travel 255 meters without air resistance. This is equivalent to 840 feet. Baseball would be quite a different sport if it were not for air resistance. Air resistance also affects tennis and football, and it is a crucial factor in table tennis and badminton. The symmetry of a trajectory disappears when air resistance is great. If an object is thrown vertically up with air resistance, will it take more time going up or coming down?

For this problem, we ascend from a simple problem of a stone falling with no air resistance to the more realistic situation where air resistance retards its motion.

A. A stone hits the bottom of a deep well and you hear the sound 3 seconds after the release. How deep is the well? Please assume that there is no air resistance and the speed of sound is a constant value of 340 m/s.

B. In this part, assume that the stone is affected by air resistance and that this resistive force is proportional to the velocity of the stone.

(1) Derive an expression for the velocity of the stone as it falls.

(2) Using 0.01 kg/s for the proportionality constant and 0.05 kg for the mass of the stone, determine the depth of the well if the sound of the stone hitting the water arrives 3 seconds after the stone is released.

Solution

Correct solutions were submitted by Zach Frazier, a senior at Ferris High School in Spokane, Washington; professor F. Y. Wu of Northeastern University, Massachusetts; John Parmon of Narbeth, Pennsylvania; and Scott Wiley, a physics teacher from Weslaco, Texas.

In part A, the total time to hear the sound is the sum of the time for the stone to fall plus the time for the sound to rise.

$$T = t_1 + t_2,$$

where

$$t_1 = \sqrt{\frac{2h}{g}}$$

and

$$t_2 = \frac{h}{v}.$$

Knowing the total time $T = 3$ s, the acceleration due to gravity $g = 9.8$ m/s^2, and the speed of sound $v = 340$ m/s, we can solve for the height of the well h by using the solver or the poly function on a TI-85 calculator or the quadratic equation. The depth of the well that satisfies these equations is 40.65 m.

Part B assumes that there is air resistance where the resistive force is proportional to the velocity of the stone:

$$mg - kv = m\frac{dv}{dt},$$
$$\int\frac{dt}{m} = \int\frac{dv}{mg - kv}.$$

Let $u = mg - kv$. We have

$$\int\frac{-k}{m}dt = \int\frac{du}{u},$$
$$\frac{-k}{m}t = \ln\frac{u}{u_0},$$
$$e^{\frac{-kt}{m}} = \frac{mg - kv}{mg},$$
$$v = \frac{mg}{k}\left(1 - e^{\frac{-kt}{m}}\right).$$

To determine the depth of the well, with air resistance, we must derive an equation for the distance traveled by the stone:

$$v = \frac{dx}{dt} = \frac{mg}{k}\left(1 - e^{\frac{-kt}{m}}\right),$$

$$\int dx = \int \frac{mg}{k}\left(1 - e^{\frac{-kt}{m}}\right)dt,$$

$$x = \frac{mgt}{k} - \frac{m^2 g}{k^2} e^{\frac{-kt}{m}} - C.$$

Since $x = 0$ when $t = 0$,

$$C = \frac{m^2 g}{k^2}$$

and

$$x(t) = \frac{mgt}{k} - \frac{m^2 g}{k^2} e^{\frac{-kt}{m}} - \frac{m^2 g}{k^2}.$$

Using the given information that $k = 0.01$ kg/s, $m = 0.05$ kg, and that the total time T to hear the splash is still 3.0 s, we can now determine the depth of the well.

The total time is the time for the stone to fall t_f and the time for the sound to rise:

$$T = t_f + \frac{x(t_f)}{v}.$$

Readers solved the equation graphically, numerically, with Mathematica, and with a spreadsheet. The depth of the well is now 34.26 m.

Doppler beats

"The most persistent sound which reverberates through man's history is the beating of war drums."
—Arthur Koestler

As a police cruiser drives by with its siren sounding, you notice that the pitch of the siren decreases. The same thing happens at the Indianapolis 500 as a race car passes you. The pitch of the engine is steady as the car approaches, decreases as the car passes by, and is steady (but lower) as the car recedes into the distance.

These are two examples of the Doppler shift. The motion of the source shifts the frequency of the sound you hear. If you are at rest relative to the air, the frequency f you hear is given by

$$f = f_0\left(\frac{v}{v \mp v_s}\right),$$

where f_0 is the frequency heard by the observer at rest, v_s is the speed of the source, and v is the speed of sound. The minus sign is used when the source moves toward the observer and the plus sign is used when the source moves away from the observer.

Doppler worked out this mathematical relationship in 1842. He pointed out that the motion of the source toward the observer causes the sound waves to reach the ears at shorter time intervals—therefore, the higher frequency. The reverse is true when the source moves away from the observer.

Doppler's formula was put to an experimental test a few years later. For two days trumpet players rode on a flat car that was pulled at different speeds. Musicians who had perfect pitch stood on the ground and recorded the notes that they heard as the train approached and receded. Their observations were in agreement with Doppler's formula.

The motion of the observer also changes the frequency. When you ride in a train, the bell at the crossing has a higher (but steady) pitch as you approach the crossing and a lower pitch as you leave the crossing behind. This effect is described by

$$f = f_0\left(\frac{v \pm v_0}{v}\right),$$

On the night side of the tracks a thief burglarizes a bank and steals the face of the clock (since time is money). He hides and listens to the approaching police car. At first he worries about the siren's steady pitch, but he's relieved when the pitch decreases. He understands the Doppler effect and knows he is out of danger. He walks to the other, sunny side of the tracks, passing all the people who also appear on the other side (*doppel* means "double" in German). Passengers on the train experience another kind of Doppler shift: while they are moving, the sound source is local though the shift is the same. Another experiment takes place down by the bridge: a moving sound source slowly passes a local idler, who probably won't notice the Doppler shift, but still enjoys the flowing tunes.

—T.B.

SAVING BANK
NEW
Doppler
RELIABLE
NOW WITH IMPROVED SOUND WAVES
POLICE
POLICE
POLICE
DOPPLER EXPRESS
TOMAS BUNK

where v_o is the speed of the observer. The plus sign is used when the observer moves toward the source and the minus sign when the observer moves away from the source.

These two effects can be combined into a single relationship

$$f = f_o\left(\frac{v \pm v_o}{v \mp v_s}\right),$$

where the upper signs refer to the motion of one toward the other and the lower signs refer to motion of one away from the other.

Another interesting sound effect occurs when two sirens produce sound waves with approximately the same pitch. The two sound waves produce a sound with a pitch halfway between the two pitches, but with an intensity that varies periodically from no sound to a sound with four times the loudness of either source. the period of this *beat* frequency is just the difference of the two frequencies.

Piano tuners use beats to tune the wires corresponding to the same note on a piano. After one string is tuned to the correct frequency, it is struck at the same time as another wire. If the two wires have the same frequency, there is o variation in loudness, that is, the beat frequency is zero. However, if the second wire has a higher or lower pitch, the loudness of the sound will vary with a frequency equal to the difference of the two frequencies produced by the wires. The piano tuner then adjusts the tension in the second wire until the beating disappears.

These two sound effects were combined in an interesting way on the second exam used to select the members of the US Physics Olympiad team that competed in the International Physics Olympiad in Reykjavik, Iceland, in July 1998.

Two sirens located on the x-axis are separated by a distance D. As heard by an observer at rest relative to the sirens, the left-hand siren has a frequency f_L and the right-hand siren has a frequency f_R. Assume that you are moving with a constant velocity v_o along the x-axis and record the following observations:

1. When you are on the right-hand side of both sirens, you hear a beat frequency of 1.01 Hz.
2. When you are on the left-hand side of both sirens, you hear a beat frequency of 0.99 Hz.
3. When you are between the two sirens, the beat frequency is zero.

A. In which direction are you moving along the x-axis?

B. What is your speed as a fraction of the speed of sound?

C. Which frequency is greater?

D. What are the numerical values of the two frequencies?

Solution

This problem was designed by Leaf Turner, one of the coaches of the US Physics Team. It was successfully solved by Zach Frazier, who graduated in 1998 from Ferris High School in Spokane, Washington, and by Stephen Hanzely from Youngstown State University in Youngstown, Ohio.

A. In the absence of any motion, the beat frequency Δf is a frequency source, just like any other type of source. Therefore, we see that the frequency is red-shifted on the left and blue-shifted on the right. This means that the observer is moving from right to left.

B. On the left-hand side, the observed frequency $\Delta f_L'$ is given by

$$\Delta f_L' = \Delta f\left(1 - \frac{v_o}{v_s}\right) = 0.99 \text{ Hz},$$

where v_s is the speed of sound. Likewise, on the right-hand side, the observed frequency $\Delta f_R'$ is given by

$$\Delta f_R' = \Delta f\left(1 + \frac{v_o}{v_s}\right) = 1.01 \text{ Hz}.$$

Dividing these two equations yields

$$\frac{\Delta f_R'}{\Delta f_L'} = \frac{1+\frac{v_o}{v_s}}{1-\frac{v_o}{v_s}} = \frac{1.01}{99},$$

from which we get $v_o = 0.01v_s$.

C. When the observer is between the two sources, there is no beat frequency, and the observer measures the same frequency from both sources. However, the source on the left is blue-shifted, and the source on the right is red-shifted. Therefore, $f_R > f_L$.

D. Numerically, we have

$$f_L' = f_L\left(1 + \frac{v_o}{v_s}\right)$$

and

$$f_R' = f_R\left(1 - \frac{v_o}{v_s}\right).$$

Setting the two shifted frequencies equal to each other, we get

$$\frac{f_R}{f_L} = \frac{1+\frac{v_o}{v_s}}{1-\frac{v_o}{v_s}} = \frac{1.01}{99}.$$

From either of the first two equations in part B and using $v_o = 0.01v_s$, we obtain

$$f_R - f_L = 1 \text{ Hz}.$$

Solving these two simultaneous equations provides us with the numerical values f_R = 50.5 Hz and f_L = 49.5 Hz.

Up, up and away

"Hands, do what you're bid:
Bring the balloon of the mind
That bellies and drags in the wind
Into its narrow shed."
—William Butler Yeats

HE'S FULL OF HOT AIR! WE all know what the expression means. Empty talk, unsubstantiated statements, pretentious verbiage, and boastful babble all come to mind when we hear the expression "full of hot air." Where did such a statement originate? O. Henry once said, "A straw vote only shows which way the hot air blows." What is there about hot air that would equate it to talking nonsense? Perhaps the hot-air diatribe is thought of as having no substance, ready to just float away.

As students of physics, we take a more substantial look at hot air. We know that hot air rises and is one means by which we can have a balloon soar above us. This month, we will ignore bees, birds, and helium-filled birthday balloons and let our minds soar with the hot air that levitates tourists on a Sunday afternoon or adventurers embarking on a 'round-the-globe expedition.

The hot-air balloon begins to rise because it is buoyant in the cooler surrounding air. It rises until the buoyant force is equal to the weight of the balloon and the air within it. To understand the rise and suspension of the balloon, we must then be reminded of the grand law of buoyancy, Archimedes' principle, and the determination of the density of the cooler air at different elevations.

Archimedes, prior to running through the streets shouting "Eureka!" realized that an object is buoyed up by a force equal to the weight of the displaced fluid. An elegant proof of this would assume that a block of water is floating amongst the rest of the water. The buoyant force, due to the pressure difference between the top and bottom of the slab of water, must be equal to the weight of the water for the static equilibrium that we observe. The pressure difference will be identical if another object replaces this slab of water. If, however, this object weighs more than the

The universal thinker in the front is thinking heavy thoughts while his brain is rising high up in the air like a hot-air balloon, carrying a bunch of other heavy thinkers, whose heads are steaming from being full of hot air. Even Archimedes, who comes running onto the stage after receiving some brilliant ideas in his bath tub and shouting with excitement "Eureka!" seems to be full of hot air. We see not only factories smoking, but also churches, minarets, and temples. The whole crooked stage doesn't seem very trustworthy, and indeed the whole human planet is floating in the air like a hot-air balloon. There seems to be nothing to hold on to, everything is hot air, and who knows, one day maybe all the great and brilliant thoughts and ideas we humans are so proud of will end up going up in smoke.

—T.B.

EVREKA!
HOT AIR
Δh = ∞
TOMAS BUNK

water it displaces, it will sink. If it weighs less than the water it displaces, it will rise.

Water is barely compressible, and the pressure differences will remain constant regardless of where the block is placed within the liquid. The atmosphere is compressible, and the pressure and density of the air varies with elevation. Assuming that the pressure and the density are proportional to one another (as they would be for a constant air temperature), we can derive the equation relating the pressure to the elevation.

A fluid element is assumed to be at equilibrium within a larger fluid. For equilibrium, the pressure pushing up from the bottom must equal the pressure pushing down from the top surface plus the weight:

$$PA = (P + dP)A + dW$$
$$= (P + dP)A + \rho gA\,dy,$$
$$\frac{dP}{dy} = -\rho g.$$

The pressure decreases with rising elevation.

Since we are assuming that the density is proportional to the pressure,

$$\frac{\rho}{\rho_0} = \frac{P}{P_0},$$
$$\frac{dP}{dy} = -g\rho_0\frac{P}{P_0},$$
$$\frac{dP}{P} = -\frac{g\rho_0}{P_0}dy,$$
$$\ln\frac{P}{P_0} = -\frac{g\rho_0}{P_0}y,$$
$$P = P_0 e^{-\frac{g\rho_0}{P_0}y}.$$

With Archimedes's principle and the derived dependence of pressure on elevation, we are now ready to embark on our journey through our problem. It is adapted from the International Physics Olympiad problem given in Germany in 1982.

A hot-air balloon, when inflated, has a constant volume $V_1 = 1.10\ m^3$. The mass of the balloon material is $m_b = 0.187$ kg, and its volume is negligible. The initial temperature of the air is $T_1 = 20.0°C$, and the atmospheric pressure outside the balloon is $P_0 = 1.013 \cdot 10^5\ N/m^2$. Under these conditions, the density of the air is $\rho_1 = 1.20\ kg/m^3$.

A. To what temperature must the air in the balloon be heated for the balloon to begin to float?

B. The balloon is tethered to the ground, and the air in the balloon is heated to a steady state temperature of 110°C. What is the net force on the balloon when it is released?

C. The balloon is tethered to the ground, and the air in the balloon is heated to a steady state temperature of 110°C and released. The balloon rises isothermally in the atmosphere, which is assumed to have a constant temperature of 20°C. Determine the height gained by the balloon under the conditions described.

D. The balloon hovers at the height calculated in part C and then is pulled from its equilibrium position by $\Delta h = 10$ m and released. Describe the subsequent motion of the balloon.

Solution

One solution to this problem soared to the top of the pile. Congratulations to Dr. H. Leverett (a medical doctor from Texas) who loves physics and *Quantum*!

Archimedes' principle informs us that the weight of the displaced air must be equal to the weight of the balloon for floating to occur:

$$m_1 g = m_a g + m_b g,$$

where m_1 is the mass of the displaced air, m_a is the mass of the air in the balloon, and m_b is the mass of the balloon material. Converting from masses to densities, we obtain

$$m_1 = \rho_1 V_1$$
$$m_a = \rho_a V_1$$
$$\rho_1 V_1 g = \rho_a V_1 g + m_b g$$
$$\rho_a = \rho_1 - \frac{m_b}{V_1}$$
$$\rho_a = 1.20\ kg/m^3 - \frac{0.187\ kg}{1.10\ m^3}$$
$$= 1.03\ kg/m^3.$$

This is the required density of air in the balloon for the balloon to float. Using the ideal gas law, we can find the corresponding temperature:

$$\rho_a T_a = \rho_1 T_1$$
$$T_a = \frac{\rho_1 T_1}{\rho_a} = \frac{(1.20\ kg/m^3)(293\ K)}{1.03\ kg/m^3}$$
$$= 341\ K.$$

In part B, the air is heated to 110°C (383 K), and the balloon rises isothermally into the atmosphere, which has a constant temperature of 20°C (293 K). The net force acting on the balloon F_U is equal to the difference between the buoyant force F_B and the weight of the balloon F_W:

$$F_U = F_b - F_W$$
$$F_b = \rho_1 V_1 g$$
$$F_W = m_3 g + m_b g = \rho_3 V_1 g + m_b g$$
$$F_U = [(\rho_1 - \rho_3)V_1 - m_b]g,$$

where m_3 and ρ_3 are the new mass and density at this temperature.

The density of the air within the heated balloon can be found, once again, from the ideal gas law:

$$\rho_3 T_3 = \rho_1 T_1$$
$$\rho_3 = \frac{\rho_1 T_1}{T_3} = \frac{(1.20\ kg/m^3)(293\ K)}{383\ K}$$
$$= 0.918\ kg/m^3$$
$$F_U = \left[\left(1.20\ kg/m^3 - 0.918\ kg/m^3\right) \times \left(1.10\ m^3\right) - 0.187\ kg\right] 9.80\ m/s^2$$
$$F_U = 1.21\ N.$$

Part C requires us to find the height to which the balloon rises. The balloon continues to rise until its buoyancy is equal to its weight:

$$F_B = \rho_y V_y g,$$

where ρ_y is the density of the displaced air at the balloon's height y. We have

$$\rho_y V_y g = \rho_3 V_3 g + m_B g$$

$$\rho_y = \rho_3 + \frac{m_B}{V_y}$$

$$\rho_y = 0.918 \text{ kg/m}^3 + \frac{0.187 \text{ kg}}{1.10 \text{ m}^3}$$

$$= 1.088 \text{ kg/m}^3.$$

Given the relationship between the air density and the height, we can now find how high the balloon will go:

$$\rho_y = \rho_1 e^{-\frac{g\rho_1}{P_0} y}$$

$$y = -\frac{P_0}{\rho_1 g} \ln \frac{\rho_y}{\rho_1}$$

$$y = \frac{-1.013 \cdot 10^5 \text{ N/m}^3}{\left(1.20 \text{ kg/m}^3\right)\left(9.80 \text{ m/s}^2\right)}$$

$$\times \ln \frac{1.088 \text{ kg/m}^3}{1.20 \text{ kg/m}^3} = 844 \text{ m}.$$

In part D, we are asked to describe the subsequent motion if the balloon were pulled from its equilibrium position by 10 m and then released. We can see that if the balloon is below the position calculated in part C, then the balloon will rise. If it goes above this position, then it will sink. The balloon will therefore undergo harmonic motion. If air resistance is taken into account, the motion will be damped harmonic motion. ◘

Warp speed

"Because we think, we think the universe is about us.
But does it think, the universe?
Then what about?
About us?"

—May Swanson

"TAKE US TO MAXIMUM warp," Captain Jean-Luc Picard orders, and the starship *Enterprise* begins to travel faster than the speed of light to avoid trouble. Warp 9.6 is the highest normal rated speed for the *Enterprise* and corresponds to a speed 1909 times the speed of light. After reaching safe haven, Captain Picard uses a subspace signal to set up a videoconference with Earth, even though the *Enterprise* is thousands of light-years from Earth.

Although such faster-than-light travel is commonplace in science fiction such as *Star Trek*, ordinary matter in our ordinary world must obey the laws of physics. The speed of light is the speed limit in the Universe. Only massless particles such as photons can travel at the speed of light; massive particles—such as those making up the starship *Enterprise*—must sluggishly travel at slower speeds.

This makes any observation of something appearing to travel faster than the speed of light rather astonishing. Such observations have been made in astronomy and will be the focus of our problem.

But first we make a digression to talk about high-speed photography. Let's assume that we have a camera with a very fast shutter speed, say, a very small fraction of a nanosecond. (Such cameras are only available in physics stores along with massless pulleys and frictionless surfaces.) We want to take a photograph of a thin rod as it passes by at a relativistic speed.

Let's assume that the camera is located at the origin and is pointing along the $+y$-direction. The thin rod is 3 m long with its ends at $y = 3$ m and $y = 6$ m as shown in figure 1.

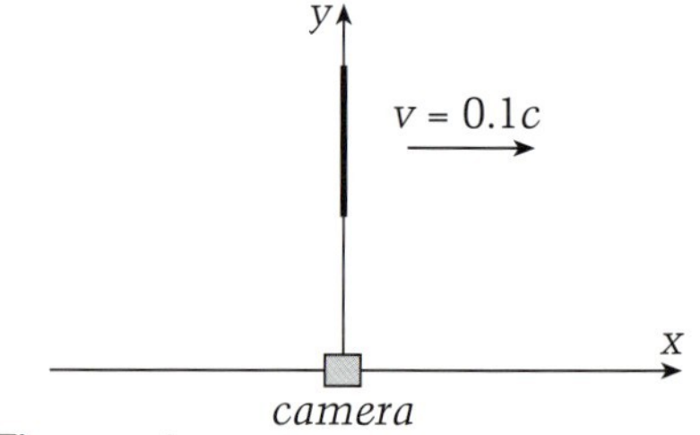

Figure 1

This illustration is different from the other ones. It is a collage, and it was done shortly after my mother passed away. On the bottom there are different images and symbols that came to mind at that time and have to do with death, the passing of time, the journey of the soul, and rebirth. At the top we see a depiction of the physics problem, showing a spaceship camera traveling at high speed in the universe and taking photographs of passing microquasars with angular speeds, one of which is faster than the speed of light. Sometimes life, too, seems to pass very quickly in front of us and we try to slow it down and save some moments by making photographs or an image like this collage.

—T.B.

1st PLAYER
2nd PLAYER
3rd PLAYER
4th PLAYER
TOM BUNK OCT '98

The rod is traveling in the $+x$-direction at a speed of $0.1c$. When the rod passes the x-axis, we consider light reflected from each end. The light reflected from the close end takes 10 ns to reach the camera and expose the film. However, the light reflected from the far end takes 20 ns to reach the camera and bumps into the closed shutter. The light from the far end that enters the camera must have been reflected earlier to allow for the extra distance it has to travel. That is, the light from the far end that arrives at the camera simultaneously with the light reflected from the near end must have been emitted a little more than10 ns earlier.

This means that the camera records the near end as being at $x = 0$, but the far end as being located at $x = -0.3$ m. The camera does not "see" the rod lying along the x-axis, but rotated through an angle of 5.7°. Who says cameras don't lie?

Our contest problem is based on one of the three theoretical problems used in the International Physics Olympiad held in Reykjavik, Iceland, on 2–10 July 1998. One of us (LDK) had the privilege of working with the Icelandic exam committee and found the problems to be very interesting and challenging.

A. In 1994, GRS1915+105 was observed to emit ejecta in opposite directions. As reported by I. F. Mirabel and L. F. Rodriguez in *Nature* (vol. 371, p. 46), the ejecta were probably produced by a neutron star or a black hole similar to the process occurring in quasars, but on a smaller scale. They call the object a microquasar.

Fits to the observations over a period of 34 days showed that the ejecta left the microquasar with angular speeds of ω_1 = 17.6 milliarcseconds/day and ω_2 = 9.0 milliarcseconds/day. If the microquasar is located at a distance $R = 3.86 \cdot 10^{20}$ m from Earth, what are the components of the velocities of the two ejecta perpendicular to the line of sight (the transverse velocities)?

B. You are probably surprised to discover that one of these velocities has a component larger than

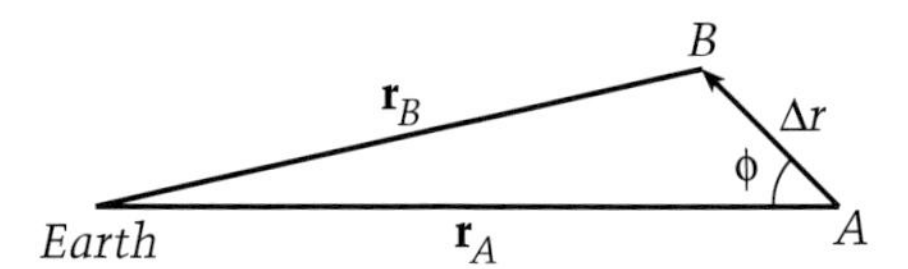

Figure 2

the speed of light. To see how this arises, let's do the following calculation. Assume that an object is traveling at a speed $v = \beta c$ at an angle $\phi < 90°$ relative to the line of sight as shown in figure 2. Denote its original position by $\mathbf{r_A}$ and its final position by $\mathbf{r_B}$. In a time interval Δt the object travels a distance Δr. What is the time interval Δt_0 between the arrival of the signal from position A to the arrival of the signal from position B as observed on Earth?

C. What is the transverse velocity observed for this motion in terms of β, R, and ϕ?

D. What is the minimum value of β for which we can observe a transverse velocity greater than the speed of light for some angle ϕ? What angle corresponds to this minimum β?

E. Draw a graph of β versus ϕ showing the region where we can observe apparent superluminal transverse velocities.

Solution

This problem proved to be quite a hit with *Quantum* readers. Art Hovey and nine students in his honors physics class at the Amity Regional High School in Woodridge, Connecticut, sent in correct solutions. The students are Steve Boyle, Stephanie Brelsford, Sid Govindan, Alex Kaloyanides, Maya Roberts, Ted Rounsaville, Samira Saleh, Mirosla Volynskiy, and Aaron Webber. Scott Wiley, a teacher at Weslaw High School in Texas, also submitted a correct solution.

A. The experimental data obtained from the photographs of the microquasar indicate that the ejecta had angular velocities of 17.6 and 9.0 milliarcseconds (mas) per day. We can obtain the transverse velocities knowing that $v = \omega r$ and converting units:

$$\begin{aligned} v &= \omega r \\ &= \left(17.6\frac{\text{mas}}{\text{day}}\right)\left(3.86\times 10^{20}\ \text{m}\right) \\ &\times\left[\frac{1\ \text{as}}{10^3\ \text{mas}}\right]\left[\frac{1^\circ}{3600\ \text{as}}\right] \\ &\times\left[\frac{2\pi\ \text{rad}}{360^\circ}\right]\left[\frac{1\ \text{day}}{24\ \text{h}}\right]\left[\frac{1\ \text{h}}{3600\ \text{s}}\right] \\ &= 3.81\times 10^8\ \text{m/s}. \end{aligned}$$

Notice that this speed is greater than the speed of light by 27 percent. The corresponding speed for the other ejectum is $1.95 \cdot 10^8$ m/s, well below the speed of light.

B. Figure 2 shows the geometry of the situation that can produce such superluminary speeds. An object is moving from point A to point B with a speed $v = \Delta r/\Delta t$. The light that leaves the object when it is located at point A takes a time

$$t_A = \frac{r_A}{c}$$

to reach Earth. The signal from point B originates a time Δt later and therefore arrives at Earth at

$$t_B = \frac{r_B}{c} + \Delta t.$$

Thus, the difference in the arrival times of the two signals on Earth is

$$\Delta t_0 = t_B - t_A = \frac{r_B - r_A}{c} + \Delta t.$$

Because the distance to the object is extremely large compared to Δr, the directions to points A and B are nearly parallel and

$$r_A - r_B \cong v\Delta t\cos\phi.$$

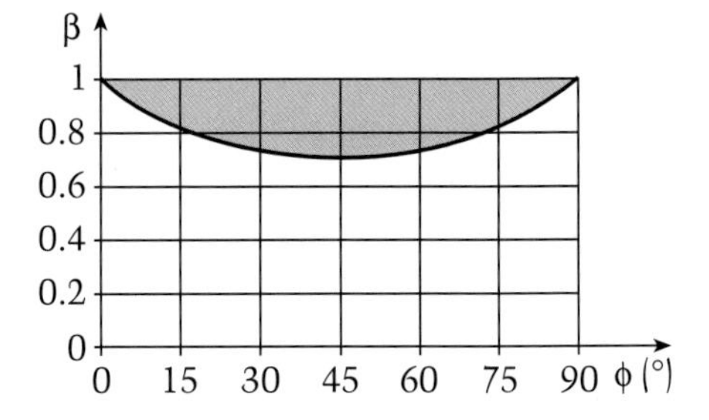

Figure 3

Therefore,

$$\Delta t_0 = \Delta t\left(1 - \frac{v}{c}\cos\phi\right)$$
$$= \Delta t(1 - \beta\cos\phi).$$

C. The observed transverse velocity for this motion is given by

$$v_\perp = \frac{\Delta r \sin\phi}{\Delta t_0} = \frac{\Delta r \sin\phi}{\Delta t(1 - \beta\cos\phi)}$$
$$= v\frac{\sin\phi}{1 - \beta\cos\phi}$$

or

$$\beta_\perp = \frac{\beta\sin\phi}{1 - \beta\cos\phi}.$$

D. We can find the minimum value of β for which we can observe $\beta_\perp \geq 1$ by looking at

$$\beta\sin\phi = 1 - \beta\cos\phi.$$

This tells us that

$$\beta = \frac{1}{\sin\phi + \cos\phi}.$$

Because $\sin\phi + \cos\phi$ has a maximum at $\phi = \pi/4$, $\beta_{min} = 1/\sqrt{2}$ for the angle $\phi = 45°$.

E. See the shaded region in figure 3. ◘

Sportin' life

"Lots of play
in the way things work,
in the way things are.
History is made of mistakes.
Yet—on the surface—
the world looks OK,
lots of play."

—Gary Snyder

MICHAEL JORDAN MAKES IT all look so easy. The ball tossed at an angle θ with an initial velocity v from a height h gracefully glides in its arc and swishes through the net. All that polish from years of practice and no formal physics.

What are we to do? Can our mathematical approach help us to replicate Jordan's skills? Definitely not. But our analysis can help us appreciate the skill of someone who can sink the jump shot. In fact, our analysis can then be used to mimic the work of the broad jumper, the javelin and discus thrower, the volleyball spiker, the football punter, the soccer midfielder, and the baseball batter.

Trajectories without air resistance follow the simple equations of kinematics for horizontal motion with no acceleration and vertical motion with the acceleration due to gravity:

$$x = v_0 t \cos\theta$$

$$y = -\frac{1}{2}gt^2 + v_0 t \sin\theta.$$

From these equations we can derive an equation for the range of a trajectory thrown from the ground and returning to the ground. For this special case, the vertical displacement y is zero. By eliminating the time t, we obtain

$$x = \frac{2v_0^2}{g}\cos\theta\sin\theta = \frac{v_0^2}{g}\sin 2\theta.$$

The maximum range of a trajectory is now proven to be 45°, since $\sin 2\theta$ is equal to 1 for this value.

If we do not restrict the vertical displacement to zero, we find that

$$y = \frac{-gx^2}{2v_0^2\cos^2\theta} + \frac{x\sin\theta}{\cos\theta}.$$

This now demonstrates that for any given initial velocity and any angle, the path of a thrown object must be a parabola!

Circus Maximus, lots of play, history is made of mistakes, but looks OK. Our story starts some years back, when some fish decided to crawl out of the water and live on dry land, initiating a chain of evolutionary developments. After we quickly arrive at *Homo erectus*, we already see the caveman inventing the wheel to get ahead faster. He is followed by the surfing Babylonian, the biking Egyptian, the rolling-in-a-barrel Greek, the hopping Roman, and so forth. The human comedy moves on and on and around, seemingly without a goal: at the finish line we slip back into the universe, where we originally came from. In the middle of the circus the soccer player is running to kick the ball into the goal. In the halls we see Dante's Belacqua sleeping—he doesn't seem to be very interested in taking part in this whole circus.

—T.B.

VIVA LA REVOLUTION
START
ONE WAY
RIP
TITANIC
SPQR
UP
BELLÀQUA
TOMAS BUNK

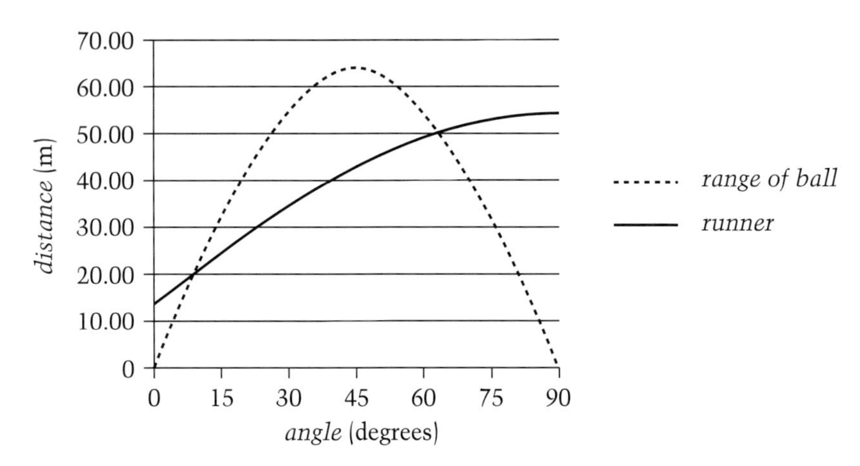

Figure 1

Using the identity relations

$$\tan\theta = \frac{\sin\theta}{\cos\theta}$$

and

$$\tan^2\theta + 1 = \sec^2\theta,$$

we can derive a new equation that will help us find the angle a ball must be thrown to reach a specific point in space:

$$y = \frac{-gx^2}{2v_0^2}\tan^2\theta + x\tan\theta - \frac{gx^2}{2v_0^2}.$$

Since the equation is quadratic, we see that we can reach any position with two different angles.

The trajectory equations can provide us with some insights to many of our sports dilemmas. The field-goal kicker can certainly use the equations to determine the range of angles that will provide his team with three points. The punter has a different job requirement. He wants to kick the ball as far as possible. But if he kicks it so far that his players can't get downfield, then the return will negate his good punt. He must maximize that distance and maximize the hang time. Maximizing the distance requires a kick at 45°, yet maximizing the hang time requires a vertical kick (with no down field component). What is a punter to do?

These decisions must be very difficult, and experience must guide the punter. Physics can provide some help. If we know the speed of the defenders as they run down the field and we know the initial velocity of the punt, we can suggest adjusting the angle of the kick so that the ball arrives when the defenders do. We will analyze this while ignoring air resistance.

The range of the punt traveling at an initial velocity v_0 and angle θ and the corresponding time of flight are given by the following equations:

$$R = \frac{2v_0^2}{g}\sin 2\theta,$$

$$t = \frac{2v_0\sin\theta}{g}.$$

If the offensive team can run downfield at a speed v_r, they can travel a distance x downfield in the time t, where $x = v_r t$. The punter kicks the ball from a position 15 yards behind the line of scrimmage. If we minimize the difference between the range of the ball R and the distance the runners travel, taking into account the 15-yard lead the runners get, we will determine the optimum angle for the punt.

The corresponding equations can most readily be solved using a solver on a calculator, a spreadsheet, or graphing the equations for R and $(x + 13.7 \text{ m})$. Figure 1 shows typical values of 25 m/s for the ball and 8 m/s for the runners. The runners will arrive as the ball arrives when the ball is kicked at an angle of 65° and travels a distance somewhat less than its maximum range.

There are some interesting trajectory problems where the physics can also provide assistance. These will make up the problems offered below.

A. (1) A free kick is being set up in soccer. The defenders form a wall with their bodies between the kicker and the goal. The ball must clear the players. The defenders are 1.8 m tall, and they set up the wall 15 m from the kicker, who kicks the ball at 35 m/s. At what locations can the ball *not* land?

(2) How does this shadow region change as the wall moves in relation to the kicker?

B. (1) A basketball player shoots a jump shot with an initial velocity v_0 at an angle θ_0. For a given basket that is h meters above the release point of the ball and L meters horizontally from the basket, determine the relationship between v_0 and θ_0.

(2) Since the ball must enter the basket during its descent, describe this constraint on the initial angle mathematically.

(3) At what angle is a minimum speed required to sink the shot?

Solution

The strategy for the solution to the first problem is to find the two angles for which the ball can be kicked at a certain velocity that will take it to the point (1.8 m, 15 m). Knowing these two angles, we then find the corresponding x values for the ball landing on the ground ($y = 0$):

$$y = \frac{-gx^2}{2v_0^2}\tan^2\theta + x\tan\theta - \frac{gx^2}{2v_0^2}.$$

Substituting the values for x, y, and $v_0 = 35$ m/s and solving the quadratic equation yields values for θ of 86.5° and 10.4°. We use the range equation to find the points where these two trajectories hit the ground.

$$x = \frac{2v_0^2}{g}\cos\theta\sin\theta = \frac{v_0^2}{g}\sin 2\theta.$$

The corresponding distances are 15.11 m and 44.04 m. The shadow region lies behind the wall and extends for 0.11 m.

Part (2) asks what happens to this shadow region as the wall moves relative to the kicker. Calculating the shadow region on a spreadsheet, we determine that the shadow region increases as the distance from the kicker increases. This makes sense because the angle required to just clear the wall will decrease and therefore the ball will not be able to fall as sharply behind the wall.

Part B shifted to trajectories in basketball. It required readers to find the relationship between the initial velocity v_0 and the initial angle θ_0 given a fixed shot position where the rim is h meters above the ball and L meters away horizontally.

Beginning with our trajectory equation

$$y = \frac{-gx^2}{2v_0^2 \cos^2 \theta} + \frac{x \sin \theta}{\cos \theta},$$

we solve for v_0:

$$v_0^2 = \frac{gL}{2\cos^2 \theta} \frac{1}{\tan \theta - \frac{h}{L}}.$$

Part (2) of this problem asks for the constraints on the initial angle if the ball is to enter the basket during its descent. The angle of entry can be defined as the angle between the horizontal and the angle of the tangent to the ball's trajectory. We will assume that we can ignore the size of the ball—a terrible assumption, but one that makes the analysis simpler.

The initial constraints should depend only on h and L, where

$$h = (v_0 \sin \theta_0)t - \frac{1}{2}gt^2,$$

$$L = (v_0 \cos \theta_0)t,$$

$$\tan \theta = \frac{v_y}{v_x} = \frac{v_0 \sin \theta_0 - gt}{v_0 \cos \theta_0}$$

$$= \frac{(v_0 \sin \theta_0)t - gt^2}{(v_0 \cos \theta_0)t}$$

$$= \frac{2(v_0 \sin_0)t - gt^2 - (v_0 \sin \theta_0)t}{(v_0 \cos \theta_0)t}$$

$$= \frac{2h}{L} - \tan \theta_0.$$

Because the ball must be falling in order to make a basket, θ and $\tan \theta$ must both be negative. This requires that the right side of the equation also be negative. Therefore, the constraint on the initial angle is

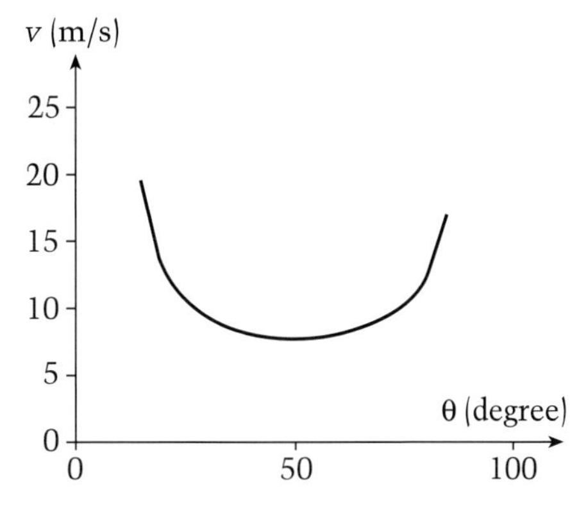

Figure 2

$$\tan \theta_0 > \frac{2h}{L}.$$

Part (3) asks for the angle where a minimum speed is required to sink the shot. To solve this, we can calculate the speed required for many different angles using the equation from part (1) and a spreadsheet, assuming values of h = 1 m and L = 5 m. The graph in figure 2 of v vs. θ indicates a minimum at approximately 50°. Alternatively, we can take the derivative of the equation from part (1) relating v to θ and set this derivative equal to zero to obtain

$$v_0 = \sqrt{\frac{gL}{2\cos^2 \theta} \frac{1}{\tan \theta - \frac{h}{L}}},$$

$$\frac{dv}{d\theta} = 0,$$

$$\tan \theta = \frac{h}{L} \pm \sqrt{\frac{h^2}{L^2} + 1}.$$

Using the sample values of h = 1 m and L = 5 m, we arrive at an optimal shooting angle of 50.7°.

Peter Brancazio, Physics Professor Emeritus of Brooklyn College, took this analysis further and then compared theory and practice on the basketball court. We highly recommend Brancazio's 1981 article "Physics of Basketball" in the *American Journal of Physics* (49), 356–365.

Elevator physics

"One man's ceiling is another man's floor."
—Paul Simon

HAVE YOU TAKEN YOUR bathroom scales to an elevator for a ride? This is guaranteed to start an interesting conversation, and you will be advancing the cause of physics at the same time.

Let's explore physics in a uniformly accelerated reference frame by considering a ball of mass m dropped from a height h while the elevator has an acceleration a in the upward direction. Let's choose the upward direction as positive, start our stopwatch at the instant we drop the ball, and assume that the elevator has a speed v_0 at this time. Using the subscripts b and f for the ball and floor, respectively, we have the following equations for their positions in the laboratory frame of reference:

$$y_\text{b} = h + v_0 t - \frac{1}{2} g t^2$$

$$y_\text{f} = 0 + v_0 t + \frac{1}{2} a t^2.$$

Notice that even though we simply drop the ball in the elevator's reference frame, the ball has an upward velocity v_0 at this time. The ball is stationary relative to the floor.

We can find the length of time t_d for the ball to reach the floor by setting $y_\text{b} = y_\text{f}$. The time is

$$t_\text{d} = \sqrt{\frac{2h}{g+a}} \equiv \sqrt{\frac{2h}{g'}},$$

where we have defined $g' = g + a$. Notice that this is exactly the same formula that we would get if we dropped a ball while standing on the ground except that the normal acceleration due to gravity has been replaced by an "effective" acceleration due to gravity g'.

Let's now calculate the velocity of the ball relative to the floor just before the ball hits the floor. The equations for the velocities are

$$v_\text{b} = v_0 - gt,$$
$$v_\text{f} = v_0 + at.$$

A single elevator is going in opposite directions at the same time. If you are going up in New York, it will look like you are going down in China, and vice versa. The Earth is a spinning ball suspended in space; there is no real up and down, but we call where gravity is pulling "down." Newton, dropping his red apple to the elevator floor, is going up (his operator is an angel going to heaven), while Galileo is going down (probably to hell, according to the Church) and dropping a red ball to the floor. Like Newton he is measuring the time it takes the dropped object to hit the floor. At the same time, at the very top and very bottom of the image, a red car is accelerating away from rest with the door slightly ajar. One wonders, at what point will the door slam shut before the driver gets a ticket for speeding?

—T.B.

...One Man's Ceiling Is Another Man's Floor

DOWN

UP

TOM BUNK

One Man's Floor Is Another Man's Ceiling

The velocity of the ball relative to the floor is equal to the velocity of the ball minus the velocity of the floor. At the time the ball hits the floor, the relative velocity v_r is

$$v_r = v_b - v_f = (g + a)t_d = g't_d.$$

Once again this is the same result we would get on the ground using the effective acceleration due to gravity.

If we assume we are standing on the ground and drop an ideal ball on an ideal floor, the ball undergoes a completely elastic collision with the floor and returns to the height from which it was dropped. Conservation of kinetic energy tells us that the velocity of the ball simply reverses direction as it rebounds from the floor. In the more general case of a two-body elastic collision, it is the relative velocity of the two bodies that gets reversed.

Let's assume that we now return to the elevator and examine an elastic collision with the floor. We assume that we reset our stopwatch and choose the height of the floor to be zero. The velocities of the ball and floor (in the ground frame of reference) are now

$$v_b = v_h + v_r - gt = v_h + g't_d - gt,$$
$$v_f = v_h + at,$$

where v_h is the velocity of the floor at the time of the collision. We can find the time t_u when the ball reaches its maximum height above the elevator floor by setting the two velocities equal to each other. (After this time the floor is moving faster than the ball, and the ball approaches the floor again. It is falling.) Therefore,

$$g't_d - gt_u = at_u,$$

or

$$g't_d = (g + a)t_u = g't_u.$$

Once again we get the same result we would get on the ground—the time to fall is the same as the time to rise.

This analysis of the ball in free fall shows that we can simplify the calculations by assuming that we are standing on the ground if we replace the normal acceleration due to gravity by the effective acceleration. Although we have only shown that this is true for one problem, this is a general statement. In fact, one of the tenets of general relativity is the equivalence principle: *A constant acceleration is equivalent to a uniform gravitational field.* In general, we find the effective acceleration by taking the vector difference of the acceleration due to gravity and the acceleration of the system.

This leads us to the following set of problems.

A. Show that the ball rises to the height from which it was dropped (relative to the floor of the elevator).

B. Argue that a ball dropped in a train moving with a constant horizontal acceleration a follows a straight line tilted at an angle with the vertical given by $\tan\theta = a/g$.

Leaf Turner developed our last problem for the first exam used to select the members of the 1999 US Physics Team that competed in Italy that summer.

C. A car accelerates uniformly from rest. Initially, its door is slightly ajar. Calculate how far the car travels before the door slams shut. Assume the door has a frictionless hinge, a uniform mass distribution, and a length L from front to back and that air resistance can be neglected.

If you are not convinced that using an equivalent acceleration due to gravity is the easiest way of solving this problem, try solving it in the inertial frame of reference attached to the ground.

Solution

Sid Govindan and Japeck Tang, two students of Art Hovey, physics teacher at Amity Regional High School in Woodbridge, Connecticut, submitted correct solutions to this problem. They reported that they particularly enjoyed the problem in part C.

Let's begin part A by writing equations for the positions of the ball and elevator after the collision with the floor using the same notation we used in developing the problem:

$$y_b' = y_h + (v_h + v_r)t - \tfrac{1}{2}gt^2,$$
$$y_f' = y_h + v_h t + \tfrac{1}{2}at^2,$$

where y_b' and y_f' are the positions of the ball and floor, respectively, y_h is the position of the floor at the time $t = 0$ of the collision, v_h is the velocity of the floor at $t = 0$, v_r is the relative velocity of the ball and the floor at $t = 0$, and a is the upward acceleration of the elevator.

The height of the ball h' above the elevator floor is given by the difference of the coordinates of the ball and floor:

$$h' = y_b' - y_f' = v_r t - \tfrac{1}{2}g't^2,$$

where $g' = g + a$ is the effective acceleration due to gravity in the elevator. Earlier we calculated $v_r = g't_d$, where t_d is the time for the ball to fall to the floor. Substituting this expression and using our results that the time t_u for the ball to rise to its highest position above the floor equals t_d, we have

$$h' = \tfrac{1}{2}g't_d^2.$$

Lastly, we use

$$t_d = \sqrt{\frac{2h}{g'}}$$

to show that $h' = h$. Once again, we obtain the same result that we would get on the ground: the ball returns to its original height.

Part B asks about dropping a ball in a train with a constant horizontal acceleration a. The effective gravity $\mathbf{g'}$ is given by the vector difference of $\mathbf{g}$ and $\mathbf{a}$ and makes an angle θ with the vertical such that $\tan\theta = a/g$. The ball dropped in the train falls along the direction of $\mathbf{g'}$ just like a ball dropped while standing on the ground falls along the direction of $\mathbf{g}$.

We use the idea of an effective gravity to solve a very interesting problem in a simple way. The door on a car is slightly ajar. If the car accelerates uniformly from rest, how far will the car travel before the door slams shut? We model the car door as a rectangle with a uniform mass distribution and a length L from front to back and treat the door as being acted on by a gravitational force in the backward direction. We can ignore the real gravitational force, because the hinges do not allow motion up and down.

The torque acting on the door is given by

$$\tau = \tfrac{1}{2} L m a \sin\theta,$$

where θ is the angle between the door and the side of the car. If we assume that the door is thin, we can think of the door as being constructed from a column of thin rods. The moment of inertia of the door about its hinges is then

$$I = \tfrac{1}{3} m L^2.$$

Newton's second law for rotational motion yields

$$\tau = I\alpha = \tfrac{1}{2} L m a \sin\theta = \tfrac{1}{3} m L^2 \alpha.$$

Therefore,

$$\alpha = \frac{3a}{2L} \sin\theta.$$

This is just the equation for a simple harmonic oscillator if we make the approximation that $\sin\theta \cong \theta$. The period for this motion is

$$T = 2\pi\sqrt{\frac{2L}{3a}}.$$

The time for the door to close is just one-fourth of this period.

Therefore, the distance traveled by the car before the door slams shut is

$$d = \tfrac{1}{2} a t^2 = \tfrac{1}{8} a T^2 = \tfrac{1}{12} \pi^2 L.$$

The eyes have it

"The history of the living world can be summarised as the elaboration of ever more perfect eyes within a cosmos in which there is always something more to be seen."
—Pierre Teilhard de Chardin (1881–1985)

SIX OF THE 30 PHYLA OF ANImals have eyes that can produce images. These mere six dominate the animal kingdom with over 95 percent of the population of animals on Earth. It is no surprise that eyes provide such a distinct advantage for survival. The blind species have to nudge up to another object to detect its presence. Does this object present itself as an obstacle, a potential food, or a potential predator? Animals with eyes have a remote sensing apparatus that allows them to avoid obstacles and predators and survey the environment for food.

The complexity of the human eye confounded Darwin. In *On the Origin of Species*, he wrote, "To suppose that the eye, with all its inimitable contrivances for adjusting the focus to different distances, for admitting different amounts of light and for the correction of spherical and chromatic aberration, could have formed by natural selection, seems, I freely confess, absurd in the highest possible degree." But Darwin's concept of adaptation and natural selection guided a steady stream of biologists who have collectively depicted a series of 40 steps, each a small advantage over the prior, which describe the evolutionary trail of the eye. You can read about this journey in Richard Dawkins's *Climbing Mount Improbable.*

A lens can be crudely modeled as a triangular prism atop a cube resting on an inverted prism. Three parallel rays of light, carefully placed, will converge at a single point. The design of a good lens requires us to "smooth" the sides of the prisms and cube so that all rays of light, undergoing refraction, will converge at a single point—the focus (figure 1 on page 190.)

Using these rays of light, we can create ray diagrams—a graphical means of determining the location and orientation of the image of an illuminated object. One ray of light emanating from the tip of the object travels parallel to the principal axis, refracts through the lens, and trav-

We are inside an eye, in the computerized control room where the brain supervises the visual enterprise. Looking out we see a desolate planet that used to be paradise, still with a Tree of Knowledge and the serpent as the evil eye, but instead of apples there are eyeballs growing on it. Whoever picks an eye will see the light, even the hopeful blind eyecatcher, who is coming to fill his bag with eyeballs. While Einstein plays his new vision blues and the wise owl continues to doze, the creator's seeing eyes approach mysteriously and our eye doctor examines the apple in our eye. Our long arm reaches for another eyeball. Maybe this one will let us see the true meaning of this enigmatic imagery, if there is any.

—T.B.

STARRY EYES
LOOK
WINDOW 2000
BLIND EYE
FOCUS
LEFT BRAIN - RIGHT BRAIN
PANVISION
THIRD EYE
EYE OPENER LIGHT
TOMAS BUNK

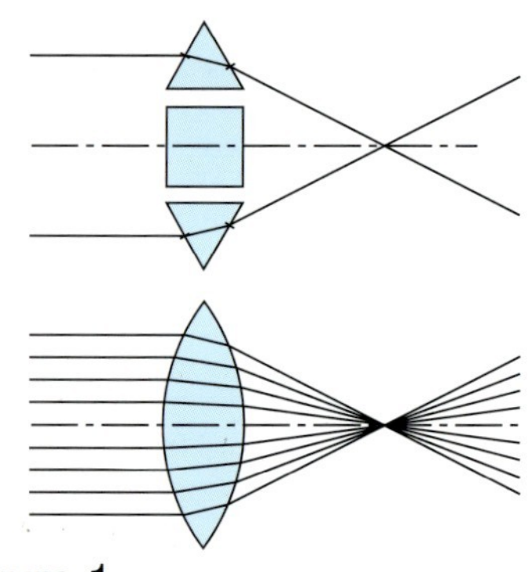

Figure 1

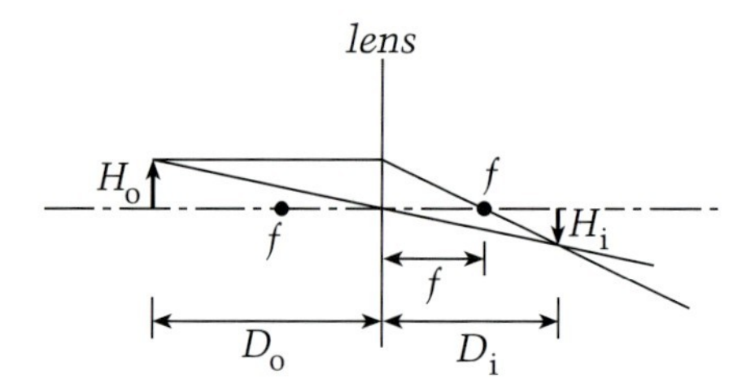

Figure 2

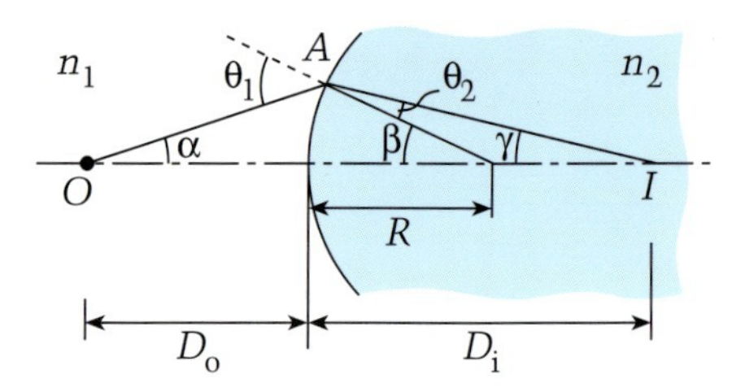

Figure 3

els through the focus. A second ray of light leaves the object and travels through the near focus emerging after refraction parallel to the principal axis. A third ray travels through the center of the lens, and if the lens is thin, is displaced negligibly (fig. 2).

A look at similar triangles yields two equations:

$$\frac{H_i}{H_o} = \frac{D_i}{D_o},$$
$$\frac{H_i}{H_o} = \frac{D_i - f}{f}.$$

Combining these yields the lens equation:

$$\frac{1}{f} = \frac{1}{D_o} + \frac{1}{D_i}.$$

A more formal proof requires us to use Snell's law to calculate the change in angle of the light as it enters (and then leaves) the glass.

Assuming that light emanates from point O and is brought to focus at point I inside the glass as shown in figure 3, we can draw a radial line from the center of curvature of the lens to the interface at point A. We can assume that the two media have indices of refraction n_1 and n_2. Thus we have the following equations:

$$n_1 \sin\theta_1 = n_2 \sin\theta_2,$$
$$\theta_1 = \alpha + \beta,$$
$$\beta = \theta_2 + \gamma.$$

Assume that all angles are small because α is small. We can then use the approximation that $\sin\theta = \theta$ in radians:

$$n_1\theta_1 \cong n_2\theta_2,$$
$$n_1(\alpha + \beta) \cong n_2(\beta - \gamma),$$
$$\beta(n_2 - n_1) \cong n_1\alpha + n_2\gamma.$$

The arc length is equal to the radius multiplied by the angle subtended (in radians). Therefore, for small angles we have

$$\alpha \cong \frac{AB}{D_o},$$
$$\beta = \frac{AB}{R},$$
$$\gamma \cong \frac{AB}{D_i},$$

and

$$\frac{n_1}{D_o} + \frac{n_2}{D_i} = \frac{n_2 - n_1}{R}.$$

Completing a parallel derivation for the ray of light leaving n_2 and refracting at a concave surface (that is, a double convex lens—convex from each side) into n_1, we emerge with the lensmaker's equation:

$$\frac{1}{f} = (n-1)\left(\frac{1}{R_1} - \frac{1}{R_2}\right).$$

We have set $n_1 = 1$ for air and $n_2 = n$ for glass. The R_2 is now negative for the double convex lens. If the lens is thick, the proof becomes a bit more tedious and yields an equation that includes the thickness d of the lens:

$$\frac{1}{f} = (n-1)\left(\frac{1}{R_1} - \frac{1}{R_2} - \frac{n-1}{n}\cdot\frac{d}{R_1R_2}\right).$$

The problems below focus on different elements in our story.

A. In the human eye, much of the imaging is due to the cornea, since the light must travel from air ($n = 1.00$) to the cornea ($n = 1.376$) before reaching the lens. The purpose of the lens is to change its focal length for accommodation. Assuming that the fixed image distance (lens to retina) is 2.50 cm, calculate the focal length of the lens/cornea system when the object distance is 20 cm and when it is 20 m.

B. If the human eye were constructed of a fixed focal length lens and moved it for accommodation (as a fish does), what distance would the lens have to move to accommodate the object distances of 20 cm and 20 m? (By the way, mollusks expand or contract the entire eye, and birds of prey change the curvature of the cornea for accommodation.)

C. Describe what would happen to the image of a candle if (1) the top half of the lens were covered, (2) the candle were much larger than the lens diameter.

D. A student completes a lens lab and records the data shown in figure 4. Calculate the focal length, graph the data with D_o on the x-axis and D_i on the y-axis, and derive the lens equation from the graph.

E. The prism from our crude lens would disperse the light into the familiar spectrum. A lens that distorts in this way is said to have chromatic aberration. In the VIII International Physics Olympiad (East Germany, 1975), students were asked to find the conditions for a thick lens such that the focal length would be the same for two different wavelengths. Please solve this problem and discuss the practical limitations of your solution with different types of lenses.

Solution

This problem inspired correct solutions from two consistent teacher contributors, H. Scott Wiley of Wescaco HS

object distance D_o (cm)	90	80	70	60	50	40	30	20	10
object distance D_i (cm)	8.8	8.9	9.0	9.2	9.5	10	11	13	40

Figure 4

in Texas and Art Hovey of Amity Regional HS in Connecticut.

A. If the human eye is modeled as a simple lens, we can use the lens equation

$$\frac{1}{f} = \frac{1}{D_o} + \frac{1}{D_i}$$

to find the focal lengths required for near and distant objects. Plugging in the values $D_i = 2.50$ cm and $D_o = 20$ cm yields $f = 2.22$ cm. Likewise, using the values $D_i = 2.50$ cm and $D_o = 2000$ cm yields $f = 2.50$ cm.

However, the eye is not a simple lens, as the rays that enter the cornea/lens from air do not return to air. Let's use the equation derived in the article for a single refracting surface:

$$\frac{n_1}{D_o} + \frac{n_2}{D_i} = \frac{n_2 - n_1}{R}.$$

Since $f = D_i$ when $D_o \to \infty$, the right-hand side of the equation must be equal to n_2/f. Therefore,

$$\frac{n_1}{D_o} + \frac{n_2}{D_i} = \frac{n_2}{f}$$

or

$$\frac{1}{f} = \frac{n_1/n_2}{D_o} + \frac{1}{D_i}.$$

Given that $n_2 = 1.376$ and $n_1 = 1.000$, plugging in the values $D_i = 2.50$ cm and $D_o = 20$ cm yields $f = 2.29$ cm. Using $D_i = 2.50$ cm and $D_o = 2000$ cm we get $f = 2.50$ cm.

B. If the human eye accommodated for distance by moving the lens and changing the image distance, we would get the following results given a constant focal length of 2.50 cm and using the simple lens equation: $D_o = 20$ cm yields $D_i = 2.86$ cm and $D_o = 20$ cm yields $D_i = 2.50$ cm. The range of lens movement would be 0.36 cm.

C. The only difference in the image after removal of half the lens will be in its lowered intensity due to some of the light not converging on the image. Research conducted with many people suggests that a misconception arises whereby these people think of the two or three rays drawn in ray diagrams as the only rays which contribute to the image. This is perhaps why people do not appreciate that removal of half the lens will not produce half an image. Similarly, the size of the object does not have an impact on the completeness of the image. Once again, students may assume that if the ray we choose to draw from the top of the object parallel to the principal axis is not able to hit the lens, part of the image may disappear. It is worthwhile to conduct experiments to see how changes in the lens shape or object size affect images.

D. Given the data in the problem and the simple lens equation, you can find the focal length for each pair of measurements and find an average of 7.99 cm.

The lens equation can be derived from a graph of D_i versus D_o shown in figure 5. We recognize this as a hyperbola that has been translated by $y = f$ and $x = f$. We also notice that in the general equation for a hyperbola ($xy = C$), the value of C appears to be f^2. Assuming this to be true, we obtain the Newtonian form of the lens equation:

$$(D_o - f)(D_f - f) = f^2.$$

This can be rearranged to obtain

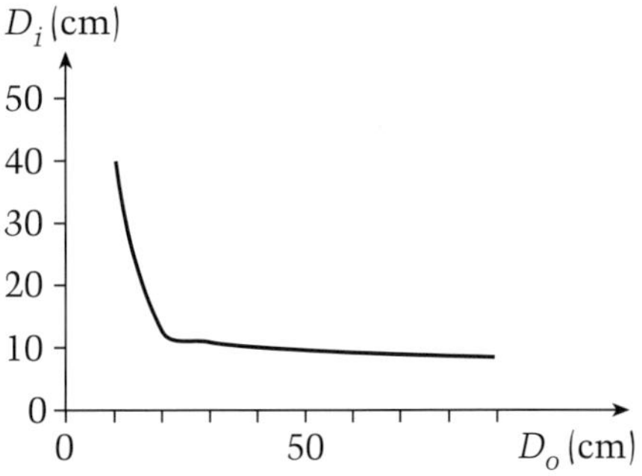

$$\frac{D_o}{D_i} - f(D_i + D_o) = 0,$$

$$\frac{1}{f} = \frac{1}{D_o} + \frac{1}{D_i}.$$

Figure 5

E. To find the conditions for a thick lens to have no chromatic aberration for two different colors we require the lens to have the same focal length for the corresponding two indices of refraction:

$$(n_1 - 1)\left(\frac{1}{R_1} - \frac{1}{R_2} - \frac{(n_1 - 1)}{n_1}\frac{d}{R_1 R_2}\right) = (n_2 - 1)\left(\frac{1}{R_1} - \frac{1}{R_2} - \frac{(n_2 - 1)}{n_2}\frac{d}{R_1 R_2}\right).$$

Multiplying both sides of the equation by $R_1 R_2 n_1 n_2$ and redistributing and canceling terms yields

$$n_1(n_1 n_2 R_2 - n_1 n_2 R_1 - n_1 n_2 d + d) = n_2(n_1 n_2 R_2 - n_1 n_2 R_1 - n_1 n_2 d + d).$$

Since $n_1 \neq n_2$,

$$n_1 n_2 R_2 - n_1 n_2 R_1 - n_1 n_2 d + d = 0$$

and

$$d = \frac{n_1 n_2 (R_2 - R_1)}{n_1 n_2 - 1}.$$

Notice that the dispersion $n_1 - n_2$ is not part of the solution—only the product of the indices appears. Since the indices of refraction are greater than 1 and the thickness of the lens is positive, we conclude that

$$R_2 - R_1 > 0.$$

The first result is that the lens cannot be plano-convex or plano-concave since those lenses require an infinite radius of curvature and would require an infinite thickness.

The second result is that a double concave (diverging) lens is possible if R_1 is negative and R_2 is positive.

The third result is that a converging lens is possible with either $R_2 > R_1$ or $R_2 < -R_1$. The converging lens cannot be symmetric ($R_2 \neq R_1$).

Image charge

"We operate with nothing but things which do not exist, with lines, planes, bodies, atoms, divisible time, divisible space—how should explanation even be possible when we first make everything into an image, *into our own image!"*

—Friedrich Nietzsche (1844–1900)

WE ARE ALL FAMILIAR WITH images formed by mirrors and lens. In fact, we often pay to see images produced in special ways such as those in fun houses or at the Haunted Mansion in Disneyland.

But what is an "image charge?" Does it have anything to do with optical images? Let's investigate this by considering the following problem.

We are given a conducting plate that is so large that we can imagine that it reaches to infinity in the plane of the plate. Alternatively, we can work close enough to the plate and far enough from the edges that the plate might as well stretch to infinity.

Let's now place a charge q a distance d in front of the middle of the plate. If the plate is grounded, is the charge q attracted to or repelled by the plate? And what is the strength of the force acting on the charge?

Experimentally, we can demonstrate that there is an attractive force between the charge and the metal plate. Run a comb through your hair and use it to pick up small pieces of aluminum foil.

We can also see this qualitatively. If we bring a positive charge near a metal plate, the positive charge will attract the electrons in the plate, causing them to concentrate in the area nearest the charge. Because these electrons have moved closer to the charge, their attractive force is larger than the repulsive force of the positive ions left behind.

Quantitatively, this looks like a complicated problem, but it can be solved rather easily using one of the "tricks of the trade." This technique relies on a uniqueness theorem for the electrostatic potential. Remember that the electrostatic potential at a point in space is the amount of work required to bring a unit positive charge to the point from a place where the potential is zero. Suppose that we are given the value of the electrostatic potential at every location on the entire boundary (surface) of a volume of space. If by hook or crook, we can find a formula that gives the correct values at all points

It's party time! And not in some tiny club but on an infinitely large metal conducting plate. Endless crowds are streaming from the subway, the attractive protons on one side and the repulsive electrons on the other. When the electrostatic dance floor, divided in squares, gets charged, the particles automatically start to rock, swing, spin, and square dance. The charge spaceships are above and under the metallic plate and make the action happen. The dance music is supplied by Einstein fiddling on a barrel, though the lady living on a small planet beneath the plate doesn't much appreciate the tumultuous activity. Why are there trains, roads, channels, bombed-out cities, the towers of Babel, and Romeo & Juliet is not clear, but maybe we'll find out when imaginary lines, planes, bodies, atoms, divisible time, and divisible space become reality. But right now it's still party time!

—T.B.

ELECTRONS! PROTONS!
COME AND JOIN THE INVERSE
SQUARE DANCE-O-RAMA
UNATTRACTIVE FORCE
ATTRACTIVE FORCE
COULOM
LAW FIRM
CON-DUCTING FLOOR
TOM BUNK

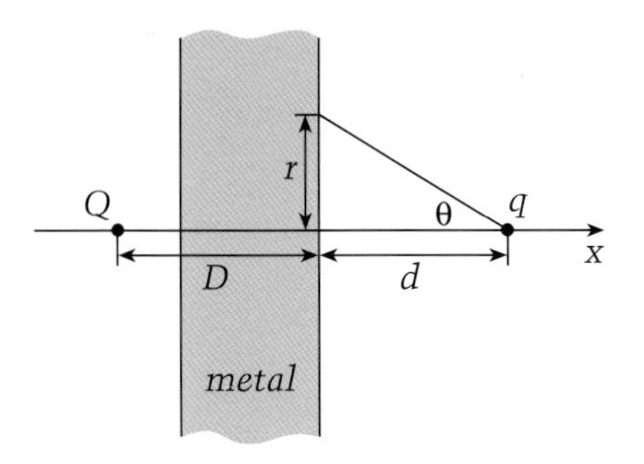

Figure 1

on the boundary, this formula also gives the values of the electrostatic potential throughout the volume. Furthermore, the formula is unique. No matter what other technique we may use, we will get the same potential. This means that we can search around for a simple way of finding the electrostatic potential.

One way is to use the method of images. We imagine that we can replace the metal plate with an image charge Q. This charge is located behind the original metal surface along the normal from the original charge to the surface as shown in figure 1. (Remember that the image of an object in front of a plane mirror is located along a similar line. In fact, we might guess by analogy that the image charge is located a distance d behind the original surface.) For the moment, let's leave the distance D of the image charge behind the plate as unknown.

If we choose the potential to be zero at infinity, the potential at a distance r from a point charge q is given by

$$V = \frac{kq}{r},$$

where k is Coulomb's constant $= 1/4\pi\varepsilon_0$. One of the reasons for introducing the idea of potential is that potential is a scalar. The potential due to a collection of point charges is just the sum of the potentials due to each charge. In contrast, we must add the electric field due to each charge as vectors to get the total electric field.

Returning to our problem, let's choose the origin of our coordinate system to be on the surface along the normal to the charge. At any location along the original surface, the potential due to both charges is

$$V = \frac{kq}{\sqrt{d^2 + r^2}} + \frac{kQ}{\sqrt{D^2 + r^2}},$$

where r is the distance along the plane measured from the normal. Because the plane is grounded, $V = 0$ and

$$\frac{q}{\sqrt{d^2 + r^2}} = \frac{-Q}{\sqrt{D^2 + r^2}}.$$

We need to find the values of two variables, Q and D, but we only have one equation. However, we also have the condition that this equation must be satisfied for all values of r. In this particular case, we can guess the solution, $Q = -q$ and $D = d$, but it is instructive to obtain the solution in a more formal manner.

Let's square the equation and multiply through by both denominators to obtain

$$(D^2 + r^2)q^2 = (d^2 + r^2)Q^2.$$

We now collect terms containing r^2 on one side of the equation and all other terms on the other side:

$$D^2q^2 - d^2Q^2 = r^2(Q^2 - q^2).$$

For this equation to be valid for all values of r, the coefficient of r^2 must be zero. Therefore, $Q = \pm q$. We choose the minus sign because this is the only way the potential can be zero at the surface. The left-hand side of the equation must also be zero, giving us $D = d$.

This tells us that the charge q can be imagined to induce a charge $-q$ a distance d behind the metal surface. According to the uniqueness theorem, we can now use this image charge to calculate results at all points in front of the metal.

Armed with this result we can answer our original questions. Because q and Q have opposite signs, the charge q is attracted to the metal plate. The strength of this attractive force is given by Coulomb's law:

$$F = k\frac{qQ}{\left(d + D^2\right)} = -k\frac{q^2}{4d^2},$$

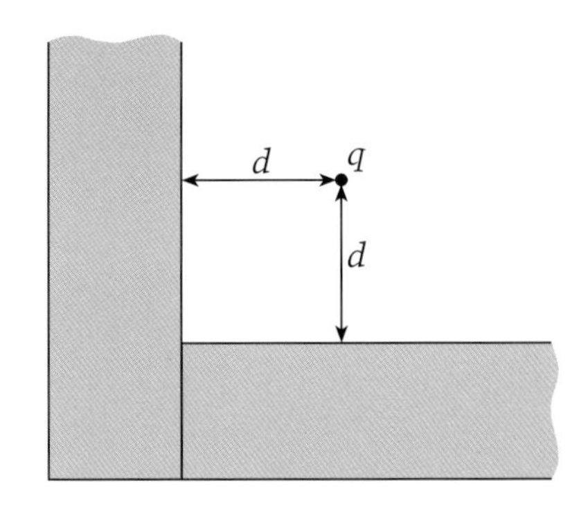

Figure 2

where k is Coulomb's constant.

The electrostatic potential is obtained by substituting our conditions back into the formula for the potential:

$$V = kq\left[\frac{1}{\sqrt{(x-d)^2 + r^2}} - \frac{1}{\sqrt{(x+d)^2 + r^2}}\right].$$

We can also use the image charge to calculate the electric field at all points in the volume and on its boundary. In particular, let's do this along the metal surface. Symmetry tells us that

$$E_r(0, r) = 0.$$

We also know this because the electric field must be normal to all metallic surfaces. The two charges contribute equal amounts to the normal component:

$$E_x(0,r) = \frac{-2kq}{d^2 + r^2}\cos\theta = \frac{-2kqd}{\left(d^2 + r^2\right)^{3/2}}.$$

One reason for calculating this electric field is that it allows us to find the actual charge distribution induced on the metal surface by the charge q. The induced surface charge density is

$$\sigma(r) = \varepsilon_0 E_x(0,r) = \frac{-qd}{2\pi\left(d^2 + r^2\right)^{3/2}}.$$

It is interesting to note that if you

integrate this charge density over the entire surface, you obtain a total charge of $-q$, the same value we obtained for the image charge.

This brings us to the statement of our problem.

A. Two large metal plates form a right-angle corner. A charge q is placed within the corner, equal distances d from both plates, and far from the edges of the plates (see figure 2). What is the force acting on the charge?

B. The second part is based on a problem given on the semifinal exam used to select members of the US Physics Team that competed in the International Physics Olympiad held in Italy in July 1999. A charge q is placed a distance d from the center of a grounded metal ball with radius $c < d$. The electrostatic potential is chosen to be zero at infinity. What is the force acting on charge q?

Solution

Using the results for a charge in front of an infinite metal plate calculated in the article and our knowledge of images formed by plane mirrors, we can guess that the image charges for part A are those shown in figure 3. You can easily verify that the electrostatic potential is equal to zero along each of the two metal surfaces and at infinity.

The electrostatic force felt by the charge is the same as that exerted by the three image charges. The two negative image charges attract the original charge toward the intersection of the two metal plates with a force

$$F_{neg} = 2\frac{kq^2}{(2d)^2}\frac{1}{\sqrt{2}} = \frac{1}{2\sqrt{2}}\frac{kq^2}{d^2},$$

while the positive image charge repels the image charge with a force

$$F_{pos} = \frac{kq^2}{\left(2\sqrt{2}d\right)^2} = \frac{1}{8}\frac{kq^2}{d^2}.$$

Therefore the net attractive force toward the corner is

$$F = \left(\frac{1}{2\sqrt{2}} - \frac{1}{8}\right)\frac{kq^2}{d^2} \approx 0.23\frac{kq^2}{d^2}.$$

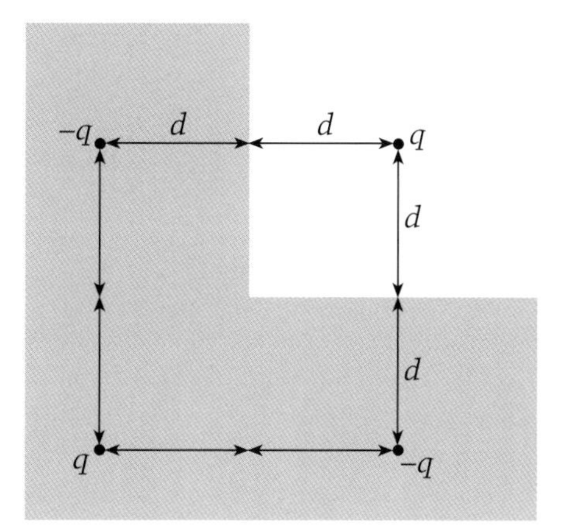

Figure 3

Note that this is very close to the force obtained for the infinite metal sheet.

In part B let's choose an image charge Q located a distance $D < c$ from the center of the ball. As shown in figure 4, let's denote the distances from the real charge and the image charge to a point on the surface of the ball as r and R, respectively. The electrostatic potential at this point is given by

$$V = \frac{kq}{r} + \frac{kQ}{R}$$

where

$$r^2 = d^2 - 2dc\cos\theta + c^2$$

and

$$R^2 = D^2 - 2Dc\cos\theta + c^2.$$

Since the electrostatic potential is zero at all points on the surface of the ball,

$$Qr = -qR.$$

We now square both sides of this relationship, plug in the values of r and R, and group terms in powers of $\cos\theta$ on each side to obtain

$$Q^2(d^2 + c^2) - 2dcQ^2\cos\theta$$
$$= q^2(D^2 + c^2) - 2Dcq^2\cos\theta.$$

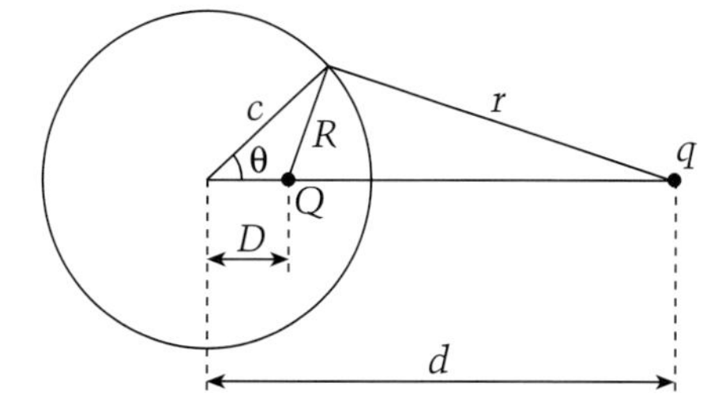

Figure 4

Because this relationship must be valid for all values of $\cos\theta$, the coefficients of each power of $\cos\theta$ must be equal on the two sides of this equation. This yields

$$Q^2 d = q^2 D$$

and

$$Q^2(d^2 + c^2) = q^2(D^2 + c^2).$$

We now solve both equations for the ratio Q^2/q^2 and equate them to obtain

$$\frac{D}{d} = \frac{D^2 + c^2}{d^2 + c^2}.$$

This is a quadratic equation in D that has two roots:

$$D = d,\ c^2/d.$$

The first root corresponds to the case where the real charge and the image charge are superimposed and the potential is zero everywhere. We are interested in the second root, for which

$$Q = -\frac{D}{c}q = -\frac{c}{d}q.$$

(Note that the location of the image is not the same as that for an object outside a spherical convex mirror.)

Using this image charge and image distance, the attractive force on the charge outside a grounded conducting sphere is

$$F = kq^2\left(\frac{cd}{d^2 - c^2}\right).$$

It is interesting to check the limiting cases. As the charge approaches the surface of the sphere, d approaches c and force becomes very large. And as the charge is moved to very large distances, the force decreases to zero. Both of these behaviors are expected and agree with the case for the infinite conducting plane. ◘

Breaking up is hard to do

"O crucified Jove, do you turn your just eyes away from us or is there here prepared a purpose secret and beyond our comprehension?"

—Dante

"THE FISSION OF THE URAnium nucleus can be considered a very interesting paragraph (but only a paragraph) in the story of physics." So stated George Gamow, a noted physicist of this century, in 1961. The technological products of the discovery of fission, notably the atomic bomb and nuclear power, have greatly elevated its importance in our culture.

The entire development of the bomb cannot be understood without a comprehensive knowledge of the events of World War II. The history of the discovery of fission includes many aspects of the politics prior to the declaration of war. We recommend that readers turn to Richard Rhodes's outstanding book, *The Making of the Atomic Bomb*, for a stirring account of the history of physics in this era. We also recommend trying to locate *Moments of Discovery, The Discovery of Fission.* This out-of-print audiotape history published in 1984 by the American Institute of Physics (AIP) includes recordings of many of the principal players including Einstein, J. J. Thomson, Rutherford, Bohr, Hahn, Frisch, Compton, Szilard, and Fermi. It is quite a thrill to hear Rutherford state that "a nucleus is a very small thing."

Our chore is somewhat modest in comparison to both the science history and the political history surrounding nuclear fission. We wish to explore the details of fission, including when it occurs and how we can explain the enormous amounts of energy that are available.

Let's begin with a recipe. The ingredients are 6 protons, 6 neutrons, and 6 electrons; the product is a carbon-12 atom:

$$\begin{aligned} \text{6 protons: } 6(1.007276\ \text{u}) &= 6.043656\ \text{u} \\ \text{6 neutrons: } 6(1.008665\ \text{u}) &= 6.051990\ \text{u} \\ \text{6 electrons: } 6(0.000549\ \text{u}) &= \underline{0.003294\ \text{u}} \\ \text{total mass} &= 12.098940\ \text{u} \end{aligned}$$

where $\text{u} = 1.66 \cdot 10^{-27}$ kg is the atomic mass unit.

Western man (and woman) wants to know everything that can be known and make anything that can be imagined. Unfortunately this is a difficult balancing act. One can improve life, but one can also destroy it. Faust has sold his soul to the devil for divine knowledge and now plays with nuclear power. The military in the front row is pleased to get their toys of destruction, while the scientists behind them are more concerned. Even Einstein forgets to play the violin for a moment, and Dante, seeing what's coming, is already writing his *Inferno.* The regular people are afraid of nuclear power but uncertain what to do, and the animals behind them worry about their own extinction. Beyond the theater we see pollution and destruction—and while impartial eyes watch from space, they do not interfere. It all may be just a dream, a delicate dance on a blue ball without any purpose except to keep everybody on her toes.

—T.B.

NUCLEAR POWER NO THANKS
NO NUKES
Niels Bohr
neutrons
TOMAS BUNK

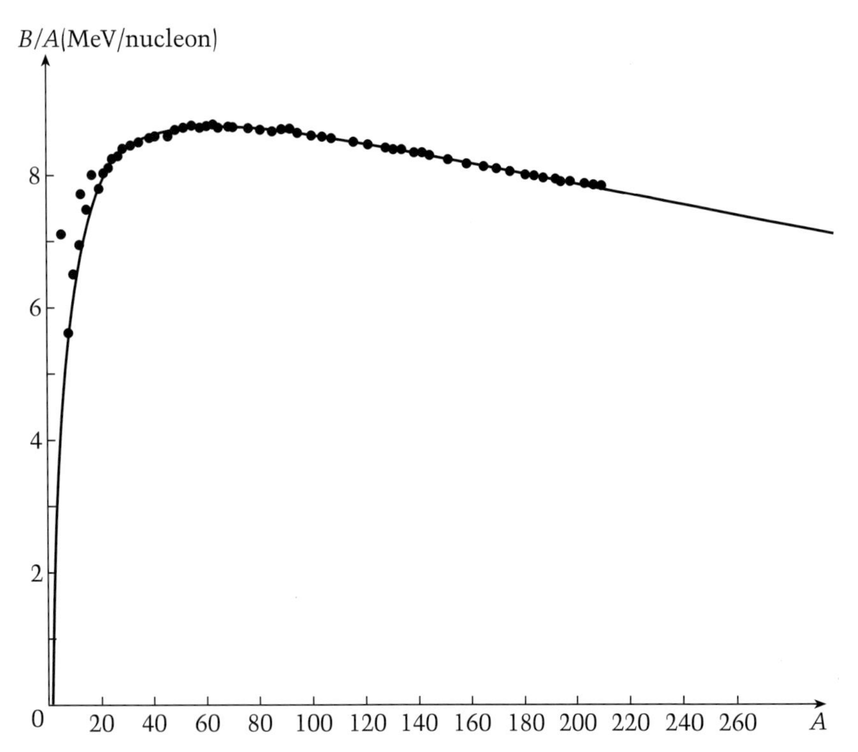

Figure 1
Average binding energy per nucleon vs. mass number, for naturally occurring isotopes. The dots are based on measured binding energies; the smooth curve is based on the liquid-drop model. The curve has a maximum at $A = 56$ for iron-56.

Surprise! The total mass of a neutral carbon-12 atom is exactly 12 u (by definition). So, where did the mass go? The missing mass, the mass defect, is actually released as energy when the nucleus is formed. From a different perspective, the carbon nucleus has a binding energy that holds the nucleus together. Removing a proton or neutron or separating all of the protons and neutrons requires an expenditure of energy. We have here a direct appreciation of Einstein's startling discovery in 1905 that mass and energy are one and the same and that the conversion factor for changing mass units into energy units is the square of the speed of light. This most celebrated equation of all of science is $E = mc^2$.

This equation shows us that a mere 1 g (10^{-3} kg) of mass yields $9 \cdot 10^{13}$ J of energy. To put this in perspective, if sold as electricity, this energy has a value of more than 2 million dollars. Applying Einstein's equation, we find that 1 u yields 931.5 MeV, where 1 MeV = $1.6 \cdot 10^{-13}$ J.

Returning to the carbon nucleus, the mass defect is 0.098940 u. This has a corresponding binding energy of 92 MeV. If we repeat the analysis for carbon-11, we find that the mass of the atom is 11.011433 u and the mass of the constituent parts is 11.090275 u. The mass defect of carbon-11 is 0.078842 u or 73 MeV. The removal of this neutron must have required an expenditure of 19 MeV. Students of introductory physics are probably more familiar with the energy required to ionize a hydrogen atom, 13.6 eV. Thus, 13.6 eV of energy must be given to the electron to free it. Most of chemistry deals with the exchange of electrons and effectively deals with energies on the order of a few eV per atom. In contrast, changes in nuclear structure have corresponding energies of millions of eV.

It is informative to compare the average binding energy per nucleon of the two isotopes of carbon. Carbon-12 has an average binding energy of 92/12 MeV or 7.7 MeV per nucleon. Carbon-11 has an average binding energy of 73/11 MeV or 6.6 MeV per nucleon. A similar calculation can be done for isotopes of all elements, and the curve that is generated is shown in figure 1. The average binding energy for most nucleons is approximately 8 MeV/nucleon. The curve has the interesting feature of having a maximum at iron-56. Nuclei with less mass and nuclei with more mass have less binding energy per nucleon. Thus, light nuclei combining to form a heavier nucleus will release energy in a process called fusion. A heavy nucleus will release energy when it splits into two lighter nuclei in a process called fission.

The binding energy curve came about after the discovery of fission. When Hahn and Strassman first recognized the fission of uranium, it startled them and the scientific world. The discovery of the neutron by Chadwick in 1932 had provided a new research tool for nuclear physics. The neutron could enter the nucleus without having to overcome the Coulomb repulsive force that a proton would experience. Enrico Fermi, soon after the discovery of the neutron, began bombarding elements with neutrons and produced many new isotopes. Since many of the isotopes Fermi produced emitted beta particles, when he bombarded uranium he thought that he had discovered a transuranic element.

Otto Hahn and his colleagues Lise Meitner and Fritz Strassman had been chemically analyzing radioactive elements for some time. Unfortunately, Meitner was forced by the Hitler regime to leave Germany in July 1938. Hahn and Strassman found that one of the elements emerging from the uranium nucleus was barium. Their first hypothesis was that it was radium, which would have required two alpha particles to leave the uranium nucleus. Even this seemed unlikely—a low energy neutron knocking two alphas from the nucleus was beyond expectation. Barium was even more unlikely, but barium it was. Meitner and her nephew Otto R. Frisch were quick to deduce that the addition of the neutron caused instability and the uranium nucleus broke into two parts. If barium was one piece, the other must be krypton. This element had indeed also been detected. Frisch mentioned fission to Niels Bohr, who was on his way to America, and en route Bohr and a colleague, Leon Rosenfeld, mapped out the liquid drop model of the atom that could predict this surprising behavior.

The liquid drop model of the nucleus treats the nucleus as a droplet of nuclear material. The nucleons on the surface are held to the drop by a surface tension. The model is quantitative and leads to an equation that can accurately predict the binding energy curve. Our explanation of the relevant equation follows that of Ohanian in his text *Modern Physics*.

To begin, we must recognize that the nuclear radius is proportional to $A^{1/3}$, where A is the atomic mass. This relationship is the result of many scattering experiments. That being the case, the volume of a nucleus ($4/3\pi R^3$) is proportional to A and the density of nuclear matter (the ratio of mass to volume) must be constant for all nuclei.

The most important term in the derivation of a binding energy equation is associated with the number of nucleons, since each nucleon attracts every other one through the short-ranged strong force. The binding energy must contain a term proportional to A. Since the nucleons on the surface do not have as many neighbors as nucleons within the interior, the second term must account for this decrease in the binding energy. Since R^2 is proportional to $A^{2/3}$, the correction term is proportional to $A^{2/3}$ and is negative. The Coulomb repulsion force between all of the protons tends to drive the nucleus apart. This term is proportional to $Z^2/A^{1/3}$, where Z is the number of protons in the nucleus.

Finally, there is a quantum-mechanical correction that takes into account the exclusion principle. Just as electrons cannot all be in the same quantum state, but fill successive shells, the nucleons must also fill shells. This leads to a term that is related to the numbers of protons and neutrons and the total number present. The constants of proportionality are found through numerous experimental data. Weizsacker's semi-empirical formula for the binding energy is

$$B = \left[15.753A - 17.804A^{2/3} - 0.7103\frac{Z^2}{A^{1/3}} - 94.77\frac{\left(\frac{1}{2}A - Z\right)^2}{A}\right] \text{MeV}.$$

The smooth curve in figure 1 is based on the liquid-drop model and can be seen to fit the data exceedingly well.

Assuming that a nucleus splits into two equal parts ($A/2$, $Z/2$ for each product), we can calculate the difference in binding energies:

$$B_{A,Z} = \left[15.753A - 17.804\,A^{2/3} - 0.7103\frac{Z^2}{A^{1/3}}\right] \text{MeV},$$

$$B_{A/2,Z/2} = \left[15.753\frac{A}{2} - 17.804\left(\frac{A}{2}\right)^{2/3} - 0.7103\frac{(Z/2)^2}{(A/2)^{1/3}}\right] \text{MeV},$$

$$2B_{A/2,Z/2} - B_{A,Z} = \left[-4.6A^{2/3} + \frac{0.26Z^2}{A^{1/3}}\right] \text{MeV}.$$

We have ignored the quantum mechanical term, which is small in comparison. If the nucleus were to undergo fission, the electrostatic force must be greater than the surface tension.

$$-4.6A^{2/3} + \frac{0.26Z^2}{A^{1/3}} > 0,$$

$$\frac{Z^2}{A} > 18.$$

In the case of ^{238}U, we have $Z = 92$, $A = 238$, and $Z^2/A = (92)^2/238 = 35.6$, which is certainly larger than 18. Fission does not proceed directly but requires an elongation of the nuclear drop. This is more likely to occur when the additional neutron is added to the uranium nucleus.

Enrico Fermi and Emilio Segre did not discover the fissioning of uranium, although fission did indeed occur during their 1934 experiments. Segre is quoted as saying, "The whole story of our failure is a mystery to me. I keep thinking of a passage from Dante: 'O crucified Jove, do you turn your just eyes away from us or is there here prepared a purpose secret and beyond our comprehension?'" (from AIP's *Moments of Discovery* audiotape). The discovery of fission in 1939 led immediately to the development of the atomic bomb effort, which included Fermi, who was then living in the United States. How might world history have been altered if the discovery of fission had occurred before the emigration of physicists to the United States and well before the start of World War II? What does this suggest about the role of chance in history?

Our problem includes analysis of a number of features of the fission process.

A. It is not hard to follow the reasoning Frisch and Meitner used to calculate the energy released in fission. Consider a typical fission reaction:

$$^{235}_{92}\text{U} + ^{1}_{0}\text{n} \rightarrow ^{140}_{54}\text{Xe} + ^{94}_{38}\text{Sr} + 2\,^{1}_{0}\text{n}.$$

The Xe rapidly decays into $^{140}_{58}$Ce and the Sr into $^{94}_{40}$Zr, with the emission of electrons of negligible mass. We now know the following masses:

$^{135}_{92}$U	235.004 u
$^{1}_{0}$n	1.009 u
$^{140}_{58}$Ce	139.905 u
$^{94}_{40}$Zr	93.906 u

Calculate the energy released in the fission reaction.

B. The discovery and the exploitation of fission did not require knowledge of $E = mc^2$. In fact, at the time the masses of the radioactive daughter nuclei were not known well enough to make a good calculation. Frisch and Meitner calculated the energy release by a second method (which was the only method Joliot used).

Calculate the radii of the Ce and Zr nuclei above using the approximate equation, $R = KA^{1/3}$, where $K = 1.0 \cdot 10^{-15}$ m. Assume that at the moment the uranium breaks into these fragments, the distance between the centers of the two fragments is equal to the sum of their radii. Calculate the electrostatic repulsion between them. Using the electrostatic potential, calculate the work done to separate these two fission products. Compare this total energy with that found in part A.

C. Surprisingly, the uranium rarely breaks into two equal products. Use the semi-empirical binding energy equation to show that the energy released is greatest for the symmetric rare fission.

Solution

For part A we find that the initial and final masses are 236.013 u and 235.829 u, respectively, for a change in mass of 0.184 u and a corresponding energy release of 171 MeV.

For part B the radii of the Ce and Zr nuclei are:

$$R_{Ce} = 5.19 \cdot 10^{-15} \text{ m}$$
$$R_{Zr} = 4.55 \cdot 10^{-15} \text{ m}$$

The distance between the centers of the two fragments is equal to the sum of their radii, $R = 9.74 \cdot 10^{-15}$ m.

We can now find the electrostatic potential energy:

$$U = \frac{-kq_1q_2}{R}$$

$$= \frac{-\left(9 \cdot 10^9 \frac{\text{N} \cdot \text{m}^2}{\text{C}^2}\right)(58)(40)\left(1.6 \cdot 10^{-19}\text{C}\right)^2}{9.74 \cdot 10^{-15} \text{ m}}$$

$$= 5.49 \cdot 10^{-11} \text{ J}$$

$$= 343 \text{ MeV}.$$

The values for the energy calculated in parts A and B are relatively close.

We built a spreadsheet to calculate the binding energies of all possible pairs of daughter nuclei using Weizsacker's semi-empirical formula. In doing this, we needed to make a decision on how to divide the neutrons between the daughter nuclei. We chose to divide them so that the ratio of nucleons was the same as the ratio of protons. For example, if the daughter nuclei had 30 and 62 protons, the number of nucleons A was taken to be (30/92) of 236 and (62/92) of 236, respectively.

After completing the spreadsheet, we graphed the energy released versus the atomic number of one of the daughter nuclei (fig. 2). The energy released is greatest for the symmetric fission with each daughter nucleus having 46 protons.

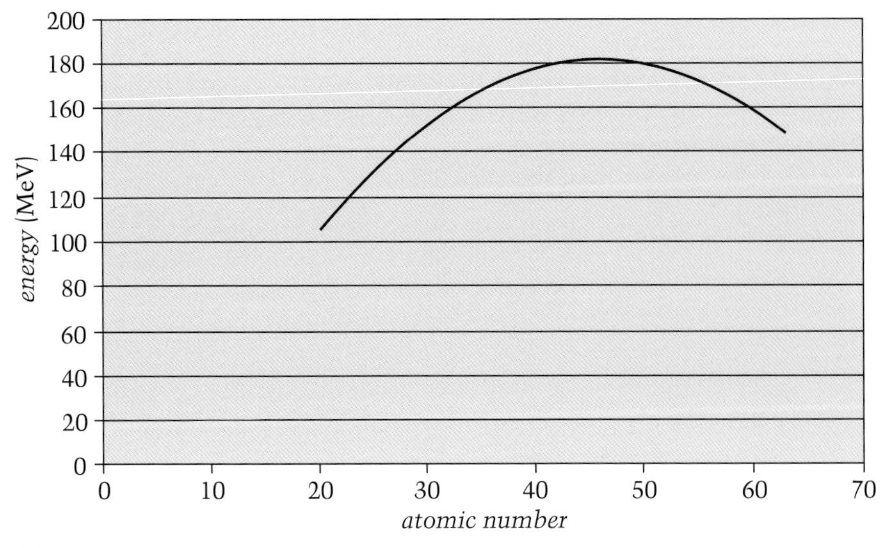

Figure 2

A question of complexity

*"Man is an over-complicated organism.
If he is doomed to extinction he will die out for want of simplicity."*
—Ezra Pound

WHAT WOULD HAPPEN IF a volcano erupted under a glacier? Can we solve such a problem? Can we do it with elementary physics? At the 1998 International Physics Olympiad (IPhO) in Iceland, the competitors were asked to solve this problem using data from an eruption that occurred in Iceland a year or two before the Olympiad took place.

Many problems at the IPhO are presented in the context of problems that physicists solve to understand the physical world. Geophysicists can generate rather elaborate three-dimensional computer codes to analyze the volcanic event, but they can get a "feeling" for what happens by simplifying the problem. For instance, they can simplify the geometry by assuming that the hot lava melts a conical cavity in the ice. They also assume that thermal conduction away from the cavity is small and that the water does not flow away. And this is what was presented to the competitors. Solving the simplified problem did not require any knowledge of physics beyond what is typically learned in the first-year university physics sequence.

What happens when you are given a problem that is either too complex to solve or one where there are crucial ingredients missing? We have observed three types of students. The first mentions that there is a piece of missing information (for example, we don't know the angle between the particle and the field) and refuses to solve the problem. The second mentions that we don't know the angle but goes on to assume that it is 90° and solves the problem. The third assumes that the angle is 90°, solves that problem, and then describes how the solution changes if the angle were not 90°. It's the third student who has the flair of a physicist.

Many problems in the real world are very complex and cannot be solved if all the complexity is included. If, however, we are able to simplify the problem without losing the key elements, we can often reduce a complex problem to a prob-

Here we are confronted with an extremely simplified history of human cultural development in three segments. On top we have the ape sitting on the tree of life and getting the divine spark by accident through a hole in space. This enables him to start thinking about his existence and making tools, like the saw he uses to cut himself off the Tree of Knowledge. In the middle part we see the modern human, who has developed tools so complex and far-reaching that he becomes dependent and loses control over his own destiny. He becomes enslaved by his own creation. In the bottom segment the planet is destroyed and the humans are replaced by self-sufficient robots, who have not much left to do but to go fishing for the divine spark that got lost on the way of human progress.

—T.B.

LAOCOON 2000
OIL
TOM
BUNK

lem (or a series of problems) that we already know how to solve.

At other times a problem may appear to be very complex because of the context in which it is presented. When we were academic directors of the US Physics Team, students would comment that the difference between the more difficult problems in an introductory physics text and problems given at the International Physics Olympiad is that you only realized that Olympiad problems are easy after you have solved them!

This is nicely illustrated by the first problem on the theoretical exam given at the IPhO that was held in Padua, Italy, in July 1999. The text of this problem runs for an entire typewritten page, but can be summarized as follows: A vertical cylinder filled with gas and capped by a moveable glass plate is illuminated for a finite time by a laser. As the gas absorbs the light, the glass plate is observed to move upward. The competitors were then asked a series of quantitative questions about the situation.

What possible assumptions about the physics could provide a pathway from complexity to simplicity? Let's begin with the friction between the glass plate and the cylinder walls. If we know nothing about the friction, we assume that the friction between the glass plate and the cylinder walls is sufficient to damp any oscillations that occur but not large enough to produce any significant loss of energy relative to the other energies involved in the problem.

What assumptions do we usually make when solving gas problems? We first assume that the gas is in thermal equilibrium. Unless otherwise stated, we usually assume that the gas is ideal and that the amount of gas is constant. In this case we assume that the cylinder does not leak gas around the piston. This allows us to use the ideal gas law

$$PV = nRT,$$

where P is the pressure, V is the volume, n is the number of moles of gas, $R = 8.31$ J/K · mol is the universal gas constant, and T is the temperature in kelvins.

Then we need to decide whether the system is thermally isolated, that is, can thermal energy enter or leave the cylinder? We have previously solved problems of both types, so if this is not explicitly stated, we will need to infer this from the context of the problem.

If we are not given any information about the coefficient of thermal conductivity between the cylinder walls and the glass plate, we simplify the problem by assuming a very low thermal conductivity and/or a very short time so that thermal losses can be neglected. In this problem, the competitors were told that the cylinder walls and the glass plate had very low thermal conductivities.

The detailed text told the competitors that light from a constant power laser was shined through the glass plate into the cylinder for a specified time interval. The radiation passed through the air and the glass plate without being absorbed but was completely absorbed by the gas in the cylinder. The molecules absorbing the radiation were excited to higher energy states and then quickly cascaded back to their ground states by emitting infrared radiation. This infrared radiation was reflected by the cylinder walls and the glass plate and absorbed by other molecules.

What was this telling the competitors? Independent of the details, the gas was being heated at a constant rate, and this energy increased the average kinetic energy associated with the chaotic motion of the molecules.

The data often provides other hints about how to simplify. In this problem, competitors were given the following values:

Atmospheric pressure:
$P_o = 101.3$ kPa
Room temperature:
$T_o = 20.0$°C
Inner diameter of the cylinder:
$2r = 100$ mm
Mass of the glass plate:
$m = 800$ g
Quantity of gas:
$n = 0.100$ mol
Molar specific heat at constant volume:
$c_v = 20.8$ J/(mol · K)
Wavelength of the laser:
$\lambda = 514$ nm
Irradiation time:
$\Delta t = 10.0$ s
Displacement of glass plate:
$\Delta s = 30.0$ mm

Even though the initial temperature of the gas is not given in the data table, the problem stated that the gas was initially in equilibrium with its surroundings. Therefore, the initial temperature of the gas is the same as the room temperature.

The pressure of the gas is not given. However, we've done piston problems before. Because the glass plate is in equilibrium, the force on the lower surface must exceed that on its upper surface by the weight mg of the glass plate. Moreover, this must be true after the heating as well as before. That is, the initial and final pressures are the same.

We are left with another problem that we have solved in other simpler contexts. What is the increase in temperature of a gas kept at constant pressure when the volume increases by a specified amount? The wrinkle in this problem is that we need to calculate the initial height of the piston first.

Notice that the power of the laser is not given. Therefore, we need a relationship between the properties of the gas and the energy added to the gas. This is provided by the first law of thermodynamics,

$$\Delta U = Q - W,$$

where U is the internal energy of the gas, Q is the heat added to the gas, and W is the work done by the gas. To use this law to obtain the heat Q, we need to know the other two quantities. Calculation of the work is straightforward, but where are we to find the change in internal energy of the system? Looking at the data gives us a hint. Because we are given the molar specific heat at constant volume for the gas, we are prompted to recall that

$$\Delta U = nc_v(T_f - T_o).$$

When we are asked about the number of photons emitted by the laser per second, we must remember that the laser beam consists of photons with energy hf and hope that the beam has a single frequency. According to the data it does.

In this problem absorption of optical energy produces a change in the gravitational potential energy of the glass plate. How efficient is this process? The word "efficiency" brings to mind its definition:

$$\eta = \frac{W}{Q},$$

where W is the part of the work done to increase the gravitational potential energy of the glass plate and Q is the energy received from the laser.

Many assumptions are required to illuminate our understanding of a gas absorbing light. Once we solve the simpler problem, we can then begin to allow the complexity to creep back in to refine our understanding.

In the problems below we present the quantitative questions from the IPhO problem that we've been discussing.

A. What are the temperature and pressure of the gas after the irradiation?

B. How much mechanical work does the gas perform? (Hint: don't forget the external pressure.)

C. How much radiant energy was absorbed by the gas?

D. What was the power of the laser and the number of photons emitted per second?

E. What was the efficiency of converting optical energy into gravitational potential energy of the plate?

F. If, after the irradiation, the cylinder is slowly rotated by 90° so that its axis is horizontal, do the temperature and pressure of the gas change? If so, what are their new values? (What simplifications do you need to make? Is the process adiabatic?)

Solution

A. In order to find the final temperature and pressure of the gas, let's begin by noting that the initial temperature is the same as room temperature, which is given to be 20.0°C. The initial and final pressures are the same. The difference in the pressures inside and outside the cylinder must be large enough to support the weight of the glass plate. Therefore,

$$P_f = P_i = P_0 + \frac{mg}{\pi r^2},$$

where P_0 = 101.3 kPa is the atmospheric pressure. With a radius r = 50 mm and a mass m = 800 g, we find that P_f = 102.3 kPa.

Now let's find the initial volume of the gas. According to the ideal gas law, we have

$$V_i = \frac{nRT_0}{P_i}.$$

Since we are only given the displacement Δs of the glass plate, we will need to find the initial height of the glass plate in order to find the final volume.

$$h_i = \frac{V_i}{\pi r^2} = \frac{nRT_0}{P_0 \pi r^2 + mg}.$$

Therefore,

$$V_f = V_i\left(\frac{h_i + \Delta s}{h_i}\right).$$

Using the ideal gas law, we have

$$T_f = T_0\left(\frac{V_f}{V_i}\right) = T_0\left(1 + \frac{\Delta s}{h_i}\right)$$

$$= T_0 + \frac{\Delta s\left(P_0 \pi r^2 + mg\right)}{nR}.$$

Plugging in Δs = 30.0 mm, we get T_f = 322 K = 49°C.

B. The mechanical work performed is given by the force exerted by the gas on the glass plate multiplied by the displacement

$$W = (P_0 \pi r^2 + mg)\Delta s,$$

which gives a value of 24.3 J.

C. The internal energy of the gas increases by

$$\Delta U = nc_V \Delta T,$$

and using the first law of thermodynamics, the heat absorbed must be

$$Q = \Delta U + W$$

$$= nc_V \frac{T_0 \Delta s}{h_i} + \left(P_0 \pi r^2 + mg\right)\Delta s$$

$$= \Delta s\left(P_0 \pi r^2 + mg\right)\left(\frac{c_V}{R} + 1\right).$$

This gives a numerical value of 85.3 J.

D. Given that the laser is on for only 10.0 s, the power of the laser is $P = Q/\Delta t$ = 8.53 W. The wavelength λ of the laser light is 514 nm, so the energy of each photon $E = hc/\lambda$, where $h = 6.63 \cdot 10^{-34}$ J·s is Planck's constant. Therefore, the number of photons emitted per unit time is

$$\frac{P\lambda}{hc} = 2.2 \times 10^{19}\ \text{s}^{-1}.$$

E. We can now calculate the efficiency of this "heat" engine for converting light energy into gravitational potential energy to be

$$\eta = \frac{mg\Delta s}{Q} = 0.28\%.$$

F. When we rotate the cylinder so that its axis is horizontal, we have an adiabatic change in pressure from P_f to P_0. We know that PV^γ is constant for an adiabatic expansion, where

$$\gamma = \frac{c_P}{c_V} = 1 + \frac{R}{c_V}.$$

This gives us

$$\frac{V_{rot}}{V_f} = \left(\frac{P_f}{P_{rot}}\right)^{1/\gamma}.$$

Because both the volume and the pressure change, we find the temperature after rotation by

$$T_{rot} = T_f\left(\frac{P_{rot}}{P_f}\right)\left(\frac{V_{rot}}{V_f}\right)$$

$$= T_f\left(\frac{P_{rot}}{P_f}\right)\left(\frac{P_f}{P_{rot}}\right)^{1/\gamma} = T_f\left(\frac{P_{rot}}{P_f}\right)^{(\gamma-1)/\gamma}.$$

This gives a temperature of 321 K, a drop of only one kelvin.

Tunnel trouble

"I wonder if I should fall right through the earth! How funny it'll seem to come out among the people that walk with their heads downwards."
—Lewis Carroll, Alice in Wonderland

HOW DO YOU LIKE THEM apples? The Garden of Eden variety brought about knowledge and led to banishment. Newton's variety extended enlightenment and led to a revolution in science and thought that echoed through politics, poetry, logic, and philosophy.

Newton was certainly not the first person to see an apple fall from a tree. He may, however, have been the first to imagine the apple and the Moon to be one and the same. The apple falls to the ground. This is a good observation, but nothing particularly special. To declare that the Moon also falls to the ground, when everyone knows it does not get any closer to the Earth, requires genius. Newton had been thinking of the gravitational force that attracts the apple to the Earth. To Newton, the Moon was merely a much larger apple that is very much further from the Earth. The Moon also falls to the Earth. It is the tangential velocity of the Moon that prevents it from getting any closer to the Earth. It is the gravitational force that holds the Moon in orbit.

Newton had begun to think of a gravitational force of attraction between the Earth and the apple. He then realized that the force is between any two masses and decreases as the square of the distance between them.

$$F = \frac{Gm_1m_2}{r^2}.$$

The proportionality constant G was not determined experimentally until a hundred years later by Cavendish in 1798. Cavendish hung pairs of 15 kilogram and 125 kilogram masses and observed the tiny attraction between them. The force of attraction is quite small indeed but it was enough to twist a tiny wire, and Cavendish was able to measure that twist and determine that $G = 6.67 \cdot 10^{-11}$ N $\cdot$ m^2/kg^2.

In middle of the Garden of Eden we have the Tree of Knowledge, which Alice has to enter to get to Wonderland. On top of the tree we recognize Galileo dropping an apple on Newton, igniting brilliant thoughts about gravitational force by comparing the apple with the Moon. Next to Newton, Einstein is fiddling for the serpent, and the police angel is following God's orders, evicting Adam and Eve and the rest of humanity from Paradise forever. Even Humpty Dumpty has a fall, not unlike Alice, who falls into the tunnel leading through hellfire to the other side of the Earth, where the banished people walk upside down and where gravitational forces work in rather peculiar ways. Newton would probably get a major headache figuring them out.

—T.B.

STAFF
In This Style 10/6
ALICE'S ADVENTURES IN WONDER= LAND
DRINK ME
EAT ME!
NEWTON
PHILOSOPHIÆ NATURALIS PRINCIPIA MATHEMATICA
TOM BUNK

Returning to Newton, one can't help but be blown away by Newton's insistence that we cannot be cajoled into agreeing that the distance between two spheres should be measured from the center of one sphere to the center of the other. Newton invented integral calculus to prove that the small attractions of each piece of Earth on the apple are equivalent to the attraction of the entire mass of the Earth if the entire mass was located at its center, one Earth radius from the apple.

The success of this treatment can be seen in the two calculations of the Moon's acceleration. In the first calculation, we look at the approximately circular orbit of the Moon about the Earth. The Moon's period is 27.3 days and its distance to the Earth is 60 Earth radii. One Earth radius $R_E = 6.37 \cdot 10^6$ meters. The centripetal acceleration of the Moon can then be calculated:

$$a = \frac{v^2}{R} = \frac{\left(\frac{2\pi R}{T}\right)^2}{R} = \frac{4\pi^2 R}{T^2}$$
$$= 0.0029\ \text{m/s}^2.$$

An inverse square relationship for gravity would predict that the acceleration of the Moon would be $(60)^2$ or 3600 times less than that of the apple.

$$(9.8\ \text{m/s}^2)/3600 = 0.0027\ \text{m/s}^2.$$

When Newton arrived at this inverse-square conclusion, it is said that he "could hear God thinking." Newton showed us that the apple is like the Moon and, simultaneously, that the Earth is like the heavens. This law of gravitation describes the planets about the Sun, the Sun about the Galaxy, and the dance of all clusters of galaxies in our Universe.

The gravitational force allows us to calculate the orbit of a satellite in low Earth orbit. The idea of a satellite orbiting the Earth first appears in Newton's landmark *Principia*, published in 1686:

$$F = \frac{GM_E m}{R_E^2} = \frac{mv^2}{R_E} = \frac{4\pi^2 m R_E}{T^2}.$$

Therefore, the orbital period is given by

$$T = 2\pi\sqrt{\frac{R^3}{GM_E}},$$

which yields a period of 88 minutes.

A more sophisticated problem is to calculate the period of an apple that travels through a hole created along a diameter of the Earth. Since on its travel through the tunnel, the apple experiences a force due to each part of the Earth, some of the mass of the Earth will be pulling inward and some will be pulling outward. A useful observation is that the force at any position is equivalent to that of all of the enclosed mass only, as if that mass was located at the Earth's center. The mass in the external shell has no contribution to the force. At an arbitrary point a distance r from the center, the attractive enclosed mass is then

$$M' = \rho V = \rho\frac{4\pi r^3}{3},$$

where ρ is the density of the Earth. Assuming that the density is constant and that there is no friction in our tunnel, we can solve for the force on the apple at any point in the tunnel:

$$F = \frac{G\rho 4\pi r^3 m}{3r^2} = \left(\frac{4\pi G\rho m}{3}\right)r = kr.$$

When the r is toward the right, the force is toward the left, and the correct form of this equation is

$$F = -kr.$$

We recognize this as the equation of a mass on a spring and as the signature of simple harmonic motion. The apple will oscillate back and forth through the Earth. The period of oscillation T is given by the equation

$$T = 2\pi\sqrt{\frac{m}{k}} = 2\pi\sqrt{\frac{3m}{4\pi G\rho m}} = \sqrt{\frac{3\pi}{G\rho}}.$$

Assuming that the density of the Earth is $5.5 \cdot 10^3$ kg/m^3, we get a period of 84 minutes. This is very close to the period of an orbiting satellite. Should it be?

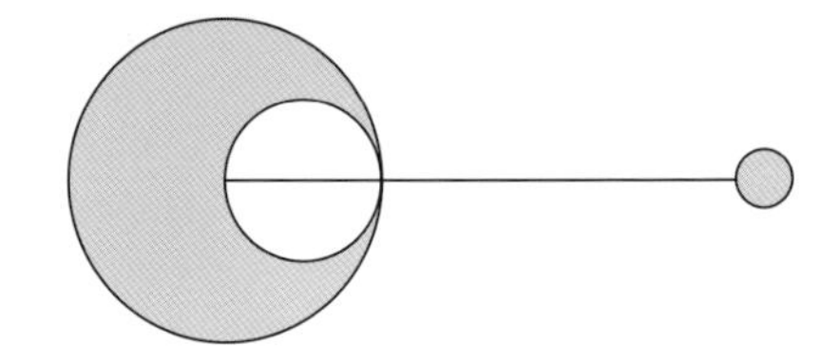

Figure 1

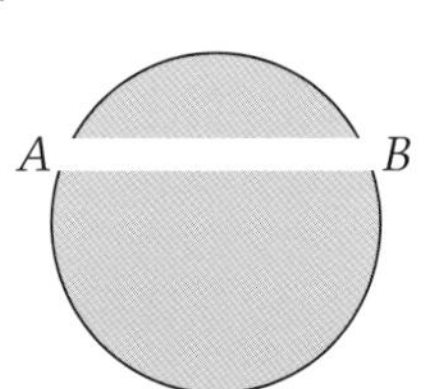

Figure 2

Our problem mixes some classic problems with some new twists.

A. Reprinted in Halliday, Resnick, and Walker is a 1946 Moscow Olympiad problem. A spherical hollow is made in a lead sphere of radius R such that its surface touches the outside surface of the lead sphere and passes through the lead sphere's center. (See figure 1.) The mass of the sphere before hollowing was M. With what force will the lead sphere attract a small sphere of mass m, which lies at a distance d from the center of the lead sphere on the straight line connecting the centers of the spheres and of the hollow?

B. A tunnel is drilled along a chord of the Earth connecting points A and B. (See figure 2.) Calculate the period for an apple to travel from A to B. Comment on the feasibility of such a tunnel for global travel.

C. Does a straight tunnel provide for the fastest journey from A to B? If not, can you find a tunnel of two straight segments that requires a smaller time?

Solution

Art Hovey of Amity Regional HS in Connecticut provided a solution to all parts, and a number of his students (Brian Chin, Alex Rikun, Josh Leven, and Victoria Buffa) were able to present solutions to parts A and B.

There are three equivalent ways of looking at the solution to part A of this problem. The first is to fill in the missing mass of the hollow in the sphere and add an equivalent mass on

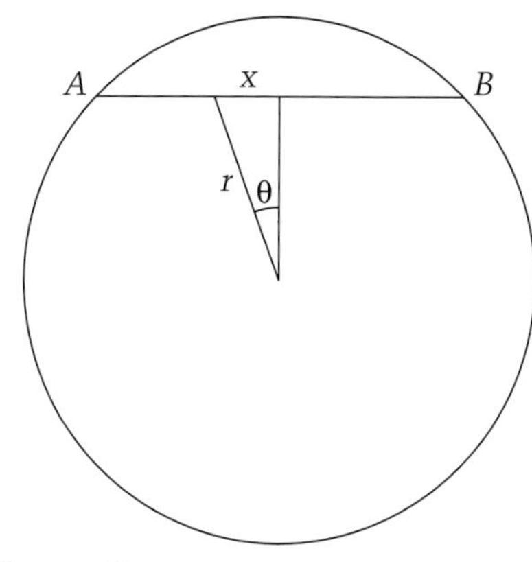

Figure 3

the opposite side of the small sphere. The difference of the two forces is the desired force. A second approach is to calculate the force of the sphere as if it were solid and subtract the force due to the mass imagined to fill the spherical hollow. The third approach is to calculate the force of the sphere as if it were solid and add a second force due to a "negative mass" filling the hollow. (The positive mass and the negative mass add together to produce the hollow.) Let's use the second approach:

$$F_1 = \frac{GMm}{d^2},\ F_2 = \frac{GM'm}{(d - R/2)^2},$$

where $M' = 1/8\,M$ because the spherical hollow has 1/2 the radius of the sphere. Therefore the force on the small sphere is

$$F_1 - F_2 = \frac{GMm}{d^2}\left(1 - \frac{1}{8(1 - R/2d)^2}\right).$$

Part B asked for an analysis of a tunnel drilled along a chord of the Earth connecting points A and B as shown in figure 3. At the position shown, there is a component of the gravitational force along the tunnel. This force is proportional to the mass that lies inside the sphere of radius r:

$$F = \frac{GM'm}{r^2}\sin\theta = \frac{G\rho\frac{4}{3}\pi r^3 m}{r^2}\sin\theta$$

$$= \frac{G\rho\frac{4}{3}\pi r^3 m}{r^2}\frac{x}{r} = \left(\frac{4\pi G\rho m}{3}\right)x = kx.$$

When the displacement is to the right, the force is to the left, so the correct form of the equation is

$$F = -kx.$$

Once again, we see that the path through the tunnel is simple harmonic motion with the same period (84 minutes) for a tunnel along an Earth diameter and also equivalent to the period of an orbiting satellite.

The tunnel would not be particularly feasible due to the difficulties of drilling through the Earth and the presence of friction, heat, and air resistance. If the Earth's molten core doesn't present enough difficulties, we will also have to worry about the walls melting and collapsing.

Part C asked if the straight tunnel provides for the fastest journey from A to B? We found that the period for any chord is 84 minutes, or a one- way travel time of 42 minutes. However, a chord is not the fastest path from A to B. It is best to travel a curved path that passes nearer the center of the Earth. Finding that curved path requires the use of the calculus of variations. Let's find a path with two straight segments that takes less time.

Consider the path from A along d and then to B as shown in figure 4. Since every cord requires 42 minutes for the trip, path w will require 1/2 that time, or 21 minutes. Path $d + x$ will also require 21 minutes, showing that path d requires less than 21 minutes.

We can find the minimum path by finding the path $d + x$ that maximizes x. First, we do some trigonometry:

$$\sin\phi = \frac{w}{R},\quad \sin(\phi + \theta) = \frac{d + x}{R},$$

$$\cos\theta = \frac{w}{d}.$$

We now write down an expression for x:

$$x = R\sin(\phi + \theta) - d$$

$$= R\sin(\phi + \theta) - \frac{w}{\cos\theta}$$

$$= R\sin(\phi + \theta) - \frac{R\sin\phi}{\cos\theta}.$$

At this point we could take the derivative of this equation and set it equal to zero, but a simple solution does not emerge. Alternatively, we can solve it numerically using a spreadsheet and finding θ for any given ϕ. As a specific example, let's choose $\phi = 20°$.

From the graph in figure 5, we obtain a maximum value for θ of 45° and a corresponding distance $x = 0.43R$. The resulting time savings can be determined by analyzing the equations for the simple harmonic oscillation... but that's another problem.

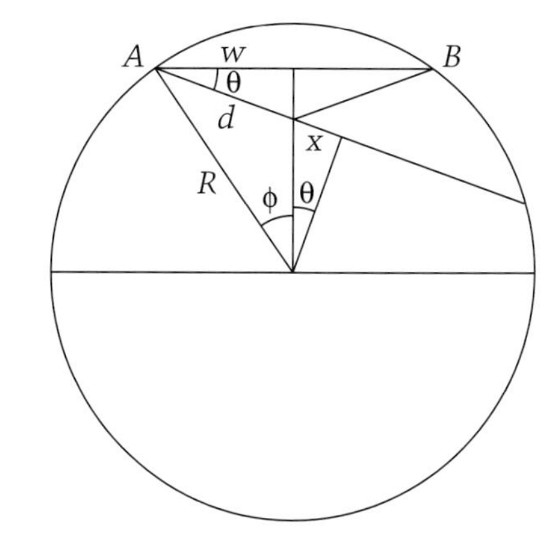

Figure 4

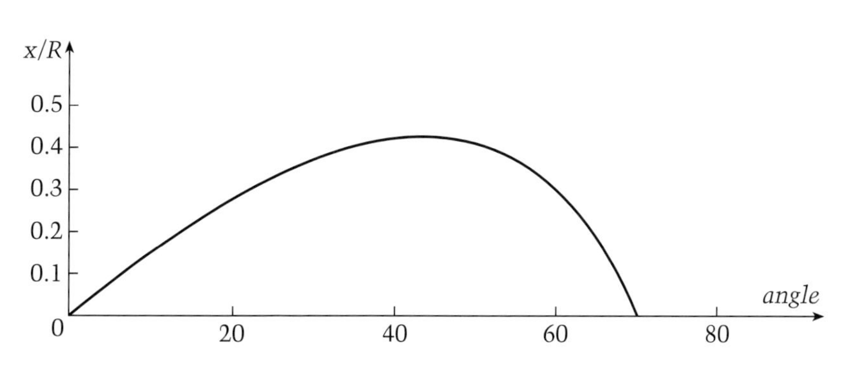

Figure 5

Magnetic vee

"The secret of magnetism, now explain that to me!
There is no greater secret, except love and hate."
—Johann Wolfgang von Goethe

One of the goals of modern day theoretical physics is to reduce all forces in Nature to manifestations of a single force. In fact, the 1999 Nobel Prize in Physics was awarded to the Dutch physicists Gerardus 't Hooft and Martinus J. G. Veltman for their contributions in unifying the electromagnetic interaction and the weak interaction. Their theoretical methods have also been applied in attempts to combine this electro-weak interaction with the strong interaction to form a grand unified theory. The biggest remaining problem is how to include the gravitational interaction in a "theory of everything."

During the early part of the 19th century, much effort went into establishing connections between electric forces and magnetic forces. Although there had been much discussion of possible connections, the first connection was established in 1819 during a classroom demonstration when the Danish scientist Hans Christian Oersted discovered that a current carrying wire deflected a compass needle. Furthermore, he discovered that the compass needle pointed at right angles to the current and that the compass needle pointed in the opposite direction when the current was reversed.

Within one week of the announcement of Oersted's discovery in 1820, the French physicist André Ampère formulated the right-hand rule: if you grasp a wire with your right hand so that your thumb points in the direction of the current, the compass points in the direction of the fingers. In modern language, the magnetic field lines are circles around the wire and your fingers point in the direction of the magnetic field.

A short time later, Ampère developed a formula for calculating the magnetic force between current-carrying wires. He also made the suggestion that all magnetic fields are due to currents, including those at

At center stage, reclining on a sofa with a control panel, we see the electromagnetic femme fatale, generating love and hate. While the repelled hate on the right can't get to her and creates war and destruction, the attracted love advances, flying to create beauty and harmony, though under the sofa we see already the serpent getting ready to cause trouble. The show is very closely observed by secret agents in the basement, where a science team is mixing up a stew of everything, with Igor bringing the prehistoric protoplasm as the basis for the mix. To this primal matter, out of which everything else comes, the scientist adds magnetism, time, space, gravity, the serpent with the apple of knowledge, and some herbs for flavor, in order to unite all known forces in a single-force stew. Whether we can digest it remains to be seen.

—T.B.

LOVE, HATE AND MAGNETS
UNIVERSE
THEORY OF EVERYTHING
ADD WATER
BIG BANG
MAGNETISM
GRAVITY
TIME
ANTI MATTER
LOVE AND HATE
SINGLE FORCE STEW.
TOM BUNK

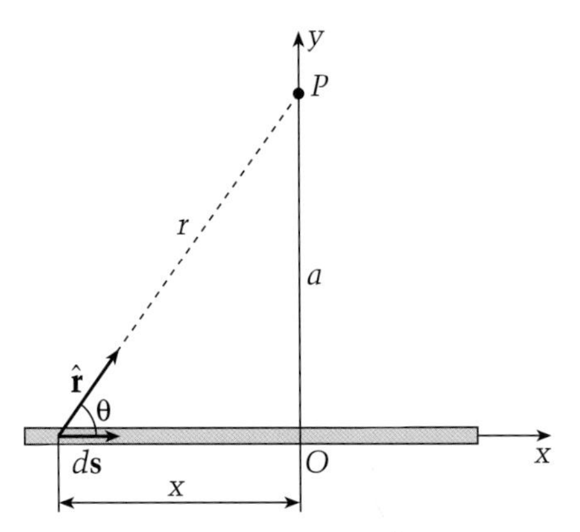

Figure 1

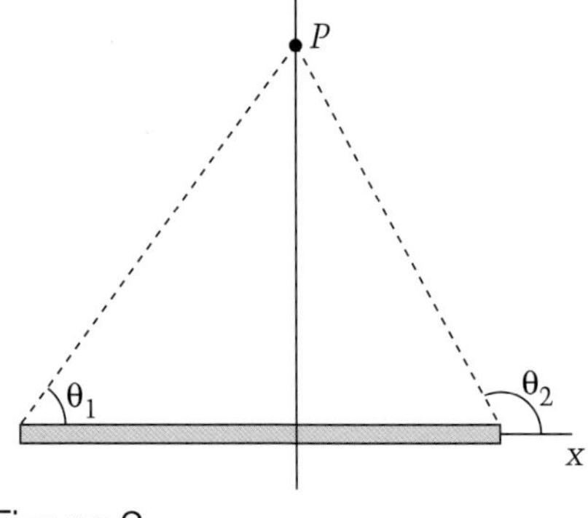

Figure 2

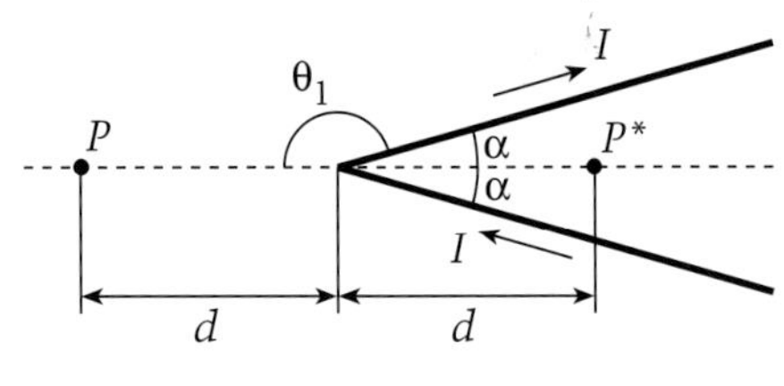

Figure 3

the atomic level. Note that this occurred three-quarters of a century before the discovery of the electron and the publication of Niels Bohr's theory of the hydrogen atom a decade later.

During this time, Jean Baptiste Biot and Félix Savart obtained a quantitative expression giving the contribution $d\mathbf{B}$ to the magnetic field at a point P due to an element of current $Id\mathbf{s}$. The full expression, now known as the Biot-Savart law, is

$$d\mathbf{B} = \frac{\mu_0}{4\pi}\frac{Id\mathbf{s}\times\hat{\mathbf{r}}}{r^2},$$

where $\hat{\mathbf{r}}$ is a unit vector directed from the current element to the point P and μ_0 is the permeability of free space with the value $4\pi \cdot 10^{-7}$ T · m/A.

To find the total magnetic field created at point P by a current of finite length, we must sum up the contributions from all of the current elements:

$$\mathbf{B} = \frac{\mu_0}{4\pi}\int\frac{Id\mathbf{s}\times\hat{\mathbf{r}}}{r^2}.$$

We must be very careful when evaluating this integral as the integrand is a vector quantity and we must take its direction into account.

Notice the similarity of the Biot-Savart law with Coulomb's law. Here a current element produces a magnetic field that varies as the inverse-square of the distance from the current element. In Coulomb's law, a point charge produces an electric field that varies as the inverse-square of the distance from the point charge.

However, the directions of the two fields are very different. In Coulomb's law the electric field is radial; it points toward or away from the point charge. In the Biot-Savart law, the magnetic field is perpendicular to both the current element and the radius vector. The magnetic field points out of the plane determined by the current element and the point P.

Let's apply the Biot-Savart law to find the magnetic field generated by a thin, straight wire carrying a constant current I. Let's set up the geometry as shown in figure 1. The wire is along the x-axis with the current in the positive x-direction. The point P is along the positive y-axis at a distance a from the origin O.

Using the right-hand rule, we see that each current element produces a contribution to the magnetic field that points out of the page. Therefore, the total magnetic field points directly out of the page and we only need calculate its magnitude. This means that we can replace $d\mathbf{s}\times\hat{\mathbf{r}}$ with $dx \sin\theta$, where θ is the angle between the direction of the current element and the direction to point P as shown in figure 1.

Before we go on, we note that we have more variables than we need. If we choose a given current element, the values of r, x, and θ are all specified. Therefore, we must express two of these three variables in terms of the third variable before we carry out the integration. Let's choose to express everything in terms of θ. We then have

$$r = \frac{a}{\sin\theta}$$

and

$$x = \frac{-a}{\tan\theta}.$$

Taking the derivative of the last expression, we obtain

$$x = \frac{-a}{\tan\theta}.$$

After making these substitutions, we are left with calculating the integral

$$x = \frac{-a}{\tan\theta}.$$

The angles are defined in figure 2.

If we look at the special case of an infinitely long, straight wire, $\theta_1 = 0$ and $\theta_2 = \pi$, and the magnetic field is given by

$$B = \frac{\mu_0 I}{2\pi a}.$$

Our problem is based on one that was given at the International Physics Olympiad that was held in Padua, Italy, in July 1999. A very long, thin, straight wire, carrying a constant current I, is bent to form a "V" of half-angle α as shown in figure 3.

A. What are the directions of the magnetic field at points P and P^*?

B. What is the magnitude of the magnetic field at point P?

C. What is the magnitude of the magnetic field at point P^*?

Solution

Part A: Using the right-hand rule, we see that the direction of the magnetic field produced by each segment of both wires is out of the page at point P and into the page at point P^*.

Part B: Symmetry tells us that the

total field at point P is twice that generated by each half of the vee. To calculate the magnetic field produced by the upper wire, we use the formula for a current segment derived in the problem. The first angle θ_1 is defined for the left-hand end of the segment. It is the angle between the current and the position vector of the point P. Therefore,

$$\cos\theta_1 = \cos(\pi - \alpha) = -\cos\alpha.$$

The second angle θ_2 is defined in the same way for the right-hand end of the segment. Because the wire is infinitely long in this direction, the angle is effectively 180° and

$$\cos\theta_2 = \cos\pi = -1.$$

Finally, the perpendicular distance between the (extended) wire and the point P is equal to $d\sin\alpha$. Remembering to multiply by 2 for the two current segments, we now have

$$B = 2\frac{\mu_0 I}{4\pi a}(\cos\theta_1 - \cos\theta_2)$$
$$= \frac{\mu_0 I}{2\pi}\frac{1-\cos\alpha}{d\sin\alpha} = \frac{\mu_0 I}{2\pi d}\tan\frac{\alpha}{2}.$$

For point P^*

$$\cos\theta_1 = \cos(-\alpha) = \cos\alpha$$

and

$$\cos\theta_2 = \cos(-\pi) = -1.$$

Therefore,

$$B = \frac{\mu_0 I}{2\pi}\frac{1+\cos\alpha}{d\sin\alpha} = \frac{\mu_0 I}{2\pi d}\cot\frac{\alpha}{2}.$$

You can also solve part B by treating point P^* as if it were outside a vee with half-angle $\pi - \alpha$ carrying current I in the opposite direction. Then

$$B = \frac{\mu_0 I}{2\pi d}\tan\left(\frac{\pi-\alpha}{2}\right) = \frac{\mu_0 I}{2\pi d}\cot\frac{\alpha}{2}.$$

As a third method, you can use the superposition principle. The problem is equivalent to two crossed infinite wires plus a vee on the left carrying current I in the clockwise

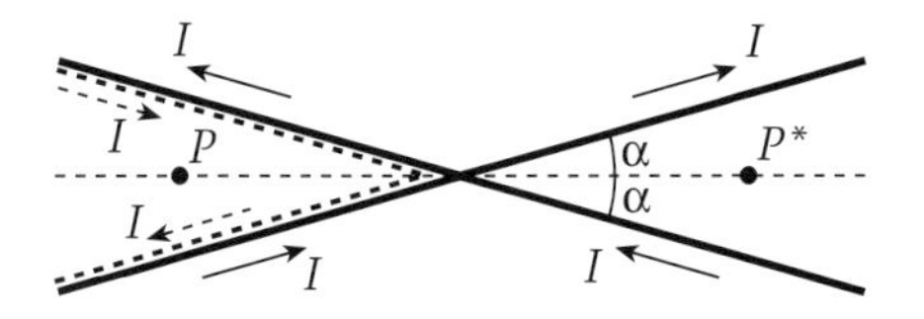

Figure 4

direction as shown in figure 4. The currents in the two vees on the left cancel each other, leaving the original situation. The superposition of the fields due to the vee on the left and the crossed infinite wires gives us the answer.

Each of the infinite wires produces a magnetic field given by

$$B = \frac{\mu_0 I}{2\pi d\sin\alpha}$$

into the page. The contribution to the magnetic field due to the vee on the left is out of the page and given by our answer for point P. Therefore, the total field is

$$B = \frac{\mu_0 I}{2\pi d}\left(\frac{2}{\sin\alpha} - \tan\frac{\alpha}{2}\right) = \frac{\mu_0 I}{2\pi d}\cot\frac{\alpha}{2}.$$

Rolling wheels

"The fly sat upon the axel tree of the chariot-wheel and said, 'What a dust do I raise!' "
—Aesop (6th century B.C.)

INVENTING THE BETTER downhill vehicle confounds engineers. If the measure of success is the speed of the car or skateboard at the bottom of an incline, how does one choose the proper wheels? Solid wheels, spoked wheels, cylinders, rotating casters, spheres, hoops, and a combination of these shapes inhabit our palette of possibilities.

The decision making certainly requires an understanding of rolling bodies. Although most people appear quite familiar with wheels, a few simple puzzles reveal the subtlety of the motion. Ask someone to trace the motion of a point on the outside of a bicycle wheel as it moves across your line of sight. Compare this with the actual trace as you roll a disk and construct the path. The path is a cycloid—not the first guess of most people!

A simpler puzzle requires you to place one quarter next to another with the faces of Washington upright. As the quarter on the right rolls without slipping on the other, what will be the orientation of Washington when the coin reaches the left-hand side of the fixed quarter? Try it and then explain it.

Finally, take a ruler and rest it on the wheels of a cart. As the cart moves forward, will the ruler move with the cart? If not, why not?

When viewing a rolling (without slipping) wheel, we notice that the bottom of the wheel is at rest with respect to the ground. There is no skidding here. The speed of the center of the wheel is equal to the speed of the wheel. When the wheel moves one meter to the right, the center of the wheel moves one meter to the right. The top of the wheel is also moving to the right, but at what speed?

An analysis of the kinetic energy of the wheel will help reveal the motion of the wheel. If the wheel is a solid cylinder, the kinetic energy of this fixed rotating cylinder is equal to the kinetic energy of each

What keeps humankind moving and rolling forward? There are many kinds of wheels to get ahead, to follow one's own bliss, inspiration, motivation, thoughts, or dreams, or just to let the good times roll. Some people like to roll in groups, some by using wind or water; some prefer mechanical constructions to move effortlessly, some push heavy rock wheels; some like to move dreaming in a barrel, others meditating on a self-painted circle; some prefer iron tanks, and some dudes choose a skateboard; Sufis like to spin, and others use their aureole, while some turn themselves into a wheel. One can get ahead by talking or resting in a coffin and being pulled all the way. The possibilities are endless, and they keep most of us busy moving forward until the end of the spiral road.

—T.B.

WALLSTREET
BLA BLA BLA BLA BLA BLA BLA BLA BLA BLA BLA BLA
UNITED STATES OF AMERICA
QUARTER DOLLAR
TOM BUNK

of the small elements

$$K = \sum \frac{1}{2} m v^2 .$$

Since each mass has the same angular velocity ω, the velocity of each small mass is equal to $r\omega$, where r is the distance from the center of rotation:

$$K = \sum \frac{1}{2} m r^2 \omega^2$$
$$= \frac{1}{2}\left(\sum m r^2\right)\omega^2 = \frac{1}{2} I \omega^2 ,$$

where I is defined as the rotational inertia of the rotating object. The rotational inertia is a measure of the difficulty of starting the rotation of an object in the same way mass is indicative of the difficulty of accelerating an object.

For a rolling wheel, the instantaneous rotation is about the point in contact with the ground. The definition of kinetic energy is unchanged but the moment of inertia about this new point must be determined. The parallel axis theorem can be used to determine this new rotational inertia:

$$I = I_{cm} + md^2,$$

where d is the distance of the rotating axis from the center of mass. In the case of a rolling body, the instantaneous point of rotation is the ground, which is one radius from the center of mass of the wheel. The kinetic energy of this rotating wheel is

$$K = \frac{1}{2} I \omega^2 = \frac{1}{2}\left(I_{cm} + mR^2\right)\omega^2$$
$$= \frac{1}{2} I_{cm}\omega^2 + \frac{1}{2} m R^2 \omega^2$$
$$= \frac{1}{2} I_{cm}\omega^2 + \frac{1}{2} m v_{cm}^2 .$$

Our analysis of the rolling wheel has become much simpler. The kinetic energy of the rolling wheel is equal to the rotational kinetic energy of the wheel about its center of mass plus the kinetic energy of the entire mass moving with the velocity of the center of mass.

That being the case, we can see that the center of the wheel moves with v_{cm}, and the wheel spins with an instantaneous angular velocity ω about the point in contact with the ground. All points have a tangential speed equal to ωr. For the point on the ground, the velocity is zero, for the central point the velocity is ωR, and for the top point the velocity is $2\omega R$. The top of the wheel moves at twice the speed of the center of mass.

We can use conservation of energy to find the speed of a wheel as it rolls down an incline. The loss in potential energy is equal to the gain in both translational and rotational kinetic energy:

$$mgh = \frac{1}{2} m v_{cm}^2 + \frac{1}{2} I \omega^2 .$$

Different wheels will have different speeds owing to their different rotational inertias.

To fully understand the rolling wheel, we must delve into the dynamics of the motion. Why would the wheel rotate—why would anything rotate? If I were to place a pen on the table, how might you apply a force to rotate the pen? A force at the center of mass will accelerate the pen, but a force at any other point is required to rotate the pen. This "off-center" force is called a torque and is defined as the cross product of the distance and the applied force:

$$\boldsymbol{\tau} = \mathbf{r} \times \mathbf{F}.$$

The force of gravity acts on the center of mass and cannot be responsible for the rotation. Similarly, the normal force of the surface on the wheel has no r (off-center component). It is the frictional force at the bottom of the wheel that creates the rolling.

Designing a wheel requires other considerations than the analysis above. The engineer must take into account the strength of materials needed, the ease of manufacturing, the cost of materials and production, the need for maintenance and repair, the durability, and the aesthetics. Maximizing the speed of a wheel and ensuring safety, low cost, and durability often requires compromise and creativity.

Part of this problem was originally given in Bucharest, Romania, in 1972.

A. Show that all solid spheres will arrive at the bottom of the incline with the same speed, independent of their radii. The rotational inertia of a solid sphere is $2/5\ mR^2$.

B. Determine the relative speeds of a cylinder, a hoop, and a solid sphere (all of the same mass) at the bottom of an incline. The rotational inertias are $1/2\ mR^2$, mR^2, and $2/5\ mR^2$.

C. Consider three cylinders of the same length, outer radius, and mass. The first cylinder is solid. The second is a hollow tube with walls of finite thickness. The third is a hollow tube with walls of the same finite thickness, filled with a liquid of the same density (both ends are closed by thin plates of negligible mass). Find and compare the linear and angular accelerations for the cylinders when they are placed on an inclined plane of angle θ. The coefficient of friction between the cylinders and the inclined plane is μ. Friction between the liquid and the wall of the cylinder is negligible.

Solution

The first two problems, considered standard fare for first-year college physics, were solved correctly by Alex Rifkin, Michelle Chung, and Victoria Butta of Amity Regional High School in Woodbridge, Connecticut. Victoria tried the more difficult and subtle third problem. Their teacher, A. Hovey, correctly solved most of this problem.

A. To show that all uniform, solid spheres arrive at the bottom of the incline with the same speed, independent of their radii and masses, we use conservation of energy. The loss in potential energy is equal to the gain in kinetic energy—both translational and rotational:

$$mgh = \frac{1}{2} m v^2 + \frac{1}{2} I \omega^2$$

or

$$mgh = \frac{1}{2}mv^2 + \frac{1}{2}\left(\frac{2}{5}mR^2\right)\left(\frac{v^2}{R^2}\right).$$

Therefore

$$v = \sqrt{\frac{10}{7}gh}.$$

B. Any object's moment of inertia can be written as kmR^2, where k is a constant that depends on the shape of the object. Comparing the relative speeds of a cylinder ($k = 1/2$), a hoop ($k = 1$), and a solid sphere ($k = 2/5$) of the same mass requires us to solve Part A for the general equation:

$$mgh = \frac{1}{2}mv^2 + \frac{1}{2}I\omega^2,$$

which yields

$$v = \sqrt{\frac{2gh}{1+k}}.$$

Notice that the mass of the objects does not appear in the answer and, therefore, does not affect the speed. All objects with the same shape have the same motion. For our three objects, we get

cylinder: $v = \sqrt{\frac{4}{3}gh}$

hoop: $v = \sqrt{gh}$

sphere: $v = \sqrt{\frac{10}{7}gh}$

C. Part C requires us to compare the linear and angular accelerations for three cylinders when they are inclined at an angle α. The cylinders all have the same length, outer radius, and mass. The first is solid, the second is a hollow tube with walls of finite thickness, and the third is a hollow tube with walls of the same finite thickness, filled with a liquid of the same density.

This problem requires an analysis of the rolling dynamics of the cylinders. We will need to find the friction required to roll without slipping since the static friction can take on any value less than or equal to μF_N.

For the linear motion we have

$$\sum F = ma$$

or

$$F_g \sin\theta - F_f = ma.$$

And for the rotational motion,

$$\sum \tau = I\alpha$$

or

$$RF_f = I\alpha = I\frac{a}{R}.$$

Therefore,

$$a = \frac{R^2 F_f}{I}.$$

Solving for the force of friction F_f and the acceleration a, we get

$$F_f = mg\sin\theta \frac{I/mR^2}{1 + I/mR^2}$$

and

$$a = g\sin\theta \frac{1}{1 + I/mR^2}.$$

The limiting angle for rolling without sliding will occur when the frictional force is equal to the normal force F_N:

$$\mu mg\cos\theta = mg\sin\theta \frac{I/mR^2}{1 + I/mR^2},$$

yielding

$$\tan\theta = \mu(1 + mR^2/I).$$

We must now find the rotational inertia of each cylinder.

1. The first cylinder has a rotational inertia $I = mR^2/2$. Using the equations above, we find that

$$a = \frac{2}{3}g\sin\theta$$

and

$$\tan\theta = 3\mu.$$

2. The second cylinder has a rotational inertia $I = m(R^2 + r^2)$, where r is the inner radius of the cylinder. We can find r by recognizing that the solid cylinder and this hollow tube have the same mass and therefore the densities ρ must be different by a factor n, where

$$\rho = \rho_{wall} = n\rho_{solid}.$$

Setting the mass of the solid cylinder equal to that of the tube

$$\rho\pi R^2 l = n\rho\pi l(R^2 - r^2),$$

we get

$$r^2 = R^2\left(\frac{n-1}{n}\right).$$

Therefore

$$I = \frac{1}{2}m(R^2 + r^2) = \frac{1}{2}mR^2\left(\frac{2n-1}{n}\right).$$

The corresponding acceleration and limiting angle are

$$a = \frac{2n}{4n-1}g\sin\theta$$

and

$$\tan\theta = \frac{4n-1}{2n-1}\mu.$$

3. The third cylinder has the same dimensions as the tube but has less mass rotating; that is, the liquid does not rotate due to a lack of friction between it and the walls:

$$r^2 = R^2\left(\frac{n-1}{n}\right),$$

$$m_{tube} = \frac{\pi R^2 l - \pi r^2 l}{\pi R^2 l}m = \left(1 - \frac{r^2}{R^2}\right)m.$$

The rotational inertia is

$$I = \frac{1}{2}m_{tube}(R^2 + r^2) = \frac{1}{2}\left(1 - \frac{r^2}{R^2}\right)m\left(R^2 + r^2\right) = \frac{1}{2}mR^2\left(\frac{2n-1}{n^2}\right).$$

The corresponding acceleration and limiting angle are

$$a = \frac{2n^2}{2n^2 + 2n - 1}g\sin\theta$$

and

$$\tan\theta = \frac{2n^2 + 2n - 1}{2n - 1}\mu.$$

Since all of the angular accelerations α are equal to a/R, the ratios of the linear and angular accelerations are

$$1 : \frac{3n}{4n-1} : \frac{3n^2}{2n^2 + 2n - 1}.$$

The ratios of the tangents for the limiting angles are

$$1 : \frac{4n-1}{3(2n-1)} : \frac{2n^2 + 2n - 1}{3(2n-1)}.$$

When the cylinders exceed the largest limiting angle and none of them have the necessary friction to roll without sliding, they all have the same linear acceleration:

$$F_g \sin\theta - F_f = ma$$

$$F_f = \mu mg \cos\theta$$

The angular accelerations are given by

$$\alpha = \frac{RF_f}{I} = \frac{R\mu mg\cos\theta}{I}.$$

Since the rotational inertia is different for each cylinder, the corresponding angular accelerations are

$$\alpha_1 = \frac{2\mu\cos\theta}{R},$$

$$\alpha_2 = \frac{2\mu\cos\theta}{R}\frac{n}{2n-1},$$

and

$$\alpha_3 = \frac{2\mu\cos\theta}{R}\frac{n^2}{2n-1}.$$

Batteries and bulbs

"The habit of analysis has a tendency to wear away the feelings."
—John Stuart Mill

THE RULES OF BASEBALL ARE the same for everyone—from the smallest Little Leaguer to the biggest Major Leaguer. However, we expect the expertise of the player to increase with age. The laws of physics are the same for everyone. We expect that the problems adults can tackle are more difficult than the ones we give children. That's usually true—but not always.

Given a flashlight battery, a flashlight bulb, and a single piece of wire, hold them together to make the bulb light. We have seen adults take more than an hour to light the bulb! And yet, this is the first activity in a lesson on circuit electricity for fifth graders. Experience has shown us that fifth graders are much more successful at this task than adults. Experience has also shown us that studying Ohm's law does not guarantee that one can successfully analyze circuits containing batteries and bulbs. Elementary education majors who have studied *Batteries and Bulbs* in physical science courses at college have often reported that their friends and spouses in electrical engineering did not have the conceptual understanding to help them with their homework.

Batteries and Bulbs was developed and written by the Elementary Science Study project in the mid 60s. Gerry Wheeler, currently the Executive Director of the National Science Teachers Association, wrote the final version of this popular unit in 1968. It stresses the development of a logical framework for understanding electric circuits and was an early example of the kind of inquiry supported by the National Science Foundation.

After unsuccessfully trying to light the bulb using arrangements such as that shown in figure 1, most students discover that they must

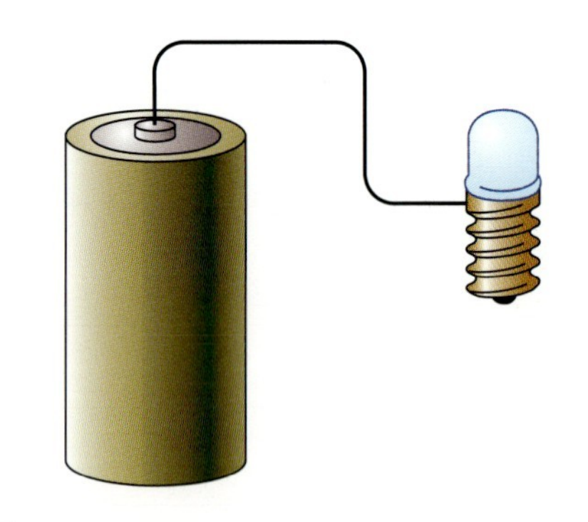

Figure 1

A lot of action in the one-ring circus called Circuit Maximus. The divine charge coming out of the heavenly battery with a bang is passed from one enlightened thinker to another. We see brilliant giants having a ball—Euclid, Confucius, Copernicus, Galileo, Buddha, Newton, Darwin, Marx, Mme. Curie, Freud, Einstein, the computer, and Death, who reconnects with the battery again to complete the circuit. Otherwise there would be no illuminated brains to enlighten us and we would still live in darkness. The very mixed audience is electrified watching the juggling clown in the center of the ring trying to keep balance on top of the world, dropping now and then a light or a bright brain.

—T.B.

DIVINECELL®
FAMILY OF MAN SIZE
1,5 VOLT
LIGHTEN UP
WHO'S IN CHARGE?
BANG!
CIRCUIT MAXIMUS

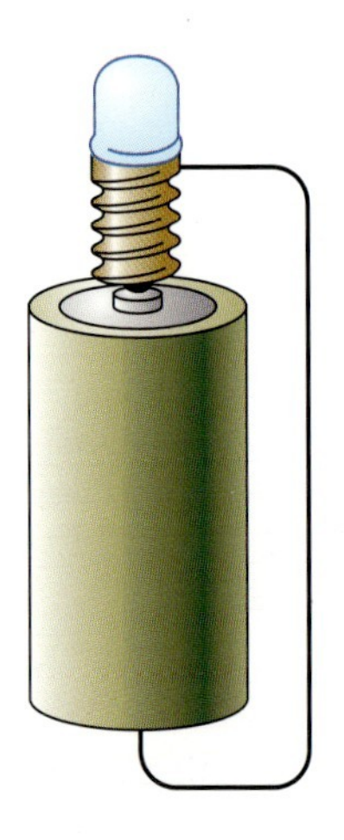

Figure 2

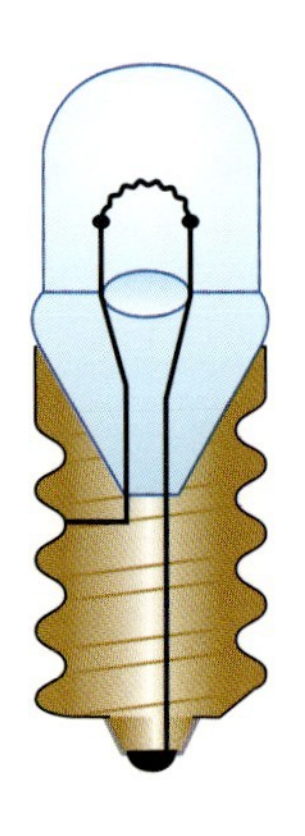

Figure 3

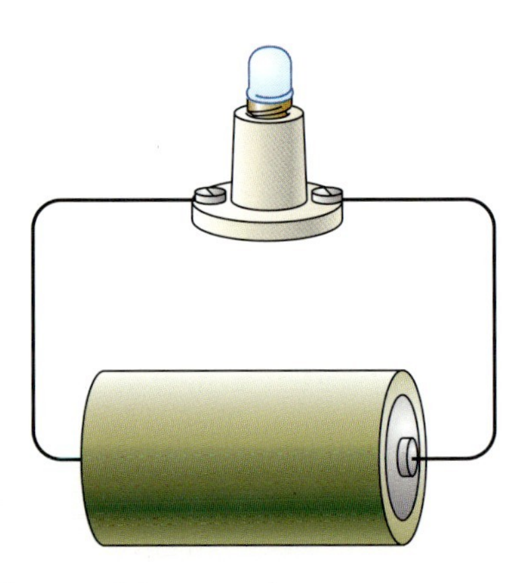

Figure 4

use two parts of each of the objects: the two ends of the wire, the two ends of the battery, and the two metal parts of the bulb. The two parts of the bulb are the metal tip and the metal around the base. Whenever all of these six parts are connected in pairs—no matter how you do it—the bulb lights. One such way is shown in figure 2. Can you find the other three ways of doing this?

All four ways have one thing in common—a continuous conducting path allows charges to flow from one end of the battery through the light bulb to the other end of the battery. This path is known as a *complete circuit*. By examining a broken flashlight bulb you can see that there is a continuous conducting path from one metal part through the light bulb to the second metal part as shown in figure 3. (The complete circuit is preserved when bulbs are screwed into sockets as shown in the rest of the figures.)

Combining the concept of a complete circuit with the law of conservation of charge leads to the conclusion that electricity flows from one end of the battery and back into the other. All of the charge that leaves one end returns to the other end. Charge does not get lost along the way.

If we light one bulb with one battery as shown in figure 4, we find that the battery runs down in something less than a day. (The actual time depends on the type of battery and the type of bulb.) However, if we connect the wire directly between the two ends of the battery, the battery runs down in less than an hour and the wire usually gets too hot to touch. We infer that the current is larger through the wire than through the bulb. We say that the bulb provides more *resistance* to the flow of charge than the wire. Let's denote the brightness of the single bulb connected to a single battery as the standard brightness.

Let's now look at what happens if we use a single battery to light two identical bulbs. We start by connecting the bulbs as shown in figure 5, an arrangement known as *series*. Because there is only a single path through the two bulbs, whatever charge flows through one of the bulbs must flow through the other bulb. If we use identical bulbs, we notice that the two bulbs have the same brightness. We also notice that these bulbs are dimmer than the standard brightness. If we leave the bulbs lit, we discover that the battery lasts longer than the battery in the standard circuit. From this we infer that there is less current in the

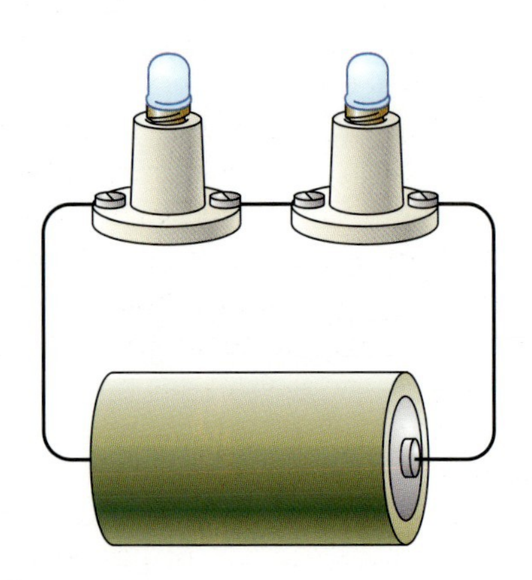

Figure 5

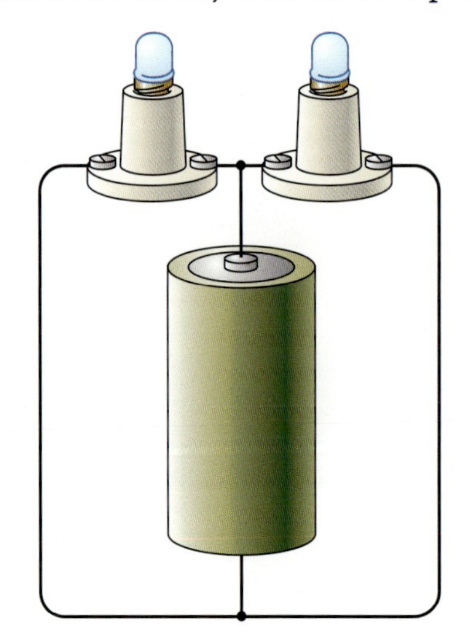

Figure 6

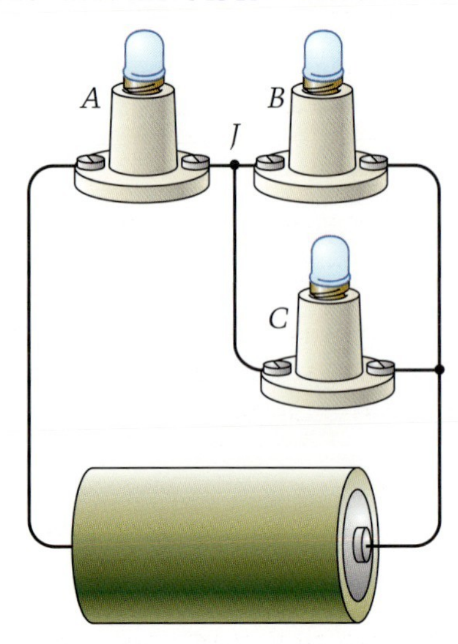

Figure 7

circuit and, therefore, the resistance of the two bulbs in series is greater than the resistance of a single bulb. (We assume that you are already familiar with the concept of resistance. If not, we would spend time developing this concept more carefully.) From these observations we infer that the brightness of a bulb is a rough measure of the current passing through the bulb. We will assume this from now on.

We can also connect two identical bulbs to the battery so that each bulb is on its own path from one end of the battery to the other, an arrangement known as *parallel*. (Note that the paths may share some of the same wires, as seen in figure 6.) In this case, each bulb has the standard brightness. The current through one bulb does not pass through the other bulb. You can check this by disconnecting either bulb and noticing that the other bulb is not affected. This means that the current through the battery must be twice that in the standard circuit and the total resistance of the combination must be one-half the resistance of a single bulb. You can verify this experimentally by letting the battery run down. It does so in approximately one-half the time. In general, adding a path in parallel always reduces the resistance of the combined paths.

Let's use these ideas to analyze the circuit in figure 7 containing three identical bulbs. Which of the three bulbs is brighter and why? How do the brightnesses of the other two bulbs compare to each other? Notice that the entire current from the battery must pass through bulb *A*. Therefore, it must be the brightest. At junction *J*, the current must split. Because each path following the junction contains a single bulb, the two paths are equivalent and the current must split equally. Conservation of charge tells us that the currents through bulbs *B* and *C* are each one half of the current through bulb *A*. Therefore, bulbs *B* and *C* are equally bright but dimmer than bulb *A*.

We can check our understanding of the model by answering the following questions about this circuit. (1) What happens to the brightness of the bulbs when bulb A is removed from its socket? (2) What happens to the brightness of the bulbs when bulb C is removed from its socket? (3) What happens to the brightness of the bulbs when a wire is connected across the two terminals of socket *A*? (4) What happens to the brightness of the bulbs when a wire is connected across the two terminals of socket *C*? Be sure to write down your answers to these questions before you read on.

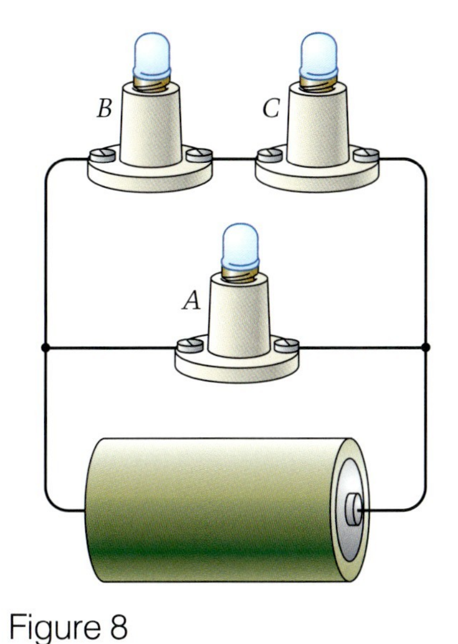

Figure 8

Figure 9

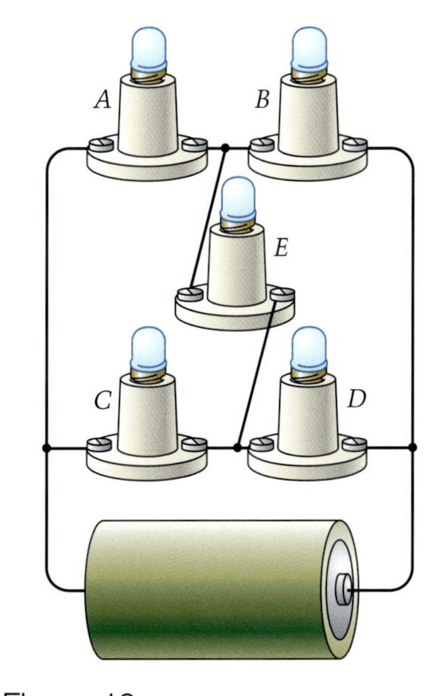

Figure 10

Now that you've committed your answers to writing, we are ready to look at the answers to these questions.

(1) Bulbs *B* and *C* will go out as the single path to the battery has been broken and there is no current.

(2) After the removal of bulb *C*, bulbs *A* and *B* are wired in series and are equally bright. Removing bulb *C* removes a parallel path to the right of the junction and therefore increases the resistance of this part of the circuit. This, in turn, reduces the current from the battery. Therefore, bulb *A* becomes dimmer. Two competing effects determine the brightness of bulb *B*. There is less current from the battery but it all passes through bulb *B*. Qualitative arguments do not tell us the answer, but observation tells us that bulb *B* gets brighter.

(3) Connecting a wire across the terminals of socket *A* provides a very low-resistance path around bulb *A*, so bulb *A* goes out. This also reduces the resistance in the circuit, so there is more current from the battery. Therefore, bulbs *B* and *C* brighten.

(4) Connecting a wire across the terminals of socket *C* provides a very low-resistance path around both bulb *B* and around *C*. Therefore, they both go out. Because this also reduces the resistance of the circuit, bulb *A* brightens.

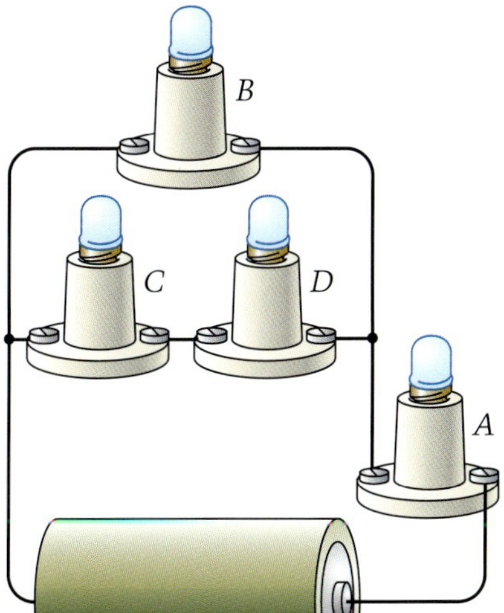

A. For the first part of our problem, examine the circuits shown in figures 8 and 9. In each case, which bulbs are the brightest and which bulbs are the dimmest? Repeat the questions asked above for each of these circuits.

B. The second part of our problem is a modification of one of the questions on the exam given to select the members of the 2000 US Physics Team. Which of the identical bulbs in the circuit in figure 10 are the brightest? Which are the dimmest? What happens to the brightness of the bulbs for each of the following? (1) Bulb *A* is removed from its socket. (2) Bulb *E* is removed from its socket. (3) Bulbs *A* and *E* are both removed from their sockets. (4) Bulbs *A* and *D* are both removed from their sockets. (5) A wire is connected across the terminals of socket *A*. (6) A wire is connected across the terminals of bulb *E*. (7) Wires are connected across the terminals of sockets *C* and *E*. (8) Wires are connected across the terminals of sockets *A* and *D*.

Solution

Our faithful contributor Art Hovey from Amity Regional High School in Connecticut and Güney Gönenç from Ankara, Turkey, submitted correct solutions for these circuits.

In figure 8, bulb *A* is the brightest; it has the standard brightness as it has its own path to the battery. Bulbs *B* and *C* are equally bright because they are wired in series and have the same current. They are dimmer than bulb *A* because their path has more resistance. For real bulbs that do not obey Ohm's law, the current through these bulbs will not be one-half of that through bulb *A*.

We now look at the questions about removing or short-circuiting various bulbs. (1) If bulb *A* is removed from its socket, the other two bulbs do not change brightness as they have an independent path to the battery. Note that we are assuming that the battery is ideal—that is, it can provide any amount of current demanded by the circuit. (2) If bulb *C* is removed from its socket, bulb *B* will go out as its path to the battery is broken. Bulb *A* is not affected. (3) If a wire is connected across the terminals of bulb *A* to short-circuit it, all of the bulbs will go out as the short-circuit across bulb *A* also shorts out the path through bulbs *B* and *C*. (4) If bulb *C* is short-circuited, it goes out. Bulb *A* is not affected. Bulb *B* gets brighter because there is less resistance on its path. In fact, it becomes as bright as bulb *A* as the two are now wired in parallel.

In figure 9, bulb A is the brightest as all of the current passes through it. Bulbs *C* and *D* are the dimmest and bulb *B* has an intermediate brightness because its path has less resistance. (1) Removing bulb *A* breaks the circuit to the battery and all of the bulbs go out. (2) Removing bulb *C* causes bulb *D* to go out as its path is broken. Bulb *A* will get dimmer because removing one of the parallel paths increases the resistance in *A*'s path. Because bulb *A* and bulb *B* are now in parallel, they will be equally bright. (3) Shorting bulb *A* causes bulb *A* to go out and converts this circuit into the circuit of figure 8. Removing the resistance of bulb *A* causes all of the other bulbs to increase in brightness. (4) Shorting bulb *C* removes some of the resistance in the path of bulb *A*, so it will brighten. Bulbs *B* and *D* are now in parallel and have the same brightness. Two competing effects determine the change in brightness of bulb *B*. There is more current from the battery but only half of it flows through bulb *B*. If we introduce the concept of voltage, we can argue that bulb *A* getting brighter requires bulb *B* to get a bit dimmer.

If we wire up the bulbs as shown in figure 10, we immediately notice that all of the bulbs have the same brightness except for bulb *E*, which does not glow. We can use symmetry to argue that there should be no current through bulb *E*. (1) If bulb *A* is removed, the circuit becomes the same as that in figure 9. The symmetry for bulb *E* is removed and it will now glow. The resistance of the circuit increases so there is less current from the battery. But all of the current now must pass through bulb *C*. The latter effect wins and bulb *C* brightens. Bulbs *B* and *D* dim. (2) Removing bulb *E* from its socket has no affect, as there was no current through bulb *E* in the original circuit. (3) Removing bulbs *A* and *E* causes bulb *B* to go out. Bulbs *C* and *D* will not be affected. (4) Removing bulbs *A* and *D* leaves us with three bulbs wired in series. Bulb *E* will brighten and bulbs *B* and *C* will dim. (5) Shorting-circuiting bulb *A* provides a direct path for bulb *B* to the battery. Therefore, it will brighten to the standard brightness. Bulbs *C* and *E* are now in parallel so bulb *E* will brighten. Bulb *C* was in series with one other bulb, but is now in parallel with bulb *E* and the combination is in series with bulb *D*. This means that bulb *C* will dim. The resistance in the path of bulb *D* decreases—the resistance of one bulb to that of two bulbs in parallel—so bulb *D* will brighten. (6) Short-circuiting bulb *E* has no affect on the brightness of the other three bulbs, as there was no current through this branch. (7) Short-circuiting bulbs *C* and *E* at the same time also short-circuits bulb *A*. Therefore, these three bulbs will go out. The remaining bulbs each have a direct path to the battery and will brighten to the standard brightness. (8) Shorting bulbs *A* and *D* leaves us with three bulbs in parallel, so each will brighten to the standard brightness. ◘

Curved reality

"There are always two choices, two paths to take. One is easy. And its only reward is that it's easy."
—Unknown

ALL CURVES ARE NOT EQUIvalent. As young babies, perhaps as young as two or three months, we can distinguish between straight lines and curves. As we enter school, we learn geometry based on straight lines and the shapes that they create—triangles, squares, hexagons, and the like. During our fascination of discovering the mundane and esoteric properties of these polygons, we forget to ask about their relevance to the natural world. Certainly, we require squares to map out our rooms and gardens. We need triangles to determine the height of a flagpole from its shadow. We need straight lines to find the shortest distance between two points.

Does nature share our enthusiasm for straight lines? How often do we find straight lines in nature? The horizon appears to be a line, but we know that it must curve if the Earth is a sphere. The rays of light piercing through the clouds provide one example of the straight line in nature. The quarter moon illuminated by the Sun at just the correct angle provides us with another natural straight line. Is the edge of a crystal a straight line? Are there other examples? We would enjoy hearing from our readers as they expand this short list of natural straight lines.

Perhaps nature's straight lines are more subtle. Perhaps the lines are there not for our visual eyes but for our mind's eye. As a ball falls, its path is a vertical line. If we could only see the ball at all places at once, how beautiful it would be. When we measure the stretch of a spring with varying weights, we discover Hooke's law, which states that the stretch is directly proportional to the force. If we write that as an algebraic equation, we succinctly state $F = -kx$. When we graph this, we find nature's straight line. Every direct relation, from $x = vt$ to $F = ma$ to $V = IR$, is a discovery of a straight line in nature.

There are many ways to experience and deal with the world around us. One is the typical western analytical scientific exploration, which is to divide nature in many little parts—measure each one, give it a shape, a formula, a name, catalog it, and then try to put it back into the whole picture again—but to no avail. During the exploration the "inquired" part has been changed and has now become separated and just one part of the universe, which as everyone knows is one and indivisible. Even our scientist, who is trying to make some geometric forms out of the chaos around him, is himself an unfinished part of the picture—he hasn't even got feet to stand securely on solid ground. Though he seems to be blind, he is growing an impressive mind's eye, replacing his analytical brain. It may help him become a wise visionary some day.

—T.B.

F = qv × B = mv²/R
mv = qBR
√2v₀² cos²θ
x-sin
cos
A
B
C
D
E
G
TOM BUNK

Visually, if not straight lines, then does nature favor curves? All curves are not equivalent. Does nature favor the circle (the Greek's symbol of perfection) over the parabola? Does the hyperbola appear more often than the ellipse? Does the cycloid make more appearances than the catenary? Let's embark on a brief tour of some simple physics with an eye toward the curves we may discover along the way. Following the earlier notion of the path of a falling ball, what would be the paths of other moving objects?

The thrown ball, without air resistance, travels along a parabola. This is simply proven with the equations of motion for horizontal and vertical motion.

$$x = v_0 t \cos\theta,$$
$$y = -\frac{1}{2} g t^2 + v_0 t \sin\theta.$$

Eliminating the time between the two equations—allowing us to see the thrown object at all times—the equation of motion becomes:

$$y = \frac{-gx^2}{2v_0^2 \cos^2\theta} + \frac{x\sin\theta}{\cos\theta}.$$

This path is identical to the general equation of a parabola: $y = ax^2 + bx$. The curve of the parabola is determined by the initial angle and velocity of the throw. But all throws result in the parabolic shape.

A charged particle shot into a region of space containing a magnetic field experiences a force that is perpendicular to both its instantaneous velocity and the magnetic field. This perpendicular force serves as the centripetal force as the particle moves in a circle.

$$F = qv \times B = \frac{mv^2}{R},$$
$$mv = qBR.$$

Knowing the charge and magnetic field, we can measure the radius of the curved path and determine the momentum of the charged particle.

An orbiting satellite also follows

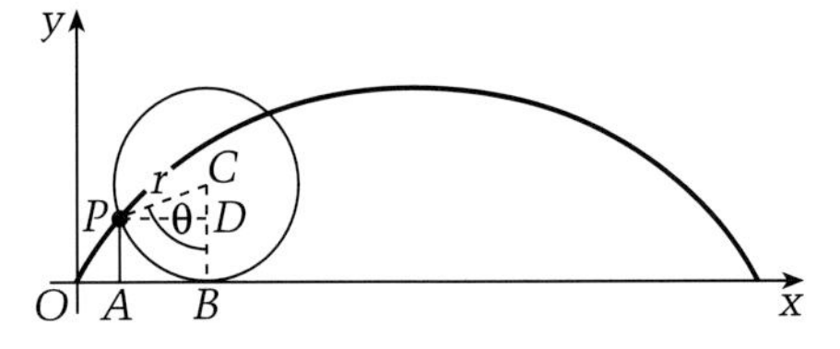

Figure 1

a curved path. In this situation, the gravitational attraction between the Earth and the satellite is the centripetal force. Since the speed is dependent on the distance, we can determine the mass of the Earth by the behavior of the satellite.

$$F = \frac{GM_E m}{R^2} = \frac{mv^2}{R},$$
$$v = \frac{2\pi R}{T},$$
$$M_E = \frac{4\pi^2 R^3}{T^2}.$$

A moon orbiting a planet provides us with the mass of the planet if we are able to measure the radius of the moon's orbit and its period.

The moon about the planet and the planets about the Sun do not travel in perfect circles. It is a testament to the experimental accuracy of Tycho Brahe's measurements that Kepler could not "settle" for the approximate circular path of the planets, but determined that they travel in ellipses. The eccentricity of Halley's comet with the Sun at one focus provides strong evidence of Kepler's mathematical insight.

We find hyperbolic paths when we observe the scattering of an alpha particle from the nucleus of a gold atom. This Rutherford scattering experiment demonstrated the existence of a nucleus, which contains (almost) all of the mass and all of the positive charge of the atom packed into an extremely small volume. Knowing the deflection angle of the alpha particle, we can determine the impact parameter (how close the particle comes to the nucleus) or vice versa. The derived equation is:

$$\tan\frac{\Theta}{2} = \frac{q_1 q_2}{msv_0^2},$$

where Θ is the scattering angle and s is the impact parameter.

A particularly interesting curve is traced out by a point on a rolling wheel (figure 1). You sometimes observe this shape at night from the reflectors on bicycle tires.

$$x = OA = OB - AB = OB - PD = r\theta - r\sin\theta = r(\theta - \sin\theta),$$

$$y = AP = BD = BC - CD = r - r\cos\theta = r(1 - \cos\theta).$$

Our problem challenges you to traverse the terrain of some of these shapes.

A. A trajectory with air resistance can be assumed to have a frictional force proportional to the velocity.

$$F_t = -bv.$$

Derive an equation for the trajectory and sketch the path for different values of b.

B. A charged particle is projected into a region of space containing crossed (perpendicular) electric and magnetic fields. It enters the region perpendicular to both fields.

(1) If the particle traverses this region without any deflection, show that the speed of the particle equals E/B. This is a way we can create a velocity selector.

(2) If the particle traverses the crossed fields from the opposite direction with the speed E/B, does it remain undeflected?

(3) If the particle traverses the crossed fields in the original direction, but with speed other than E/B, determine the curved path.

Solution

In part A of the problem, we asked you to derive an equation for a trajectory that has a frictional force proportional to the velocity and to sketch the paths for different values of the proportionality constant b. It is probably best to begin with the one-dimensional problem of an object falling with a retarding force. The total force on the ball is

$$F = mg - bv,$$

where the constant b depends on the size and shape of the object and the

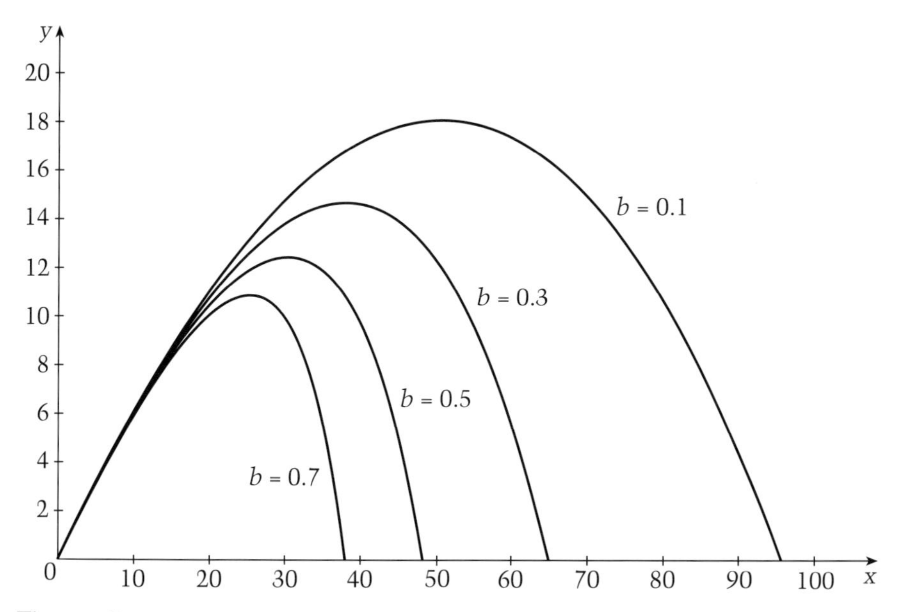

Figure 2

viscosity of air. Putting this force into Newton's second law, we have

$$m\frac{dv}{dt} = mg - bv.$$

Separating the variables and assuming that the initial velocity is zero, we obtain

$$\int_0^v \frac{dv}{v-(mg/b)} = -\frac{b}{m}\int_0^t dt.$$

We now integrate and solve for v:

$$v = \frac{mg}{b}\left(1-e^{-bt/m}\right).$$

As t gets large, the velocity approaches mg/b, which is called the *terminal velocity*.

In the two-dimensional problem, we realize that the initial velocity will not be zero and that when the object is ascending, the air resistance opposes the rise and when the object is descending, the air resistance opposes the fall. Since the velocity has a horizontal component, there will be an air resistance in the horizontal direction as well as in the vertical direction.

The appropriate equations for an object traveling in the xy-plane are:

$$m\frac{d^2x}{dt^2} = -b\frac{dx}{dt}$$

and

$$m\frac{d^2y}{dt^2} = -mg - b\frac{dy}{dt}.$$

Assuming that the object starts from the origin, the solutions of these equations for velocity and position are:

$$v_x = v_{x0}e^{-bt/m},$$
$$x = \frac{mv_{x0}}{b}\left(1-e^{-bt/m}\right),$$
$$v_y = \left(\frac{mg}{b} - v_{y0}\right)e^{-bt/m} - \frac{mg}{b},$$
$$y = \left(\frac{m^2g}{b^2} + \frac{mv_{y0}}{b}\right)\left(1-e^{-bt/m}\right) - \frac{mg}{b}t.$$

In the same way that we eliminate t and combine the equations to find that trajectories with no air resistance travel in parabolas, we find that the equation for the trajectory with air resistance is

$$y = \left(\frac{mg}{bv_{x0}} + \frac{v_{y0}}{v_{x0}}\right)x - \frac{m^2g}{b^2}\ln\left(\frac{mv_{x0}}{mv_{y0} - bx}\right).$$

We can graph this equation for different values of b using a graphing calculator or a spreadsheet.

As we can see from the graph in figure 2, the trajectory approximates a parabola for low air resistance ($b = 0.1$) but for larger air resistance ($b = 0.7$), the path falls off more rapidly than a parabola. If you crumple a piece of paper and toss it at an angle, you will see that the path is quite similar to that of our theoretical calculation.

In part B the net force on the particle must be zero. Therefore, the magnitude of the electric force must equal the magnitude of the magnetic force:

$$qE = qvB,$$

and therefore

$$v = \frac{E}{B}.$$

In part (2), we asked what would happen if the particle entered the region of crossed fields in the opposite direction. In this case the particle would no longer travel undeflected since the direction of the electric force would be the same but the direction of the magnetic force would be reversed.

Part (3) asks what the path of the particle would be if it traveled at a speed other than E/B. Qualitatively, we can look at a laboratory frame that moves at a speed of E/B. In this frame, the particle will move in a circle. We can show this by looking at the general force equations for the particle, changing reference frames, and analyzing the resulting equations.

Using the orientations indicated in figure 3, we have

$$a_x = \frac{q}{m}E + \frac{q}{m}v_yB$$

and

$$a_y = -\frac{q}{m}v_xB.$$

Changing to a reference frame that moves with speed E/B in the negative y-direction requires the following equations:

$$v_x = v_x'$$

and

$$v_y = -\frac{E}{B} + v_y',$$

where the primed variables refer to the moving reference frame.

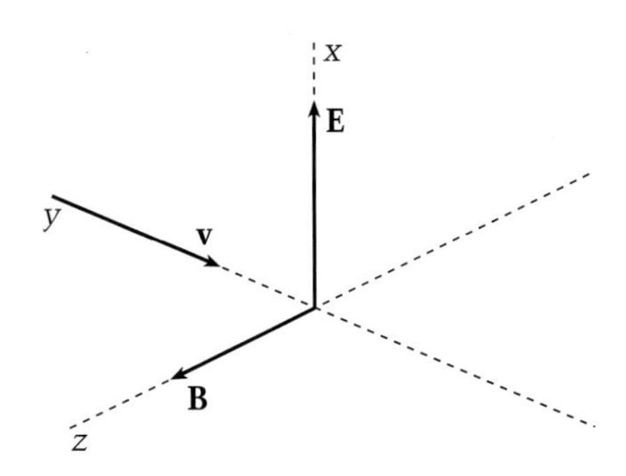

Figure 3

Substituting these expressions into the acceleration equations, we get

$$a_x' = \left(\frac{qB}{m}\right)v_y',\ a_y' = \left(-\frac{qB}{m}\right)v_x'. \quad (1)$$

These are the equations for a circle (the acceleration is perpendicular to the velocity and the velocity is constant).

If we choose a particle having a velocity other than E/B, the path of the particle will be a cycloid, the same curve as a point on a bicycle wheel rolling with uniform speed E/B. This can be shown by assuming that the particle is at the origin and at rest in the laboratory frame. The particle will begin to accelerate due to the electric force in the x-direction. Once it has a velocity in the x-direction, it will then be deflected by the magnetic force. It will travel in a cycloid finally coming to rest again at a point on the y-axis where its motion will be repeated.

We can prove this mathematically. By "guessing" the solutions to equation 1, we can then transform the solution back to the laboratory frame and analyze the final equations. Our "guess" at the solutions of equation 1 satisfying the initial condition that the speed in the x-direction is 0 and the velocity in the y-direction is E/B are

$$v_x' = \frac{E}{B}\sin\left(\frac{qB}{m}\right)t,$$

$$v_y' = \frac{E}{B}\cos\left(\frac{qB}{m}\right)t.$$

We can take the derivative of these equations and substitute back into equation 1 and see that they are solutions. Transforming these equations back into the laboratory reference frame, we arrive at

$$v_x = \frac{E}{B}\sin\left(\frac{qB}{m}\right)t,$$

$$v_y = \frac{E}{B}\cos\left(\frac{qB}{m}\right)t - \frac{E}{B}.$$

Integrating these equations, substituting $\omega = qB/m$ and $R = E/\omega B$, and using the initial conditions that $x = y = 0$ at $t = 0$ yields:

$$x = r(1 - \cos \omega t),$$

$$y = r(-\omega t + \sin \omega t).$$

These are the equations of a cycloid as shown in the article. The resulting path of the particle will be a cycloid having loops, cusps, or ripples, depending on the initial conditions and on the magnitude of E.

Relativistic conservation laws

"In mathematics you don't understand things. You just get used to them."
—John von Neumann

CONSERVATION LAWS ARE everywhere! Conservation of energy is one of the most useful laws in all branches of science. Other conservation laws in physics include charge, momentum, angular momentum, and those associated with the more esoteric baryon and lepton attributes. But conservation laws are much more pervasive. They even apply to poker. Certainly, the number of cards is conserved in any one game. The amount of money can also be conserved if the game is carefully constructed.

The World Series of Poker is held in Las Vegas each year. It is not a tournament for the poor or the faint of heart. In the final game each player buys in for $10,000! In 2000 there were 512 players, so the total amount of money in the game was $5,120,000. The game is Texas Hold 'em and the rules require that no money be taken from the game or added to the game—so-called "table stakes."

As the game proceeds, some players accumulate lots of money while others lose. However, at all times the total amount of money in the game is a constant—5.12 million dollars. When players lose all of their chips, they must exit the game. So the number of players is not conserved, but the amount of money is.

After all but one of the players have lost their chips, the winner has accumulated $5,120,000. You can imagine the size of the bets when there are only two players left at the table. No, the winner does not get to keep all of the money; the money is divided among the players according to when they left the game. Of course, those leaving early get smaller amounts and the last player gets the most. In 2000 the winner pocketed $1,500,000 and the second place player walked away with $896,500. Even the players who finished in 37th through 45th place won $15,000.

Poker may not interest some

Since *Homo sapiens* doesn't have to spend 24 hours a day hunting for food to survive, he has a lot of time for games and other entertainment, called history. The most popular game was and still is war and making money. On the left we see the king playing war with his army as if it's a chess game, and on the right people play the stock market. In the middle we have a chessboard floor with military men playing poker. On the right are judges more or less engaged in surveying the conservation laws, while the chess game is out of hand. There is a pawn delivering the king to the guillotine, while another king (Ubu) is knocking down any figure in its way, etc. Apparently there seems to be a never-ending supply of conserved energy, so that the show will go on for a long, long time to come.

—T.B.

ONE DAY...
LET THE GAMES BEGIN!
BANG!
THE GAME THEORY SHOW
IN LIFE WE DON'T UNDERSTAND THINGS
WE JUST GET USED TO THEM.... OR NOT.
K. DOSE
IGNORAMUS ET IGNORABIMUS
THE TRAVELING PHOTONS
PICK ANY NUMBER!
Cogito ergo sum.
RESTROOM
CONSERVED ENERGY
RULES
RULES
MORE RULES
EXIT
TOM BUNK

readers as such, but it is actually one small example from a branch of mathematics called game theory. Games (in mathematics) are situations that involve people or machines with conflicting interests. Simple games can have complete "solutions" but serve as insights into more complex games like checkers and chess and more serious games like politics, warfare, and property law. One type of game is the zero-sum game, where one player's loss is another player's gain. Not all games are zero-sum games. The stock market or the nation's economy can gain value over time.

We are very familiar with the conservation laws of energy and momentum in classical mechanics. Although in special relativity the sizes of time intervals, lengths, energies, momenta, angular momenta, and so on are not the same in different inertial reference systems, the conservation laws are still valid. However, we must modify our classical expressions for energy and momentum to make them work in relativistic situations.

The relativistic momentum p is given by

$$p = \gamma m v,$$

where m is the (rest) mass of the particle, v is its speed, and

$$\gamma = \sqrt{\frac{1}{1-\beta^2}}$$

with $\beta = v/c$, the ratio of the speed of the particle to the speed of light. Notice that the relativistic momentum reduces to the classical value for slow speeds—that is, when $v \to 0$. This must be true because we know that Newton's laws of motion work very well for ordinary speeds.

The photon has no rest mass but it does have momentum. Therefore, we must use a different expression for the momentum of a photon:

$$p = \frac{h}{\lambda} = \frac{hf}{c},$$

where $h = 6.63 \cdot 10^{-34}$ J·s is Planck's constant and λ and f are the wavelength and frequency of the photon, respectively.

The relativistic energy of a particle with nonzero mass is given by

$$E = \gamma m c^2.$$

When the particle is at rest $\gamma = 1$, so the rest-mass energy is $E = mc^2$. This is Einstein's famous equation giving the equivalence of mass and energy. The difference between the total energy and the rest-mass energy is equal to the kinetic energy of the particle. We can show that this reduces to the classical formula for the kinetic energy as $v \to 0$:

$$KE = \gamma m c^2 - mc^2 = (\gamma - 1)mc^2.$$

We now use the binomial expansion

$$(1 + x)^n \approx 1 + nx,$$

when $x \ll 1$. This gives us

$$KE = \left(\frac{1}{\sqrt{1-\beta^2}} - 1\right)mc^2$$

$$\approx \left(1 + \frac{1}{2}\beta^2 - 1\right)mc^2 = \frac{1}{2}mv^2.$$

The relativistic energy of a photon is

$$E = hf = pc.$$

With these new expressions, conservation of energy and linear momentum work the same way they do in classical physics. The mathematics can be more difficult because the factor γ depends on the speed of the particle.

The present problem comes to us from the second exam used to select the 2000 US Physics Team. It was created by Leaf Turner, who worked at Los Alamos National Laboratory and was a senior coach of the team.

A relativistic particle decays into two photons. One of the photons travels along the positive x-axis with frequency f_1, while the second photon travels along the negative x-axis with frequency $f_2 < f_1$.

A. What is the velocity **v** of the particle?

B. What is the rest mass of the particle?

C. What are the frequencies of the photons in the rest frame of the particle?

You are given the formula

$$p'_x = F_1 p_x + F_2 \frac{E_\gamma}{c},$$

where p'_x is the x-component of the momentum of either photon in the laboratory reference frame and E_γ and p_x are the energy and x-component of the momentum, respectively, of the photon in the rest frame of the particle.

D. What are the functions F_1 and F_2 in terms of β?

Solution

A. We apply energy and momentum conservation to the decay process:

$$\gamma m c^2 = hf_1 + hf_2, \qquad (1)$$

$$\gamma m v = \frac{hf_1}{c} - \frac{hf_2}{c}. \qquad (2)$$

Let's now divide equation (2) by equation (1), cancel common factors, and solve for the velocity:

$$v = c\frac{f_1 - f_2}{f_1 + f_2}. \qquad (3)$$

B. We solve equation (1) for the mass:

$$m = \frac{h}{\gamma c^2}(f_1 + f_2) = \frac{h}{c^2}(f_1 + f_2)\sqrt{1-\beta^2}$$

and substitute for β from equation (3) to obtain

$$m = \frac{2h}{c^2}\sqrt{f_1 f_2}. \qquad (4)$$

C. In the rest frame of the particle the two photons must travel in opposite directions with the same size momentum. Therefore, the frequency f will be the same for both photons. Energy conservation requires

$$mc^2 = 2hf$$

or

$$f = \frac{mc^2}{2h}. \qquad (5)$$

D. In this part we are to determine the functions F_1 and F_2 in the expression

$$p'_x = F_1 p_x + F_2 \frac{E\gamma}{c},$$

where the unprimed variable refers to the rest frame of the particle and the primed variable refers to the laboratory frame. Writing this equation for the first photon, we obtain

$$\frac{hf_1}{c} = F_1 \frac{hf}{c} + F_2 \frac{hf}{c} = (F_1 + F_2)\frac{hf}{c}.$$

Using equation (5), this simplifies to

$$F_1 + F_2 = \frac{2hf_1}{mc^2}. \qquad (6)$$

For the second photon we obtain

$$-\frac{hf_2}{c} = -F_1 \frac{hf}{c} + F_2 \frac{hf}{c} = (-F_1 + F_2)\frac{hf}{c},$$

$$F_1 - F_2 = \frac{2hf_2}{mc^2}. \qquad (7)$$

Adding equations (6) and (7) and dividing by 2, we get our expression for F_1:

$$F_1 = \frac{h}{mc^2}(f_1 + f_2).$$

Comparing this to equation (1), we see that

$$F_1 = \gamma.$$

Subtracting equation (7) from equation (6) and dividing by 2 yields

$$F_2 = \frac{h}{mc^2}(f_1 - f_2).$$

Comparing to equation (2) shows that

$$F_2 = \beta\gamma.$$

A good theory

"In my work, I have always tried to unite the true with the beautiful; but when I had to choose one or the other, I usually chose the beautiful."
—Hermann Weyl

CREATION CAPTIVATES THE mind. We view creation as a series of miracles. The birth of a human being is an act of creation. The first simple thought of a young child is another act of creation. We create and we admire those individuals whose creations extend our own perceptions. As a culture, we struggle with the creation of the world by inventing stories and theories as to how everything we know came into being. "In the beginning, God created the Heavens and the Earth" is one such story. The Big Bang is another such story. "Non-being existed not, nor being" from the Upanishads is still another story.

The painter, the writer, and the composer share their personal conceptions of the world and allow us to peer into their minds. The scientist must also create a personal conception. The scientist, however, bears the pressure of a large constraint. As fertile as the imagination may be, the scientist's creation must be consistent with measurements of the physical world.

The greatest scientists create their world views and help us to see our world through their lenses. As noted in the quote above, Weyl describes how the inner vision of the scientists guides their work. The great scientists have such a superb intuition about the world that when they conceive of a world different from the one we know, we often discover that the "real" world embodies manifestations of their vision and that society's earlier view was myopic. Of course, that vision is further corrected centuries later by a new set of corrective lenses, which further clarifies the blur.

How does a good theory get judged? It must first be able to explain what the prevailing theory has successfully explained. It must also be able to explain some known phenomenon that the prevailing theory is unable to explain. When that

Being a NON-BEING can get pretty boring after a while, that's why a long time ago our Creator decided to BE and fill the void with some action. He ordered a big box of elements to be assembled in 7 days. He then carved a ladder out of the Tree of Life, climbed to the top, and kicked off the creation with a big bang! This event, with all its wonders, is still unfolding today and is full of powerful energy. Its inexplicable mysteries are being closely watched by admiring fans, scientists who are trying to decipher the secret laws of the divine enterprise, and the artists who themselves create, each according to their inner calling. But to make things not too perfect, the Maker's Adversary enters the stage to add even more excitement to the cosmic show.

—T.B.

BANG!
Z
EMERGENCY EXIT
TAO
NON BEING
OUT TO LUNCH
GENESIS
CREATE YOUR OWN WORLD IN 7 DAYS!
BATTERIES NOT INCLUDED
INTUITION
1 STEIN
N. BOHR
ACTIVE PHYSICS
ARTHUR
TOMAS BUNK

theory is able to predict something that nobody has perceived, and that something is then discovered, we realize that we have a very good theory.

One such good theory is Newton's theory of gravity. When Newton proclaimed that every mass attracts every other mass with a force that is inversely proportional to the distance between them, he was able to explain the motion of the planets about the Sun. Kepler had described this quite well, but Newton's synthesis did more. It was able to explain why the ratio of the square of the period to the cube of the distance was a constant for all planets orbiting the Sun. Newton provided a means to weigh the Sun. He was also able to describe the motion of the tides. What was Newton's surprise prediction? Long after Newton's death, slight perturbations were measured in Neptune's orbit. If Newton was correct, these perturbations were signaling the existence of a planet that nobody knew about. Uranus' discovery was an enormous support for Newton's theory of gravitational attraction.

Einstein's theory of gravity supplanted Newton's 200-year success story. Einstein's warping of a space-time continuum was able to explain the motions of the planets and was as successful as Newton's explanation. Einstein's creation was also able to explain the precession of Mercury's orbit about the Sun. The precession was well known but it could not be explained using Newton's theory. Einstein succeeded in shedding light on this puzzle. What did Einstein predict that nobody knew about? He predicted the bending of light as the light passed by a large mass—a bending that was larger than one might expect from Newton's gravity and Einstein's earlier $E = mc^2$. When Arthur Eddington viewed the eclipse of 1919, his experimental team found that Einstein had indeed predicted a phenomenon that nobody had envisioned.

Niels Bohr and all of his colleagues were aware that hydrogen had a distinct spectrum. Rydberg found a mathematical means of calculating the wavelengths of the emitted light. Bohr's theory of electron orbits about the nucleus was able to do a better job than Rydberg. Bohr postulated that the electrons were restricted to orbits whose angular momentums were whole number multiples of Planck's constant divided by 2π:

$$mvr = \frac{nh}{2\pi}.$$

Combining this with the notion that the electrons move in circular orbits due to a Coulomb force

$$\frac{mv^2}{r} = \frac{kq_1q_2}{r^2},$$

Bohr was able to find both the radii and the energy levels of the different orbits. The energy levels could be succinctly described in terms of the lowest energy assigned to the tightest orbit:

$$E_n = \frac{E_1}{n^2},$$

where E_1 is the ground state energy of -13.6 eV and n is the number of the orbit.

Bohr was able to explain the spectrum of hydrogen by noting that electrons jumping from one energy level to a lower energy level would emit photons with an energy corresponding to the difference in energy of the two levels. If the electron jumps from the $n = 3$ orbit to the ground state $n = 2$,

$$E_3 - E_2 = hf,$$

where $h = 6.63 \times 10^{-34}$ J·s is Planck's constant and f is the frequency of the light.

The visible spectrum of hydrogen emerged from jumps of the electron from the $n = 3$ level to the $n = 2$ level, the $n = 4$ level to the $n = 2$ level, and the $n = 5$ level to the $n = 2$ level. What if the electrons could also jump from higher levels to the $n = 1$ level? Bohr could calculate those frequencies of light—frequencies in the ultraviolet part of the spectrum. When explorations of the ultraviolet took place, Bohr's predictions of the frequencies were right on.

Bohr also predicted frequencies of light for the hydrogen spectrum, which would be emitted if an electron jumped to the $n = 3$ level. These infrared frequencies were also right on. Some of these frequencies had been known. Some had not. The frequencies corresponding to jumps to the $n = 4$ and $n = 5$ levels had never been observed. As humans, our eyes limit our world view to the visible spectra. Bohr, with the assistance of his very good theory, helped us peer into the invisible world.

What else did Bohr's theory tell us that we never expected? Bohr was able to create an understanding of the periodic table and predict the chemical properties of element 72, which had previously been a blank in the table. The discovery of element 72 (named hafnium after Bohr's home town) came just hours before Bohr was to accept his Nobel Prize.

Our problem draws from Newton's and Bohr's theories.

A. This problem comes from a wonderful, slim book *Thinking Like a Physicist*, edited by N. Thompson. A small moon of mass m and radius a orbits a planet of mass M while keeping the same face towards the planet. Show that if the moon approaches the planet closer than

$$r_c = a\sqrt[3]{\frac{3M}{m}},$$

loose rocks lying on the surface of the moon will be lifted off.

B. This problem was first given at the 7th International Physics Olympiad in Warsaw, Poland, in 1974. A hydrogen atom in its ground state collides with another hydrogen atom in its ground state at rest. What is the least possible speed for which the collision is inelastic? If the speed is greater, a photon is emitted that can be observed in the direction of the initial velocity or in the opposite direction. How much do the frequencies of these photons differ from the frequency they would have if emitted from an atom at rest? The mass of the hydrogen atom is 1.67×10^{-27} kg and its ionization energy $E = 13.6$ eV $= 2.18 \times 10^{-18}$ J.

Solution

Once again, Art Hovey from Amity Regional High School of Connecticut submitted solutions that were basically correct.

In the first problem, the gravitational and normal forces acting on a rock of mass μ lying on the surface of the moon provide the centripetal force that keeps it in orbit about the planet. Let's assume that the rock is located on the side of the planet facing the moon, that the radius of the moon is r, and that the distance between the moon and the planet is d. We also assume that the mass of the planet is very much larger than the mass of the moon:

$$\frac{GM\mu}{(r-a)^2} + F_N - \frac{Gm\mu}{a^2} = \mu\omega^2(r-a). \quad (1)$$

The gravitational force of the planet acting on the moon causes the moon to revolve about the planet with the same angular velocity:

$$\frac{GMm}{r^2} = m\omega^2 r. \quad (2)$$

Solving equation (1) for ω and substituting into equation (2), we get

$$\frac{GM\mu}{(r-a)^2} + F_N - \frac{Gm\mu}{a^2} = \mu(r-a)\left(\frac{GM}{r^3}\right).$$

The loose rock will leave the surface whenever the normal force is negative. The limiting case is found by setting the normal force equal to zero:

$$\frac{GM\mu}{(r-a)^2} - \frac{Gm\mu}{a^2} = \mu(r-a)\left(\frac{GM}{r^3}\right).$$

Recognizing that the distance between the planet and the moon is much greater than the radius of the moon ($r \gg a$), we can ignore the terms of a that are added to much larger values of r:

$$Ma^2r^3 - mr^5 = Ma^2(r^3 - 3r^2a),$$

$$r = a\sqrt[3]{\frac{3M}{m}}.$$

The second problem described an inelastic collision between two hydrogen atoms. For this inelastic collision, the two hydrogen atoms "stick" together, forming a diatomic molecule. Momentum must be conserved:

$$m_1 v_0 = 2m_1 v_f.$$

The loss in kinetic energy can now be calculated:

$$\Delta K = K_f - K_0$$
$$= \frac{1}{2}(2m)\left(\frac{v_0}{2}\right)^2 - \frac{1}{2}(m)v_0^2$$
$$= -\frac{1}{4}mv_0^2 = -\frac{E_0}{2}.$$

Where did the energy go? With billiard balls, the energy of an inelastic collision may be transformed into sound, deformation of the objects, or heat, but these don't make sense at the atomic level. The energy must have raised an electron to a higher energy state. The smallest energy change for a ground state electron in hydrogen can be calculated:

$$\Delta E = E_2 - E_1 = \frac{E_1}{2^2} - E_1 = -\frac{3E_1}{4}$$
$$= -\frac{3(2.18 \cdot 10^{-18}\ \text{J})}{4} = 1.63 \cdot 10^{-18}\ \text{J}.$$

Setting these energy differences equal to one another, we can solve for the initial velocity of the hydrogen atom:

$$v_f = 3.13 \cdot 10^4\ \text{m/s}$$

Since the diatomic molecule is moving with a speed $v_0/2$, the frequency of the emitted photon will be Doppler shifted. For speeds that are small relative to the speed of light, the fractional change in the frequency is approximately equal to the ratio of the speed of the molecule to the speed of light:

$$\frac{\Delta f}{f} \cong \frac{v}{c} = \frac{6.26 \cdot 10^4\ \text{m/s}}{3 \cdot 10^8\ \text{m/s}} = 0.021\%.$$

The frequency is larger if the photon is emitted in the forward direction and smaller if emitted in the backward direction.

The fundamental particles

"For we do not think that we know a thing until we are acquainted with its primary conditions or first principles, and have carried our analysis as far as its simplest elements."
—*Aristotle,* Physics

THE SEARCH FOR THE FUNdamental building blocks in nature has gone on for more than two thousand years. Aristotle felt that all the materials around us were composed of varying quantities of four basic *elements*—earth, fire, air, and water. In hindsight, this may seem simplistic. How can the myriad of properties of these materials arise from only four basic elements that don't share these properties? Is this any more strange than believing that everything is composed of chemical elements? For instance, hydrogen—a very flammable substance, and oxygen—a gas required for combustion, combine to form water, which is used to put out fires!

With the number of chemical elements exceeding one hundred, it was quite comforting to discover a hundred years ago that atoms were composed of three more basic particles—electrons, protons, and neutrons. For instance, the most common neutral atom of carbon is composed of six electrons, six neutrons, and six protons. Combinations of only three basic particles determine the myriad of chemical properties that exist in nature.

Beginning in 1932, scientists discovered many new "elementary particles" such as the muon, the pion, and the kaon. In 1932 the first of the "antiparticles" was discovered. The positron is just like an electron, except that it has a positive charge. Over the next few decades more than a hundred new particles were discovered, leading to renewed efforts to find a simpler set of fundamental building blocks.

The elementary particles are grouped into two families—the *leptons* and the *hadrons*. The lepton family has six members: the electron, a heavy electron known as the muon, a still heavier electron known as the tau, and three neutrinos, one associated with each of these "electrons." It is currently be-

This ancient document shows the human being as a collection of animals, vegetables, mysteries, elements, and symbols. His chimneylike finger, after being touched by the Maker's spark of life, suggests factory production, while the fish inside represents evolution, his right foot stands for the elements earth and air, the left for water and fire, the guts are the enlightened serpent coming from Hell and snapping at the rabbit, the light-bringer of spring. The soul, embodied by a blue bird, is captured in a cage and is watched ominously by a black cat. The human is shown on a permanent phone line with his ape ancestor, who still lives in the Tree of Knowledge. His ribs are removed for further use, and the hectic construction activities around him suggest that he obviously needs some more work.

—T.B.

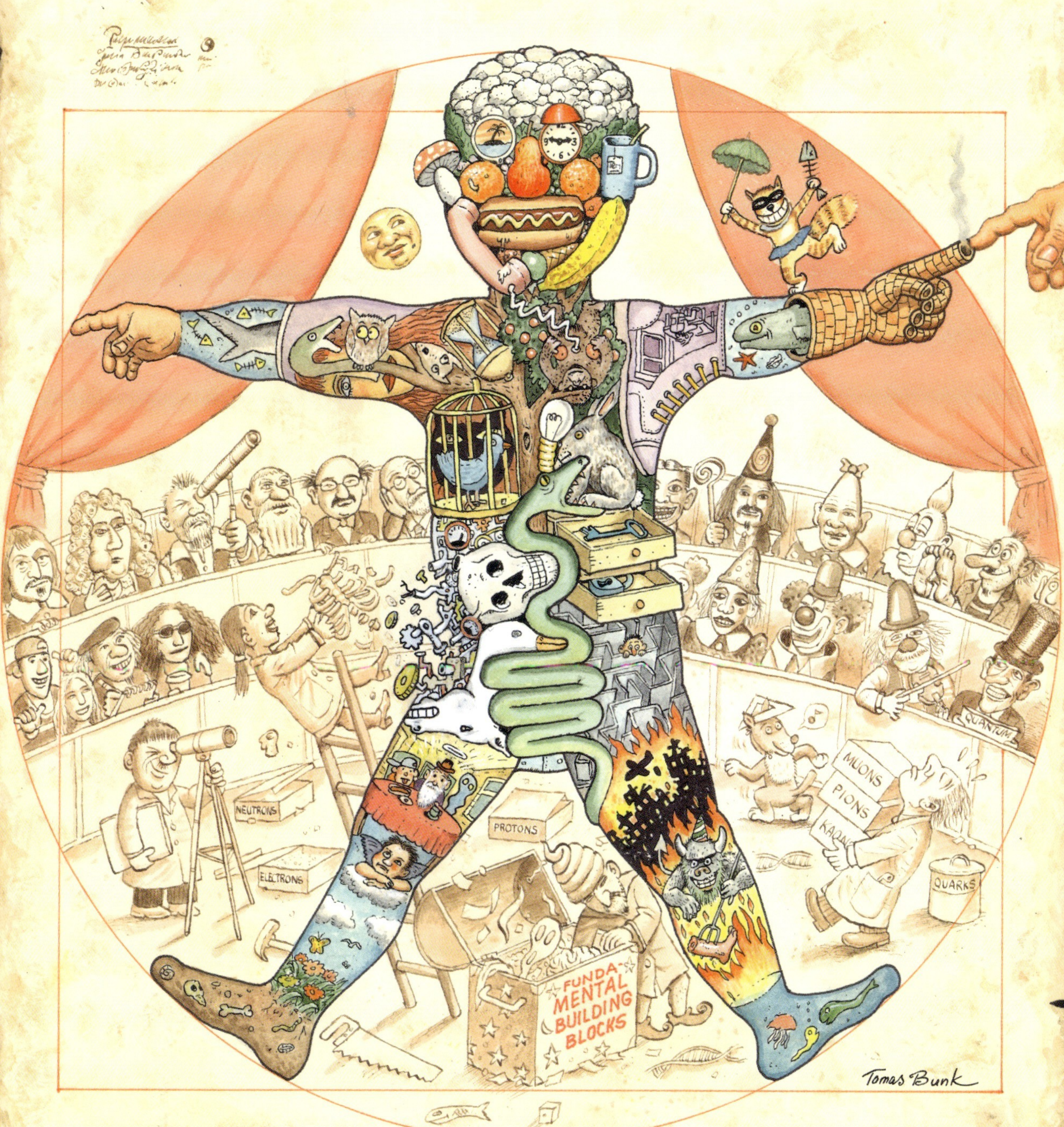
NEUTRONS
PROTONS
ELECTRONS
MUONS
PIONS
QUARKS
QUANTUM
FUNDA-
MENTAL
BUILDING
BLOCKS
Tomas Bunk

LE MUSEE
QUANTUM
PATAPHYSIQUE

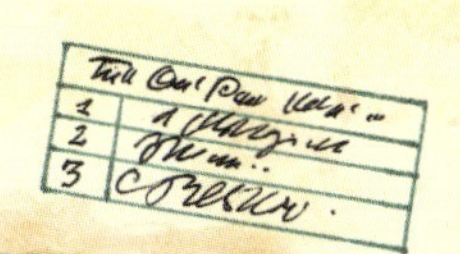

NEW ROCHELLE
1-11-01
NEW YORK

Name	Symbol	Charge	Other
Down	d	$-^1/_3$	
Up	u	$+^2/_3$	
Strange	s	$-^1/_3$	Strangeness = −1
Charm	c	$+^2/_3$	Charm = +1
Bottom	b	$-^1/_3$	Bottomness = +1
Top	t	$+^2/_3$	Topness = +1

Table 1

lieved that these six particles are truly fundamental, as they do not show any internal structure.

The other elementary particles are composites. Murray Gellmann and George Zweig hypothesized that these particles are combinations of more fundamental particles known as *quarks*. Theorists believed (and still believe) that there should be one quark for each lepton. Therefore, it was very comforting to discover six quarks. In order to account for the known hadrons, the quarks must have baryon number $^1/_3$, spin $^1/_2$, and the properties given in table 1.

Let's look at an example of how this works. The proton is a *baryon* with a spin of $^1/_2$ and charge of +1. All baryons are composed of three quarks—hence the assignment of baryon number $^1/_3$ to the quarks. Each quark has a spin of $^1/_2$. If two of these are aligned in one direction and the third is aligned in the opposite direction, they can combine to form a particle with a spin of $^1/_2$ if the quarks have no orbital angular momentum. This leaves us with obtaining the correct charge. We see that the combination uud has a charge of +1. We could conceivably use some of the other quarks but the proton is the lightest baryon and we expect it to be composed of the lightest quarks, the up and down quarks.

The antiproton is an antibaryon and is composed of three antiquarks. Antiquarks have the same properties as the quarks, but many of these properties have the opposite sign, in particular, the baryon number, the charge, and the "other" properties in table 1. The composition of the antiproton is just like that of the proton except that all of the quarks are replaced by the corresponding antiquark. Therefore, the composition of the antiproton is $\overline{u}\overline{u}\overline{d}$, where the overbar indicates an antiquark. This yields a baryon number of –1, a spin of $^1/_2$, and a charge of –1.

The members of the other subfamily of hadrons are known as *mesons*. Mesons are composed of a quark and an antiquark, which yields a baryon number of zero. For example, a positively charged pion has a spin of a zero and a charge of +1. Verify for yourself that $u\overline{d}$ has the correct properties. To get a spin of zero, the spins of the two quarks must be aligned in opposite directions.

The positively charged kaon is a meson with a spin of zero, a strangeness of +1, and a charge of +1. Its composition is $u\overline{s}$. The neutral pion is an interesting case as it is composed of two combinations, $u\overline{u}$ and $d\overline{d}$.

A. What combinations of up, down, and strange quarks make up the particles in table 2?

Name	Symbol	Baryon	Spin	Charge	Strangeness
Neutron	n	+1	$^1/_2$	0	0
Pi minus	π^-	0	0	−1	0
K zero	K^0	0	0	0	+1
Lambda zero	Λ^0	+1	$^1/_2$	0	−1
Antineutron	$\overline{n}$	−1	$^1/_2$	0	0
Xi minus	Ξ^-	+1	$^1/_2$	−1	−2

Table 2

Name	Symbol	Baryon	Spin	Charge	Strangeness
Neutron	n	+1	$^1/_2$	0	0
Antineutron	$\overline{n}$	−1	$^1/_2$	0	0
Pi minus	π^-	0	0	−1	0
Xi minus	Ξ^-	+1	$^1/_2$	−1	−2
Delta plus plus	Δ^{++}	+1	$^3/_2$	+2	0
Antilambda	$\overline{\Lambda^0}$	−1	$^1/_2$	0	+1

Table 3

There is one problem that we've not mentioned. The omega minus (Ω^-) is a baryon with a spin of $^3/_2$ and a strangeness of –3. The only combination of three quarks that gives the correct properties is sss. In order to get the spin of $^3/_2$ all three of the spins need to point in the same direction. This is the root of the problem. The Pauli exclusion principle says that no two quarks in a hadron can have the same set of properties—that is, none of the quarks can have the same set of *quantum numbers*. But, in the Ω^-, the three quarks have the same set of quantum numbers.

This led to the idea that there must be another quantum number that describes the quarks. This quantum number is known as *color*, and has three values. But, in this case, the values are not numbers. They are called *red*, *green*, and *blue*. If we think of these quantum numbers as combining like colored lights, all hadrons must be *white*. Therefore, one of the strange quarks in the Ω^- must be red, another green, and the third blue.

The antiquarks have the complimentary colors; the antired quark is *cyan*, the antigreen quark is *magenta*, and the antiblue quark is *yellow*. Therefore, the positive pion is a combination of

$$u_{red}\bar{d}_{cyan} + u_{green}\bar{d}_{magenta} + u_{blue}\bar{d}_{yellow}.$$

B. What combinations of up, down, and strange quarks make up the particles in table 3?

Solution

The only two properties that are needed to build each of the hadrons in table 2 are its charge (in units of the electronic charge) and its strangeness.

The neutron, as its name implies, is neutral. Because it is a baryon, it must be composed of three quarks, and because it does not have any strangeness, it must be composed of only down and up quarks. Since the down quark has a charge of −1/3 and the up quark has a charge of +2/3, the neutron is composed of two down quarks and an up quark (n = ddu).

The negative pion has a charge of −1 and a strangeness of zero. Once again it must be composed of only down and up quarks. However, the pion is a meson and consists of a quark–antiquark pair. Because the antiquark has the opposite charge of a quark, we need a down quark and an up antiquark to get a charge of −1 $\left(\pi^- = d\bar{u}\right)$.

The neutral kaon is a meson with a strangeness of +1. The only way we can get this strangeness is by using a strange antiquark, which converts the strangeness of −1 to +1. This gives us a charge of +1/3. Therefore we need a down quark to give us an overall charge of zero $\left(K^o = d\bar{s}\right)$.

The lambda baryon only comes with a zero charge. Its strangeness of −1 requires that we use a strange quark, giving us a charge of −1/3. The other two quarks must be down and up quarks. The only combination of two of these that gives us a charge of +1/3 is one of each. Therefore $\Lambda^0 = dus$.

The antineutron is an antibaryon and must be composed of three antiquarks. We simply use the antiquark version of each quark in the neutron to build the antineutron. Therefore $\bar{n} = \bar{d}\bar{d}\bar{u}$.

The cascade minus has a strangeness of −2 and a charge of −1. This requires two strange quarks for a charge of −2/3. We need a down quark to give us the additional charge of −1/3. Thus, we have $\Xi^- = dss$.

Up to this point we have not had to worry about the Pauli exclusion principle. However, to complete the picture, we need to add a new quantum number. Color comes in three varieties: red, green, and blue. Let's use the subscripts r, g, and b, respectively, to represent these three values, and the subscripts c, m, and y to represent the complimentary colors cyan, magenta, and yellow, respectively. In this scheme all hadrons must be white if we imagine the colors to combine as lights; that is, additively.

Four of the particles in table 3 are repeats. All we need to do is to make sure that we include all combinations of the color quantum number.

$$n = d_r d_g u_b + d_g d_b u_r + d_b d_r u_g,$$

$$\bar{n} = \bar{d}_m \bar{d}_c \bar{u}_y + \bar{d}_c \bar{d}_y \bar{u}_m + \bar{d}_y \bar{d}_m \bar{u}_c,$$

$$\pi^- = d_r \bar{u}_m + d_g \bar{u}_c + d_b \bar{u}_y,$$

$$\Xi^- = d_r s_g s_b + d_g s_b s_r + d_b s_r s_g.$$

The delta plus plus is a baryon without strangeness. Therefore, it is composed of only down and up quarks. The only combination that gives us a charge of +2 is three up quarks. Thus, we have

$$\Delta^{++} = u_r u_g u_b.$$

Finally, the antilambda is the antiparticle of the lambda. All we need to do is to change each quark to the antiquark and each color to the complementary color.

$$\overline{\Lambda^0} = \bar{d}_m \bar{u}_c \bar{s}_y + \bar{d}_c \bar{u}_y \bar{s}_m + \bar{d}_y \bar{u}_m \bar{s}_c.$$